# GENETICS, ETHICS AND THE LAW

George P. Smith, II

ASSOCIATED FACULTY PRESS, INC.

# New Studies on Law and Society

*Advisor:* J. Myron Jacobstein
Law Librarian and Professor of Law
Stanford Law School

Horrow, *Sports Violence*
Liu, *Sino-American Juvenile Justice System*
Riga, *Death of the American Republic*
Rubenstein and Fry, *Of a Homosexual Teacher*
Kushner, *Apartheid in America*
McClintock, *NLRB General Counsel*
Laitos, *A Legal-Economic History of Air Pollution Controls*
Smith, *Genetics, Ethics and the Law*
Gordon, *Crime and Criminal Law*
Rosenfield, *Labor Protective Provisions in Airline Mergers*
Riga, *Right to Die or Right to Live?*

Copyright © 1981
GEORGE P. SMITH, II
All Rights Reserved

Published by
Associated Faculty Press, Inc.
Gaithersburg, MD

Exclusive distributor:
University Publications of America, Inc.
Frederick, MD

Library of Congress Catalog Number: 81-68027
ISBN: 0-86733-005-8

# TABLE OF CONTENTS

Acknowledgments

INTRODUCTION .................................................... 1

CHAPTER 1: CHANGING VALUES AND PERCEPTIONS. ................... 8
    Uncertainties On The Spiral Staircase. ............................ 9

CHAPTER 2: IMPLEMENTING A NEGATIVE EUGENICS PROGRAM ....... 18
    Amniocentesis. ................................................ 18
    Genetic Screening and Counseling Programs. ....................... 19
    Restrictions On Marriage ........................................ 19
    Restrictions On Reproduction. .................................... 20

CHAPTER 3: THE VAGARIES OF INFORMED CONSENT ................ 32
    The Historical Perspective ....................................... 32
    Implied Consent And Eugenic Sterilization. ........................ 35
    Voluntary Sterilization .......................................... 37
    Control ....................................................... 39
    Consent In Human Experimentation .............................. 39
    Fetal Consent. ................................................ 40
    Structuring A Right Of Refusal—A Synthesis. ...................... 41

CHAPTER 4: WRONGFUL LIFE v. WRONGFUL BIRTH—AN IN DEPTH CONSIDERATION ................................................ 70
    Elements Of A Cause Of Action .................................. 70
    Park v. Chessin: A Study In Conflicts. ............................ 72
        The Sixth Cause of Action ................................. 73
        The Appeal. .............................................. 74
        A Suggested Resolution .................................... 75
    Wrongful Conception ........................................... 76
    Wrongful Birth And Allowable Recovery .......................... 77
    Allowing Suits In Tort For Pre-natal Injuries. ....................... 80
    Recapitulation ................................................. 81
    Epilogue: A Long Awaited Resolution?. ............................ 82
        Complications ............................................ 85

CHAPTER 5: THE NEW BIOLOGY AND A PROGRAM FOR POSITIVE EUGENICS ........................................... 104
    Artificial Insemination ......................................... 104
    In Vitro Fertilization And Embryo Implants. ....................... 104
    Asexual Reproduction: Cloning And Parthenogenesis ............... 105

CHAPTER 6: THE LEGAL RESPONSE .............................. 118
    Confidentiality. ............................................... 120

CHAPTER 7: THE SCIENTIFIC METHOD AND THE NEW
BIOLOGY: AN OVERVIEW........................................... 129
New Avenues Of Regulation: The Federal Government Acts .............. 131
Ethical Tribunals In Nongovernmental Areas............................ 132
Finding An Equilibrium................................................ 133

CHAPTER 8: THE BIOETHICAL CONUNDRUM: A CONSIDERATION
IN MICROCOSM.................................................... 145
The Interaction Of Science And Ethics................................. 145
The Metaethical Quagmire.............................................. 145

CHAPTER 9: SCIENCE AND RELIGION: COMPATIBILITIES
AND CONFLICTS.................................................... 153
Theological Considerations ........................................... 155
The Roman Catholic View............................................. 155
The Protestant View................................................. 156
The Jewish View..................................................... 157

CONCLUSIONS.................................................................. 164

APPENDICES:

A — Model Informed Consent Law............................................ 167
B — Proposed Voluntary Sterilization Act.................................. 168
C — Model Statute For Artificial Insemination............................. 170
D — Guide To Clinical Research: The Declaration of Helsinki............... 172
E — The Nuremberg Code Of Ethics In Medical Research...................... 174
F — Research Involving Prisoners: Recommendations Of The National Commission For The Protection Of Human Subjects Of Biomedical And Behavioral Research......................................................... 176
G — Decision Of The Director, National Institute of Health, To Release Guidelines For Research On Recombinant DNA Molecules (1976) And Pertinent Parts Of The Guidelines And The Amended Guidelines of 1980; Statement By Joseph A. Califano, Jr., Former Secretary of HEW Regarding The Promulgation Of The Revised Guidelines Of 1978............... 193
H — Research Involving Children: Recommendations Of The National Commission For The Protection Of Human Subjects Of Biomedical And Behavioral Research......................................................... 216
I — United Nations General Assembly: Declaration On The Rights Of Mentally Retarded Persons (1971)............................................ 225
J — The Belmont Report Of The National Commission For The Protection Of Human Subjects: Ethical Principles For Treating Humans................... 227

TABLE OF PRINCIPAL CASES ................................................... 236

INDEX........................................................................ 238

To My Mother
Louise Barrett-Smith Cornell

*"We know how human nature may be degraded; we do not know how by artificial means any improvement in the breed can be effected."*

PLATO

# About the Author

Mr. Smith received both his B.S. degree in Business-Economics, together with his J.D. in law (with honor) from Indiana University in Bloomington. He is a member of the American Law Institute as well as the bars of the District of Columbia, Indiana, and The United States Supreme Court. He earned his LL.M. degree at the Columbia University School of Law in New York City, where he held a University Fellowship, and was a Commonwealth Fellow in Law, Science and Medicine at Yale University Law School during the 1976-1977 school year. In December, 1980, he was a Residential Scholar at the Rockefeller Foundation's Bellagio Study in Italy. A prolific author, whose works have been cited by The United States Supreme Court, The United States Congress, The New York Times, The Wall Street Journal and TIME Magazine, Mr. Smith is currently Professor of Law at The Catholic University in Washington, D.C., where one of his principal teaching areas is Law, Science and Medicine. He is an occasional lecturer at The School of Medicine, Uniformed Services University of the Health Sciences in Bethesda, Maryland, and has been a Distinguished Visiting Scholar in Bioethics at The Kennedy Institute of Ethics at Georgetown University since 1977.

# ACKNOWLEDGEMENTS

I am indebted to such a large number of people for aid—moral, spiritual, philosophical, intellectual and otherwise—in not only the preparation of this book but for their stimulation and encouragement to develop and pursue scholarship over the years that I risk embarrassing omission by naming but a few. Indeed, it is exceedingly difficult to begin to list all of these friends and colleagues who have influenced me. Perhaps by mentioning at least some, I gesture toward all.

To my mother who—through her initial guidance, encouragement and love—engendered a spirit of intellectual curiosity in me and buttressed it with countless hours of "on the job training" in English grammar and composition, I acknowledge my lifetime gratitude. My intellectual maturation, if you will, was nurtured at the secondary level by three teachers whom I, in turn, express my great debt of appreciation: Mildred E. Hipskind, Martha J. Jones and the late Ruth A. Jones. In college, law school and graduate school, I was exceedingly fortunate to have a number of great teachers whose legacy of intellectual perspicacity and literary expressiveness I willingly and eagerly fell heir: Professor Ross Allen, Dr. Mary Elizabeth Campbell, the late Dr. Charles Frankel, Professor Edward W. Gass, Professor John S. Grimes, Professor Walter Gellhorn, the late Dr. Clifford Kirkpatrick, Professor W. Howard Mann, Dr. Ernest Nagel, Dr. Lester B. Orfield and the late Dr. Ross Robertson. To these teachers go my enduring thanks and appreciation for their steadfast dedication to excellence.

I wish to acknowledge with particular pleasure the kind and consistent encouragement, faith and solid support of two particular friends and colleagues, Dr. John L. Garvey, Dean Emeritus of The Catholic University Law School, and Professor Raymond B. Marcin, during the continuing research and re-writing of this book.

Additional thanks for their generous support through the years are owing to: Rev. Jonathan Briggs Appleyard; Professor James F. Bailey, III; Gary H. Baise, Esquire; A. James Barnes, Esquire; Dean Emeritus Ralph C. Barnhart and Mary; Hon. Judge Betty Barteau; Hon. Judge John W. Beauchamp; Professor William W. Bishop, Jr.; Professor David R. Bookstaver; Robert E. Bostwick, Esquire; Barry E. Bretschneider, Esquire; Dr. Charles R. Britton and Jana; Gregory D. Buckley, Esquire, and Judy; Dean Thomas Buergenthal; Rev. Professor Harold A. Buetow; Leewell H. Carpenter; Michael F. Colligan, Esquire; Professor Alfred F. Conard; Nancy A. Cowgill; Professor Charles Davenport; Michael K. Diamond and Pamela; D. Jane Drennan, Esquire; James L. DeArmond and Aunt Mildred; Elizabeth C. Dymond; Professor Cyril A. Fox, Jr.; Dean Steven P. Frankino; Jonathan Z. Friedman, Esquire; Professor W. Samuel Furlow; Lane R. Gabler, Esquire; Professor Daniel J. Gifford and Ann; Rev. Professor David J. K. Granfield; Mabel Gray; the late Dr. Adolf Homburger; Dr. Edmund G. Howe, III; Dr. Kenneth B. Hughes and Mae; Josh T. Hulce, Esquire, and Crol; Allan E. Kaulbach, Esquire and Linda; Professor William A. Kaplin and Babs; Barbara A. King; Brenda J. King; Dr. Josephine Y. King; Professor Joseph Laufer and Lily; Agnes C. Lee; Professor Urban A. Lester; Henry R. Lord, III, Esquire; Professor A. Leo Levin and Doris; Gerald A. Malia, Esquire; Olive Manley; Robert R. McCallen, Esquire; the late Judge Donald R. Mote; Joseph H. Nixon; Dean Emeritus Charles B. Nutting; Henry P. Stetina, Esquire; Michael W. Owen, Esquire, and Sharon; Professor John Oxley-Oxland; Professor Robert L. Potter and Linda; Hon. Chief Judge Edward D. Re; Professor Robert I. Reis and Ellen; Dr. W. Taylor Reveley, II, and Marie; Professor Ralph J. Rohner; William D. Ruckelshaus, Esquire; Ambassador Rudolf E. Schoenfeld; Dr. David L. Scott and Kay; Professor Philip Shuchman and Hedvah; Nora C. Skilliter; Dr. Susan A. Skilliter; Hon. Chief Judge William E. Steckler; Professor Edward L. Symons, Jr.; Della A. Tillman; Professor Oscar Trellas; Lawrence R. Velvel, Esquire; Dr. G. Graham Waite and Mary Ellen; Faye Wakefield; Thom Wertenberger; Professor Welsh S. White; Dr. David A. Williams and Linda; the late Pearl A.

Williams; Professor George G. Winterton and Rosalind; Dr. Robert R. Wright, III, and Jackie, and Nicholas C. Yost, Esquire.

Finally, I must acknowledge the special kindness and gracious hospitality of my good and valued friend, William D. Mett, Esquire, who provided me with a "Pu'uhonua-o-Honaunau," or place of refuge, at his estate, Hale Enukai, In Kailua, Hawaii, for the completion of this book.

A number of students whom I have taught at The Catholic University of America Law School and elsewhere were of inestimable assistance; some by participation in seminars and courses; others as research assistants and still others by example as a well-spring for both maintaining and, in some cases, expanding my intellectual curiosity. My profound thanks, then, go to: Mitchell J. Birzon, Peter M. Brizon, Dr. Ann M. Brooks, Helen W. Clark, Roger J. Dodd, Vincent Gelardi (my Platonian sleuth), Patricia Kauffman, Richard F. McManus, Joseph A. Mecca, William H. Miller, Marc S. Moskowitz, Dr. Barry S. Reed, Stephen F. Riley, Stephen J. Stabler, Paul B. Taylor, Maureen A. Thompson and—especially—Garry A. Williams, Patricia A. O'Leary and Robert B. McLaughlin.

Helpful criticisms of parts of an earlier draft of the manuscript from which this book evolved were made by Professor R. Kent Greenawalt and the late Dr. Charles Frankel of the Columbia University School of Law together with Professor William T. Vukowich of the Georgetown University School of Law. Dr. Margery W. Shaw, Director of the Medical Genetics Center at the University of Texas Health Science Center at Houston, made significant comments as the book took final form. Of course, any errors of omission or commission are acknowledged to be my own.

The writings of Guido Calabresi Sterling Professor of Law at Yale, together with his personal spirit of humanism and Christian example have influenced me greatly in my work and provided me with marked opportunities for intellectual growth.

During the 1976-77 school year, I was a Fellow in the Law, Science and Medicine Program at the Yale University Law School and was enabled with financial support from the Commonwealth Fund to continue my researching and writing of this book. It is difficult to express the full measure of my gratitude to the Librarian and Associate Dean for Library Operations at the Law School, Arthur A. Charpentier, and Mr. Gene Coakley, the Circulation Supervisor, for their assistance. Never in the course of my career, have I had the opportunity to work with two such outstanding professionals. They lightened my research tasks immeasurably and provided me with treasure trove not only at the Sterling Law Library but at the Yale Medical School Library and the Yale Divinity School Library.

I must also acknowledge the kind assistance of Helen Y. Zimmerberg, Librarian at the Guyot Biological Library at Princeton University, Beverly Zink of the Speer Library at the Princeton Theological Seminary, and the staff of the Andover-Harvard Theological Library at the Harvard Divinity School.

The research facilities at the Kennedy Institute of Ethics at Georgetown University, Washington, D.C. have richly aided me over the years. To the late Dr. Andre E. Hellegers, the Director of the Institute, Fr. Richard A. McCormick, S.J., the Rose Kennedy Professor of Christian Ethics, Doris Goldstein, the Institute Librarian, and Patricia Schifferli, the most able Administrator of the Institute, I express my very genuine gratitude for their assistance and innumerable kindnesses.

I appreciate the permission of the Board of Editors of the Georgetown University Law Journal to draw upon and develop certain themes and ideas for this book which were part of an article that I published in volume sixty-four of the Journal. Permission to reproduce pertinent parts of the Cincinnati Law Review and the University of Pennsylvania Law Review in the Appendices of this book is also acknowledged.

# INTRODUCTION

Substantial scientific evidence exists which indicates man's genetic inheritance acts as a major influence not only upon his behavior but on his health.[1] In the United States, it is estimated that one out of every twenty babies is born with a discernible genetic deficiency;[2] of all chronic diseases, between twenty and twenty-five per cent are predominantly genetic in origin.[3] At least half of the hospital beds in America are occupied by patients whose incapacities are known to be of a genetic origin.[4] Because modern medicine can alleviate the symptoms of some genetic disease syndromes through sophisticated treatment, many who are afflicted with genetic disease and who would not have survived in the past now survive. Medicine is unable to do much by way of curing genetic defects,[5] however, and those afflicted with genetic diseases who are kept alive by modern technologies can reproduce and--thus--may increase the number of defective genes in the genetic profile of the human population.[6]

Considerable research into techniques for perfecting genetic engineering has been undertaken in an attempt to develop new, effective treatment for individuals with inherited diseases. Under the rubric of the "New Biology," scientists are both investigating and developing many procedures, including gene deletion surgery, splicing and transplantation, cloning, in vitro or test tube fertilization, embryo implantation, parthenogenesis, amniocentesis and experimentation with the scope and application of DNA.[8] Genetic engineering utilizes some of these procedures to reorganize human genes to produce varied, particular characteristics.[9]

In order to combat genetic disease, genetic engineering may--and frequently does in fact--rely upon eugenics, the science that deals with the improvement of heredity. Stated simply, a positive eugenics program seeks to develop superior qualities in man through the propagation of his superior genes;[10] with the positive eugeneticist seeking to produce a "new breed" of many with keener and more creative intelligence.[11] Contrariwise, a negative eugenics program attempts only to eliminate genetic weaknesses.[12] When seen in application, positive eugenics programs encourage the fit and proper individuals to reproduce while negative eugenics programs discourage the less fit and those with inheritable diseases from procreating.[13] Abortion is one way of implementing a program of negative eugenics after earlier measures of regulation have failed.[14]

Although genetic research has expanded in recent years, the motivating force behind the New Biology has been basic to human society. Since the time of Plato, people have attempted to improve the human race,[15] and research and experimentation in genetics have followed this tradition, seeking to relieve or totally alleviate human suffering that is genetically determined. These research efforts reflect the belief that society as a whole would prosper from methods to make humans more fit because it would be populated by the best physical specimens who, in turn, would beget superior offspring.[16] Some individuals have been motivated to undertake genetic experiments by the power of possible scientific creation and manipulation.[17] Adolf Hitler directed his nation toward the achievement of "Master Race" status, not only by attempting to rid Germany of what he considered negative genetic strains, but also by promoting a positive program of eugenics, Lebensborn or Fountain of Life, designed to create a German "super race" through selective breeding.[18] The current quest to manipulate the human genetic code results from both the traditional desires to increase scientific knowledge and rid the world of disease and infirmity, and the more modern objectives to limit or contain population growth and to provide children to infertile marriage partners that desire a family. The desire to remake man physically might also be attributed to what one scholar has termed the modern idea of social and political revolution.[19]

Despite the traditional origins of genetic research, genetic engineering presents serious legal, ethical, and social questions that remain unanswered. The misapplication of scientific knowledge may have adverse and sometimes ruinous effects.[20] The central problem posed by current efforts to manipulate the genetic code concerns man's reaction to self-knowledge about his future. As man begins to induce and manipulate life, he must also begin to question the limits of free will and of self-determination.[21] As man acquires these godlike powers, he must endeavor to execute them with a rational purpose and in a spirit of humanism;[22] he should seek to minimize human suffering.[23] Genetic engineering that contributes to the social good should be utilized fully.[24]

Under this ethical standard, society still must carefully define the social good. The quality of life that genetic manipulation promises must be evaluated and weighed against the sanctity of life. Genetic manipulation provides a perilous opportunity that either may threaten freedom or may enhance it, depending upon the balance struck between its use for individual need satisfaction and societal good.[25] Discussion of several possible genetic engineering programs will highlight the important legal and social choices that society must confront as these programs become possible to implement.

Notes--Introduction

1. A. Hellegers, Problems of Bioethics: A Report to the Sacred College on the Doctrine of Faith, at 43 (1974) (unpublished monograph in Kennedy Institute of Ethics Library, Georgetown University).

2. Gorney, "The New Biology and the Future of Man," 15 U.C.L.A.L. REV. 273, 291 (1968).

See generally, J. HALDANE, DAEDALUS OR SCIENCE AND THE FUTURE (1924).

3. Robinson, "Genetics and Society," 1971 UTAH L. REV. 487.

It has been estimated that 15 million Americans suffer today from the consequences of birth defects at varying degrees of severity. Interestingly, not all of these disorders are genetic: 20% are estimated as not involving a heritable component--but represent the effects of agents such as infection, drugs, and physical injury to the fetus. The remaining 80% or roughly 12 million Americans, carry the true genetic diseases due wholly or in part to defective genes or chromosomes. There are further estimates of the severity of genetic diseases which indicate that: 36% of all spontaneous abortions are caused by gross chromosomal defects (amounting to more than 100,000 per year in the United States); at least 40% of all infant mortality results from genetic factors; genetic defects are present in 4.8% to 5% of all live births; of the 3% of the population in the United States who are mentally retarded, about four-fifths are believed to carry a genetic component; about one-third of all patients admitted to hospital pediatric wards are there for genetic reasons; each of us carries between 5 and 8 recessive genes for serious genetic defects and, hence, stands a statistical chance of passing on a serious or lethal condition to each child; each married couple stands a 3% risk of having a genetically defective child; from 10% to 12% of the American population is estimated to have enzyme abnormalities or other genetically determined deficiencies which make them likely to react adversely to one or another of many commonly used drugs.

Largely because of incomplete data, no comprehensive assessment of the number, variety and distribution of genetic and partially genetic diseases has yet been made. Every year, the list of genetic disorders grows as "new" ones are recognized. The list of disorders, each of which is caused by a single defective gene (in single or double dosage), now numbers nearly 2,000 and is growing at a rate of 75 to 100 yearly. The list does not include diseases caused by multiple genes or by chromosomal defects.

Approximately $100 million is spent each year by the National Institute of Health on research related to genetic disease. WHAT ARE THE FACTS ABOUT GENETIC DISEASE? at 6, 9, 29, U.S. Dept. of H.E.W., Public Health Service, N.I.H., DHEW Public. No. (NIH) 75-370 (1975).

4. A. Hellegers, supra note 1, at 41.

5. See Waltz & Thigpen, "Genetic Screening and Counseling: The Legal and Ethical Issues," 68 NW.U.L. REV. 696, 698 (1973).

6. See Waltz & Thigpen, supra note 5, at 698.

See also, Schmeck, "Scientists Now Can Make Human Genetic Maps," N.Y. Times, Sept. 14, 1975, at 18, col. 1; "First Finding of Organisms Gene Makeup Recorded," N.Y. Times, Feb. 26, 1977, at 1, col. 2.

7. Kass, "The New Biology: What Price Relieving Man's Estate?," 174 SCIENCE 779, 780 (1971).

See also, C. HEINTZE, GENETIC ENGINEERING: MAN AND NATURE IN TRANSITION (1973).

8. See generally, "Symposium--Reflections on the New Biology," 15 U.C.L.A.L. REV. 267 (1968); TIME Mag., April 18, 1977, at 32.

Creative, scientific impulses for research and investigation should be neither systematized nor controlled, "Some part of life--perhaps the most important part--must be left to the spontaneous action of individual impulse, for where all is system, there will be mental and spiritual death." B. RUSSELL, THE IMPACT OF SCIENCE ON SOCIETY 89 (1952).

9. Waltz & Thigpen, supra note 5 at 696. Some researchers maintain that this type of biomedical engineering is not qualitatively different from toilet training, education, and moral teaching--asserting that all these acts of social engineering are designed to shape man in a particular way in present as well as future generations. See Kass, supra note 7, at 779; M. FRANKEL, GENETIC TECHNOLOGY: PROMISES AND PROBLEMS (1973).

See generally, Sullivan, "Genetic Decoders Plumbing Deepest Secrets of Life," N.Y. Times, June 20, 1977, at 1, col. 1; "Ingenious Chemical Techniques for 'Reading' the Messages of DNA," N.Y. Times June 20, 1977, at 54, col. 1.

DNA, deoxyribonucleic acid, is a molecule. It is the substance from which genes are made of and, thus, contains the chemical record in which hereditary information is encoded. New methods of gene experimentation allow DNA molecules to be cut and then recombined to scientific specifications. Using new cutting and splicing techniques, scientists can attach any genes to certain carrier genes. These linked genes, termed "recombinant DNA" are then conveyed inside a host cell, which virtually adopts the new gene as one of its own. Because of scientific familiarity with E. coli (Escherichia coli), a common benign bacterium found within the intestine of every living person, experimentation has largely focused to date on use of this with DNA.

The rather simple object of today's experiments with DNA is to transplant genes from animals or plants to bacteria. If foreign genes for insulin (possibly cancer) or blood clotting factors could be made to work--instead of merely copied--inside their bacterial hosts, these so-called "bugs" might then become efficient chemical factors producing substances to meet the needs of diabetic, cancer victims or hemophiliacs.

Recombinant DNA is quite easy to make. Any high school student having either a course in physics or chemistry could successfully undertake to conduct the necessary experiments needed to make it. Restrictions enzymes are available commercially and may be used to split DNA molecules from any source--man, cancer, viruses, bacteria, plants, insects. There is a very real concern that absent carefully observed safety precautions, experiments with cancer gene hosts, for example, could easily escape the confines of a laboratory and avoid capture--thus giving rise to the spread of "experimental cancer." Bennet & Gurin, "Science That Frightens Scientists--The Great Debate Over DNA," ATLANTIC MONTHLY, Feb. 1977, at 43; Cavalieri, "New Strains of Life--or Death," N.Y. Times Mag., Aug. 22, 1976, 8 at 9, 58.

See Fletcher, "Ethics and Recombinant DNA Research," 51 SO. CAL. L. REV. 1311 (1978) and Appendix G, infra.

Senator Ted Kennedy of Massachusetts introduced Senate Bill 1217 in 1977 designed to regulate federal research activities in DNA, but subsequently withdrew it before it was considered by the full Senate.

See also, Friedman, "Health Hazards Associated With Recombinant DNA Technology: Should Congress Impose Liability Without Fault?" 51 SO. CAL.L. REV. 1355 (1978); J. LEAR, RECOMBINANT DNA: THE UNTOLD STORY (1978).

10. See Vukowich, "The Dawning of the Brave New World--Legal, Ethical and Social Issues of Eugenics," 1971 U. ILL L. F. 189, 222.

See generally, Davis, "Prospects for Genetic Intervention in Man," 170 SCIENCE 1279 (1970); Hirschorn, "Human Genetics," 224 J.A.M.A. 597 (1973).

11. Frankel, "The Specter of Eugenics," 57 COMMENTARY 25, 30 (1974).

See also, P.B. & J.S. Medawar, "Revising the Facts of Life--A Framework for Modern Biology," HARPERS, Feb. 1977, at 41. "Eugenics is the political arm of genetics." Id. at 46.

12. Id.

To be justifiable, the acceptance or rejection of eugenic policies should be based upon more than one criterion. The following requisites should be a part of every eugenic program: scientific validity (e.g., a demonstration of sufficient genetic variation to allow for selection of the attribute in question); moral acceptability (i.e., a demonstration that the attributes chosen for selection are properly considered socially desirable); and ethical acceptability (i.e., a demonstration that the programs needed to institute a eugenic program do not compromise individual rights and liberties presently sanctioned by both public policy and the law). Lappe, "Why Shouldn't We Have a Eugenic Policy?," in GENETICS AND THE LAW 421 at 425 (A. Millunsky, G. Annas eds. 1976).

See Osborn, "Qualitative Aspects of Population Control: Eugenics and Euthenics," 25 LAW AND CONTEMP. PROBLEMS 406 (1960).

13. See Smith, "Through a Test Tube Darkly: Artificial Insemination and the Law," 67 MICH.L. REV. 127, 147 (1968).

14. See generally Green, "Genetic Technology: Law and Policy for the Brave New World," 48 IND. L. J. 559 (1973).

15. See T. DOBZHANSKY, MANKIND EVOLVING 245 (1962); M. HALLER, EUGENICS 3 (1963).

Today's biological revolution may be properly viewed as--in the words of Marx and Engels--the "transformation of quantity into quality." J. FLETCHER, THE ETHICS OF GENETIC CONTROL 3 (1974).

See generally, MAN'S FUTURE BIRTHRIGHTS (E. Carlson ed. 1973); STUDIES IN GENETICS--THE SELECTED PAPERS OF H.J. MULLER (1962); RECENT ADVANCES IN HUMAN GENETICS (L. Penrose ed. 1961); CLASSIC PAPERS IN GENETICS (J. Peters ed. 1959); GENETICS, MEDICINE AND MAN (H. Muller, C. Little, L. Snyder ed. 1947); D. HALAEY, JR., GENETIC REVOLUTION (1974).

16. See Dobzhansky, "Comments on Genetic Evolution," 90 DAEDALUS 451, 470-73 (1961); Kass, "New Beginnings in Life," in THE NEW GENTICS AND THE FUTURE OF MAN 18 (M. Hamilton ed. 1974).

17. Davis, "Ethical and Technical Aspects of Genetic Intervention," 285 NEW. ENG. J. MED.799, 800 (1971).

18. See TIME Mag., Oct. 28, 1974, at 33-36. Phase one of Lebensborn involved a program of selective sexual intercourse between carefully screened German women and SS men, who were regarded as racially and politically perfect German men. Once a successful pregnancy resulted, the prospective mothers were taken to one of twelve special maternity centers and given both special medical attention and lavish personal care. The second phase of Lebensborn involved the kidnapping of hundreds of thousands of children from Poland, Czechoslovakia, Yugoslavia, Norway and France in order to improve the "breeding stock" of the Fatherland. The children were taken to indoctrination centers and then put up for adoption to racially pure and ideologically trustworthy German families. Id.

Minthorn, "Nazi Babies: What has become of the 'Lebensborn' Children," Wash. Post, Sept. 10, 1978, at 4, cols. 1,2.

See also, "Biomedical Ethics and the Shadow of Nazism" (a conference), 6 THE HASTINGS CENTER REPORT 1 passim (Aug. 1976). See generally, G. STINE, BIOSOCIAL GENETICS (1977).

19. Frankel, supra note 11, at 33.

20. McElheny, "'Genetic Engineering,' an Advancing Stage," N.Y. Times, Dec. 15, 1975, at C37, col. 1.

It has been asserted that the ultimate goal of genetic engineering is not directed toward the amelioration of patients' ills--both prenatally or post natally--but rather the initiation of healthy people free of disease through the practice of medicine preconceptively. "It is a matter of directed and rational mutations, over against the accidental mutations now going on blindly in nature. It aims to control people's initial genetic design and constitution..." J. FLETCHER, THE ETHICS OF GENETIC CONTROL 56 (1974).

Dr. Martin J. Cline of the UCLA Medical School became the first scientist known to have used the new techniques of genetic engineering on human subjects. Dr. Cline attempted to treat two patients suffering from a fatal blood disease by placing normal genes in their defective bone marrow cells. See Jacobs, "Doctor Tried Gene Therapy on 2 Humans," Wash. Post, Oct. 8, 1980, at 1, col. 1.

See also, Lederberg, "Genetic Engineering, or the Amelioration of Genetic Defect," 34 THE PHAROS 9 (Jan. 1971); L. KARP, GENETIC ENGINEERING: THREAT OR PROMISE? (1976).

21. A. Hellegers, supra note 1 at 44.

See generally, Smith, "Theological Reflections and the New Biology," 48 IND. L. REV. 605 (1973).

22. See K. Vaux, BIOMEDICAL ETHICS 111 (1974); J. RAWLS, A THEORY OF JUSTICE, 284-293 (1971).

See generally, L. AUGENSTEIN, COME LET US PLAY GOD (1969); R. DUBOS, MAN ADAPTING (1965).

The essentials of the good life for which we all strive were listed by Bertrand Russell as being knowledge and love. B. RUSSELL, WHAT I BELIEVE (1925).

23. See Pauling, "Foreword to Symposium," supra note 8 at 270.

See generally, Fried, "The Value of Life," 82 HARV. L. REV. 1415 (1969).

"Human Life," said Alfred North Whitehead, "is a flash of occasional enjoyments lighting up a mass of pain and misery, a bagatelle of transient experience." ALFRED NORTH WHITEHEAD--THE INTERPRETATION OF SCIENCE 182 (A. Johnson ed. 1961).

24. Greenawalt, "A Contextual Approach to Disobedience," 70 COLUM. L. REV. 48, 50, (1970).

25. Shinn, "Genetic Decisions: A Case Study in Ethical Methods," 52 SOUNDINGS 229, 309 (Fall 1969).

See Callahan, "Symposium--The Law and the Biological Revolution," 10 COLUM. J. L. & SOC. PROBS. 47, 70 (1973); Canavan, "Genetics, Politics and The Image of Man," 4 THE HUMAN LIFE REV. 50 (1978).

A culture assumes a unique characterization by the way in which it shapes or--if you will-fabricates the image of man. E. Pellegrino, "Medicine and Philosophy--Some Notes on the Flirtations of Minerva and Aesculapius," Annual Oration, Soc. Health, Human Values, Washington, D.C., Nov. 8, 1973, at 5 (1974).

# CHAPTER 1

## CHANGING VALUES AND PERCEPTIONS

Values--even those regarded as fundamental--evolve and change with time. An evolutionary standard structures and determines what values are to remain stabilized and what ones are to be totally changed or merely modified.[1] Moral conflict within a society serves as a basis for determining morality itself--unless, for example, rigid and inflexible laws set rules of obedience the violation of which constitutes right and wrong.[2] As two astute commentators have so eloquently observed, "We do not live in the timeless days of a dog or sparrows. As we become aware of what we, as a society are doing, we bear responsibility for those allocations that will be made as well as for what has been done in our names."[3]

Life, as a human resource, should be preserved and developed in such a manner to achieve its fullest potential for total economic realization, maximization or productivity. Indeed, human life at whatever stage of development, is both a precious and a sacred resource. Its initial advancement or abrupt curtailment should be guided always by a spirit of humanism. As such, the quality of purposeful, humane living becomes an ultimate co-ordinate to total economic utility. By considering a case of life nearing its conclusion, one may develop a construct which--in turn--may be applied with relative ease to problems of genetic development.

Recently, a Probate Court in Massachusetts ruled--and was subsequently affirmed on appeal by the Supreme Judicial Court of the state--that Joseph Saikewicz, a severely retarded sixty-seven year old patient in the Belchertown State School, a facility of the Department of Mental Health, who suffered from acute myeloblastic monocytic leukemia, would not be required to undergo painful chemotherapy in order to sustain his life.[4] In taking this posture, the court made a landmark decision by endorsing quality of living as a valid legal standard in decision-making.

Mr. Saikewicz had an I.Q. of ten and a mental age of approximately two years and eight months. He was ambulatory, but was unable to communicate verbally--often left to making gestures and grunts in order to make his desires known. He had lived in state institutions since 1923 and been at Belchertown State School since 1928. When it was discovered in April, 1976, that Mr. Saikewicz had leukemia, the Superintendent of Belchertown and a staff attorney at the School petitioned the Probate Court for the appointment of a guardian ad litem whose duty it would be to make all necessary decisions relative to the care and treatment of the patient. The judge appointed the guardian who, in turn, in early May, 1976, filed his report which noted: the terminal nature of Saikewicz's illness and that while chemotherapy would normally be the course of medical treatment, it would cause the patient more significant adverse effects and discomfort and, thus, should not be administered. The inability of the patient/ward to understand the treatment and its nature coupled with the fear and pain he would suffer as a consequence, outweighed the very limited prospect of any positive benefit from such treatment--namely, the possibility of an uncertain but limited extension of life.[5] The court ruled that it was in the "best interests" of Mr. Saikewicz that chemotherapy be withheld.[6] Thus, by balancing the gravity of the harm of commencing the treatment against the utility of the good of prolonging and making more painful an admittedly wretched life by commencing the treatment, the court concluded--both wisely and humanely--to withhold treatment and thereby allow Saikewicz the right to die with a modicum of dignity.[7] Surely, the basic societal standard of justice is in no way affronted by such judicial action.[8]

What uneasiness that occurs here is the nature and the extent to which the medical profession removes from its direct control--or attempts to share with the courts--the standard of societal morality and consciousness vested in the profession, itself. "The nature, extent and duration of care by societal standards is the responsibility of a physician."[9] One can understand, perhaps, the need of the medical profession for a sharing of heretofore non-delegable decision-making processes with the judiciary--particularly in light of the growing (and justifiable) fears of malpractice litigation. But, at the most a sharing--not an abdication of responsibility should be encouraged. A doctor's expertise in medical matters can never be fully shared or totally understood by one untutored in the medical arts.[10]

Extraordinary measures undertaken to prolong a life of suffering is unjust to the individual in question and to the societal standard of decency and humanity. In part, perhaps a realization of this prompted the Saikewicz court to embrace a doctrine of some considerable precedence--with over one hundred and fifty years of application: the doctrine of substituted judgment.[11] Applied as such over the years in the area of the administration of estates of incompetent persons, the doctrine attempts to ascertain the incompetent person's actual interests and preferences. Consequently, the decision in cases of the nature of the instant one "should be that which would be made by the incompetent person, if the person were competent, but taking into account the present and future incompetency of the individual as one of the factors which would necessarily enter into the decision making process of the competent person,"[12] The Massachusetts Supreme Judicial Court concluded that had Saikewicz been competent, and considering all consequences of the acts of chemotherapy to be administered to him, he would have declined the therapy.[13] As shall be seen later, the doctrine of substituted judgment may also be employed most effectively in the area of human experimentation.

Uncertainties On The Spiral Staircase

The doctrine of substituted judgment, as stated in Saikewicz, when combined with the underlying purpose in cases of genetic improvement, namely to avoid suffering and promote the social and economic good, becomes a useful reference standard in the consideration of specific problems in human genetics. The central question to be raised, then, in the following illustration is: would the individual in question, if either presently alive or yet to be born, and being possessed of a rational reasoning process, choose a course of action in resolving the particular problem which would avoid for it personal suffering and/or, at the same time, promote the social and economic good?[14]

A clear illustration of how advances in modern surgery present correspondingly complex ethical problems in genetics is to be found in the treatment of spina bifida--or failure of the spinal canal to close. Unless neurosurgical treatment is followed, a goodly number of children born with this defect and related disorders die. Interestingly, even though "successfully" treated, a significant proportion of the children remain paralyzed, mentally retarded, or both--thus necessitating continual and often complex medical treatment. Although upon reaching adulthood, many do not procreate--thereby presenting no deleterious effect on the extended genetic composition of the population--the basic ethical problem remains: whether an initial rescue of such afflicted individuals should be undertaken, when knowledge is such that their quality of life will be seriously compromised over the years.

There has been a general consensus as of recent years, to withhold heroic or surgical treatment in order to save such a person's life. These decisions are normally made by teams of neurosurgeons and pediatricians in consultation with the families of the patients and based upon empirical criteria which shows the patient will be seriously retarded or paralyzed.[15] The decision making process here presents clearly the uses of a balancing test by weighing the gravity of the harm and suffering to the patient in initially undertaking heroic treatment to sustain its life against the utility of the social good in allowing it to die an early death; obviously a humane application of the substituted judgment doctrine.[16]

Retinoblastoma, or tumor of the retina, is no longer fatal. With timely operation or radiation treatment, most of the lives of the victims of this genetic defect can be saved. Tragically, the cure inevitably results in either blindness in one or, more usually, both eyes, the geneticist will have little compunction in strongly advising the cured persons against having children.[17]

Assume that a man who, as a child, had retinoblastoma had the condition operated on and was one of the seventy per cent who in turn successfully survived surgery, although both his eyes were removed. Assume, further, that this man subsequently married and fathered two children--each of who developed retinoblastoma and that they too were operated successfully, with one child suffering the loss of one eye and the other, both eyes. What actions should be taken in order to prevent further transmission of the retinoblastoma gene: refuse to perform surgery on the two children in question and let nature take its course; operate on their eyes--but also sterilize them; or, operate without sterilization, thereby leaving it to the afflicted carriers to eventually decide whether they, in turn, wish to have children with the disease?[18] Should any greater emphasis be placed upon the need for sterilization if the father and his sons were unemployed, not educated and living on public assistance? Does the social or economic good "demand" greater consideration in resolutions of cases of this nature?

Another type of hereditary blindness termed, juvenile cataract or cataract of the young, is pertinent to the consideration. Here, the disease affects the lens of the eye and typically results in blindness. In most families, the disease is inherited as a dominant abnormality--without about half the children of affected persons being affected again. Cataract is, however, operable. If the operation is successful, the patients can see well with appropriate glasses and become adjusted members of society. Should efforts be undertaken to prevent or dissuade carriers from having children--approximately half of whom will have to submit to a surgical intervention and wear glasses? In some countries, juvenile cataract is listed as being among those hereditary defects for which sterilization is recommended.[19]

Huntington's chorea is unquestionably a dominant gene with one hundred per cent penetrance.[20] The average age of onset of the defect is about thirty-five years, but it may not become apparent in many individuals until the age of fifty or later. Those afflicted by it are characterized by disordered and involuntary movements and by progressive mental deterioration.[21] Although ultimately fatal, those who possess the gene can produce a family before the visible onset of the disease occurs.[22] The offspring of a parent with a dominant gene have a fifty-fifty chance of inheriting the gene from that parent.[23] Until recently, no treatment was available, but now drugs are being used with some success in controlling the disease.[24] Given all of this information, would it be advisable and in the best interests of society, future generations, and of the individual carrier to advocate sterilizing all such identifiable carriers of Huntington's chorea genes before they reach maturity?

Assume a middle-income couple has four sons, all hemophiliac. The special burdens associated with the care of the children have made the family virtually dependent upon their community for not only financial assistance but for donations of several thousand pints of blood needed for transfusions. The children are under the age of ten years. Once the first hemophiliac son was born, it was certain there was a one-in-two risk appearing in any succeeding son. Should pressure have been properly exerted on the mother to refrain voluntarily from having any more children, or, coercively to be sterilized? Should, in the alternative, the four hemophiliac sons, themselves, be sterilized?[25]

In all of the illustrations, one can see the dynamics of sociocultural, economic, genetic, philosophical, legal and political vectors of influence in focus. The social costs of maintaining genetic variation of the types seen previously is arguably so great and the personal suffering so intense that artificial selection against them is ethically, as well as economically, the wisest, most humane and most equitable solution.[26]

Sterilization of individuals suffering from certain severe dominant disabilities will--without question--reduce the frequency of the responsible genes in future generations, thereby validating a program of negative eugenics. Yet, the frequency of the most harmful dominant genes will hardly be affected by sterilization--since these genes, by killing or sterilizing their carriers, are self-eliminating, thus resulting in more new occurrences of such disease being not due to inheritance, but rather to mutation.[27]

Even though these problem areas discussed are genetic in origin, they also present a social etiology of one degree or other. Indeed, the argument for either restraining or limiting reproduction because of presumed unfitness and genetic deficiencies is strengthened when the very real social and psychological handicaps that might be suffered by their offspring are evaluated. Today, more and more emphasis is placed on the environment, itself, as a tool for positive or--as the case may be--negative influence and conditioning. And so, the proper question to be posited is: what chance do children have to be "normal" if reared by parents who are psychotic, mentally retarded, habitually criminal or afflicted with serious physical ailments or deformities?[28] Efforts designed at preventing the deterioration of human quality, whatever the causes, made under the rubric of negative eugenics will, perhaps, in the short run, be more effective and stand a greated likelihood of wider public adoption than moves to sustain positive eugenics.[29]

As biology becomes more and more a branch of politics--what with experimentations in genetic engineering, especially DNA and human experimentation in the prolongation of life--[30]individual freedom is diminished as government influence predominates.[31] Painful decisions of an individual and societal nature, cannot be postponed indefinitely. Utilization of scarce resources for sustaining life processes, as an example, inevitably forces choices to be made of an oftentimes tragic nature.[32]

Notes--Chapter 1

1. G. CALABRESI, P. BOBBITT, TRAGIC CHOICES 197 (1978).

2. Id. at 198.

3. Id. at 199.

4. Superintendent of Belchertown State School, et al v. Joseph Saikewicz. 379 N.E. 2d 417 (1977).

5. Id. at 419.

6. The Probate Court Judge then entered findings of fact and an order that in essence agreed with the recommendation of the guardian ad litem. Two questions were subsequently certified by the judge to the Supreme Court: 1.) Does the Probate Court under its general or any special jurisdiction have the authority to order, in circumstances it deems appropriate, the withholding of medical treatment from a person even though such withholding of treatment might contribute to a shortening of the life of such person? 2.) On the facts reported in this case, is the Court correct in ordering that no treatment be administered to said Joseph Saikewicz now or at any time for his condition of actual myeloblastic monocytic leukemia except by further order of the court? Id. at n. 2.

7. Id. at 432.

The Probate Court considered six factors weighing against chemotherapy--with the sixth being, "the quality of life possible for him even if the treatment does not bring about remission." The Supreme Court chose instead of reading the lower court judge's "formulation in a manner that demeans the value of life of one who is mentally retarded, the vague, and perhaps ill-chosen, term 'quality of life' should be understood as a reference to the continuing state of pain and disorientation precipitated by the chemotherapy treatment." Viewing it within this context, the court affirmed the decision to withhold treatment. Although the Supreme Court endeavored to qualify the concept of quality of life, it is inescapable that its attempted qualification in actuality was but a definition of non-quality, if you will.

See Baron, "Assuring Detached But Passionate Investigation and Decision: The Role of Guardians Ad Litem in Saikewicz-type Cases," 4 AM J. L. & MED. 111 (1978).

See generally, A. STONE, MENTAL HEALTH AND THE LAW: A SYSTEM IN TRANSITION (1976); R. VEATCH, DEATH, DYING AND THE BIOLOGICAL REVOLUTION (1976).

8. An affront to a community's basic sense or feeling of justice, fairness or sanctity of life occurs when certain courses of action that are undertaken are perceived widely as unjust or unfair. G. CALABRESI, THE COST OF ACCIDENTS 294 (1970).

Society should allocate its resources in such a way as to recognize and give validity to the importance of avoiding death attended by great suffering. Fried, "The Value of Life," 82 HARV. L. REV. 1415, 1435-1437 (1969).

9. In re Quinlan, 70 N. J. 10, 44, 355 A.2d 647, 665 (1976).

See also, P. RAMSEY, ETHICS AT THE EDGE OF LIFE, Ch. 7 (1978).

10. "...Questions of life and death seem to us to require the process of detached but passionate investigation and decision that forms the ideal on which the judicial branch of government was created." Supra note 4 at 435.

11. Id. at 431.

12. Id.

Using the same substitute judgment test, a Franklin County Probate Court Judge in Massachusetts ruled that Earle Spring, a senile 78 year old man, would rather die than prolong his life with kidney dialysis treatment which meant he would spend the major part of it sleeping. TIME Mag., Feb. 11, 1980 at 95. See Barbash, "Judge Orders Life Support in 'Death-With-Dignity' Case," Wash. Post, Jan. 24, 1980, at A2, Col. 1; "Subject of 'Death-With-Dignity Hassle Dies in Nursing Home," Indianapolis Star, April 7, 1980, at 3, col. 5.

See generally, McCormick, "The Quality of Life, The Sanctity of Life," THE HASTINGS CENTER REPORT 30 (Feb. 1978).

13. For an excellent analysis of the Karen Quinlan case, compared with Saikewicz, see Brant, "Beyond Quinlan and Saikewicz: Developing Legal Standards for Decisions Not to Treat Terminally Ill Patients," 21 BOSTON BAR J. 5 (1977).

See also, Curran, "Law-Medicine Notes: The Saikewicz Decision," 298 NEW ENG. L. MED. 499 (1978); RAMSEY, supra note 9, ch. 8.

See also Eichner v. Dillon, 426 N.Y.S. 2d 517 (App. Div. 1980) where it was held that a guardian of an 83 year old patient (Catholic priest) who was terminally ill and comatose had a right--under certain specific circumstances--to terminate extraordinary measures that keep the patient alive. See Ch. 3, note 128, infra, for a fuller discussion of this case.

Dr. Sissela Bok presents a most enlightening analysis of the consequences of lying to terminally ill patients about their actual condition in her brilliant book, LYING: MORAL CHOICE IN PUBLIC AND PRIVATE LIFE, Ch. 15 (1978).

14. A series of case studies, compiled over a six year period and rooted in real experiences concerning problems in medical ethics serves as interesting reading. See R. VEATCH, CASE STUDIES IN MEDICAL ETHICS (1977).

15. Motulsky, "Brave New World?" in HUMAN GENETICS: READINGS ON THE IMPLICATIONS OF GENETIC ENGINEERING 280, 287-288 (1975).

16 A physician in Sheffield, England, Dr. John Lorber, has gained international prominence for his work with newborns having spina bifida. For some of the babies, they are heavily sedated and let to die peacefully--without period feedings. Others undergo a series of operations over many years. Although they lead impaired lives, Dr. Lorber judges them to be worth living. P. RAMSEY, ETHICS AT THE EDGE OF LIFE 193 (1978).

17. C. AUERBACH, THE SCIENCE OF GENETICS 95 (Rev. ed. 1969).

18. A. SCHEINFELD, YOUR HEREDITY AND ENVIRONMENT 709, 710 (1965).

19. Supra note 17 at 95.

20. Supra note 18 at 78; G. HARDIN, NATURE AND MAN'S FATE 153 (1959).

A dominant gene is one which, singly, passed on by just one parent, can produce a given defect in a child. The presence of a dominant condition in either parent implies a fifty-fifty risk of its reappearing in any offspring.

A recessive gene is one which must be coupled with a matching gene--that is, there must be a pair of such genes, one coming from each parent--in order for the effects to show in a child. If both parents are afflicted by a simple recessive condition (and each therefore carries two of the genes), every child of theirs will receive the two recessive wayward genes and can develop the disease or defect.

A sex-linked recessive gene is one found in the X chromosome, and when wayward in its action, it strikes most often against males. The reason is that a male has only one X, so if any of its genes are bad, the effects will show in him. But in a female who has two X's, any wayward gene in one X will in most cases have its effects blocked by a corresponding normal gene in her other X.

Chromosomes are highly sustainable thread-like structures in cell nuclei. The chromosomes have the ability to duplicate themselves identically.

Whether wayward genes are dominant, recessive or sex-linked, they may be qualified or governed in their action by environmental factors or by certain other genes. This accounts for the fact that wayward genes sometimes appear to be capricious and unpredictable in their behavior. In order to describe the degree of regularity with which a gene works, geneticists use the term "penetrance." If a gene--a single dominant, or a recessive in a double dose--manifests its effects in all individuals who carry it, the gene is said to have "100 per cent penetrance"; if its effects show in 75 per cent of the cases, "75 per cent penetrance"; and so forth. Thus, a single dominant gene with 100 per cent penetrance may cause the same defect (or quirk) to appear in successive descendants of a family, generation after generation. Yet, if a gene has only partial penetrance--75 per cent, 50 per cent, or whatever--its effects may be discontinuous and may not show through in all persons who inherit it. In other words, the trait (but not the gene) will sometimes skip a generation. G. RODERICK, MAN AND HEREDITY 225 (1968); A. SCHEINFELD, YOUR HEREDITY AND ENVIRONMENT 189 (1965).

21. H. PAPAZIAN, MODERN GENETICS 77 (1967).

22. G. RODERICK, MAN AND HEREDITY 86 (1968).

23. Id. at 208.

See Table 1, "Risks of a Repeat Defective Child."

Table 1

Risks of a Repeat Defective Child

Referring Mainly to Congenital Abnormalities and Mental Defects Apparent in Early Childhood, Whether Hereditary, Environmental, or a Combination of Both*

| If a couple has had one child with this defect: | The chances of having another child with the same defect are: |
|---|---|
| Albinism | 1 in 4 |
| Cleft palate (alone) | 1 in 7 |
| Clubfoot | 1 in 30 |
| Dwarfism, achondroplastic (when hereditary) | Up to 1 in 2 |
| Extra fingers and/or toes | 1 in 2 |
| Hands or feet malformed (hereditary types only) | Up to 1 in 2 |
| Harelip (with or without cleft palate) | 1 in 7 |
| Heart, malformed | 1 in 50 |
| Idiocy | |
|    Mongoloid | |
|    (a) If mother has normal chromosomes: (Chances slightly increased over average for mother of same age. But still low risk) | 1 in 50 |
|    (b) If mother has a chromosomal translocation (an extra No. 21 attached to another chromosome): | Up to 1 in 3 |
|    Amaurotic | 1 in 4 |
|    Phenylpyruvic amentia | 1 in 4 |
|    Cretinism | Uncertain |
| Intestines, pyloric stenosis | 1 in 17 |
| Kidney, polycystic congenital | 1 in 4 |
| Spina bifida | 1 in 25 |

As a rule, for any simple recessive condition the repeat risk is 1 in 4; for a simple dominant, 1 in 2. But in no case should any determinative judgment be made regarding these matters without consulting a qualified doctor or a medical geneticist. *A. SCHEINFELD, YOUR HEREDITY AND ENVIRONMENT 657 (1965).

---

    Cystic fibrosis, for example, presents yet a different challenge for the genetic counselor. This genetic condition affects about one birth in each two thousand, the frequence of heterozygous persons in the population (one in twenty-two), and signifies that only one in four hundred eighty-four matings--random with respect to the possession of the gene--will, in turn, run the high genetic risk (one in four) of producing an affected child. This frequency is regarded as not enough to require all marrying persons be screened in order to determine the carriers among them. Perhaps all first cousins who marry whould be tested since the rarer the condition, the greated the probability of consanguinity among the parents. Glass, "Human Heredity and Ethical Problems," 15 PERSPECTIVES IN BIO. & MED. 237, 240 (Winter, 1972).

24. Supra, note 22 at 86.

25. A. SCHEINFELD, YOUR HEREDITY AND ENVIRONMENT 710 (1965).

26. Id. at 711 (quoting Dobzhansky).

27. C. AUERBACH, THE SCIENCE OF GENETICS 87 (rev. ed. 1969).

    For recessive abnormalities the situation is entirely different. From the point of view of the individual family, sterilization may still be desirable--as when for example, parents of one child with amaurotic idiocy wish to avoid having a second one like it. Yet, from the standpoint of the greater population, sterilization against recessive abnormalities is not considered very effective by some authorities; principally because recessive genes, by their very nature may be carried hidden in what appears to be outward "normal" individuals. These wayward genes may be recessives in the single state--or they may be "incomplete" or "qualified" dominants--or they may be other genes largely inactive in the carriers but which could act decisively in their offspring. The mere thought of a wholesale curbing of the reproduction of these carriers would be ludicrous; especially when it is appreciated--as has been observed previously--that each person carries some unrevealed wayward genes (estimated to be about eight). Perhaps the best approach at containment is to discourage matings of unrelated individuals known to carry matching wayward genes, of cousins with above average chances of carrying matching recessives, and a heightened use of genetic screening at early pre-school ages. Id. at 87, 88; Supra, note 5 at 682, 683.

    So it is seen, then, the recessive, or other multiple-gene, conditions present the principal eugenic predicament. These are largely responsible for not only the greater part of the common serious hereditary mental diseases and defects--but the very worst of the inherited neurological, ocular, muscular, skeletal and blood afflictions. Consider genetically produced imbeciles and idiots. These defectives pass but a very small percentage of their own genes due in large part because most are sterile or are institutionalized. The morons provide the real problem. They are individuals with an I.Q. of from 50 to 70. If in the United States one assumes that there are 600,000 low-grade morons produced by recessive genes--simple or multiple--a further assumption can be made: namely, that all valid figures point to an estimate that there would be at least ten times as many, or 6 million normal or close to normal persons, each carrying a hidden one of these genes. Thus, matings of these persons would continue to replenish the supply of morons. If not a single one of the existing morons had offspring, their counterparts in the next generation might possibly be reduced by no more than ten per cent. At this rate it could well take ten generations to cut the incidence of these defectives to one fourth their present number and twenty-two generations to lower it to one tenth (not allowing for any possible new mutations). Supra, note 25 at 681.

28. Supra, note 25 at 712.

29. Id. at 709.

30. Prof. Lawrence H. Tribe, Rose F. Kennedy Lecture, "Clones, Cyborgs and Chimeras," April 5, 1978, Georgetown University, Washington, D. C.

    See also, Dobzhansky, "On Genetics and Politics" in HEREDITY AND SOCIETY 30 (A. Baer ed. 1973).

31. Morrison, "Misgivings about Life-Extending Technologies," 107 DAEDALUS 211 (1978); Swazey, "Protecting the 'Animal of Necessity:' Limits to Inquiry in Clinical Investigation," 107 DAEDALUS 129 (1978).

32. See G. CALABRESI, P. BOBBITT, TRAGIC CHOICES 178 passim for an illuminating analysis of the allocation of renal dialysis units, as a paradigmatic example of the utilization of scarce economic resources

which inevitably require a tragic choice of uses between limited users.

The extent or duration of human resources was tested again by the Massachusetts courts in two cases: In re Dinnerstein, 380 N.E. 2d 134 (Mass. App. Ct. 1978) and In re Spring, 405 N.E. 2d 115 (Mass. Sup. Ct. 1980). The High Court, ruling in Spring, re-affirmed its decision in Saikewicz although--admittedly--when the Dinnerstein rule is factored into the consideration, some confusion results in determining the ratio decidendi of the case. Saikewicz was viewed as an incompetent with a choice to make. Because he had a chance, although slight, of prolonging life with chemotherapy, it was determined that this was a choice concerning the form of treatment. Dinnerstein's prognosis--to the contrary-- was hopeless. His disease was progressive, unremitting, and no medical breakthrough was anticipated. Thus, the doctrine of substituted judgment although integral in Saikewicz, was held inapplicable to Dinnerstein because there was no treatment of any nature available that could improve Dinnerstein's condition.

The case holding of In re Spring states that an incompetent's right to treatment is the basis for the distinction between Saikewicz and Dinnerstein. Spring concerned the decision regarding continuance of hemodialysis treatment for an incompetent. Spring's condition was similar to Dinnerstein's in that it was terminal in nature--irreversible--and the medical care required custodial. Yet the same Court which decided Saikewicz considered Spring--based on its facts--to be a RIGHT to treatment case and therefore concluded that the decision-making responsibility should remain with the Probate Court. The Spring Court distinguished between a patient for whom treatment is useless, and one for whom further treatment is useless, and one for whom further treatment is a viable alternative by condiseration of such factors as: the extent of impairment of the patient's mental faculties, the prognosis without (or) with the proposed treatment, the complexity, risk, and novelty, its possible side effects, the urgency of decision, the consent of the patient, spouse or guardian, the good faith of those who participate in the decision and the clarity of professional opinion as to what is sound medical practice.

CHAPTER 2

IMPLEMENTING A NEGATIVE EUGENICS PROGRAM

In seeking to eliminate genetic weaknesses from the society, a negative eugenics program necessarily requires some process to determine genetic composition. Genetic screening and counseling accomplish this objective by identifying carriers of genetic diseases and advising couples whether reproduction is biologically desirable.[1] That screening and counseling may occur at both preconceptual and postconceptual stages.[2] A simple preconceptual screening procedure consists of withdrawing and analyzing a blood sample to determine if an individual possesses any recessive traits for a genetic disease.[3] Postconceptual screening and counseling procedures are more medically complicated and also pose more complex legal issues.

Amniocentesis

A recently developed postconceptual screening procedure, amniocentesis, has emerged as a principal element of negative eugenic programming. The procedure consists of inserting a needle through the abdominal wall of a pregnant woman into the amniotic sac containing the fetus, withdrawing a sample of the sac fluid, and analyzing it.[4] Since the sac contains cells from different parts of the fetus, analysis of this sample reveals the sex of the fetus and also whether it will be affected with certain genetic disabilities.[5] By permitting a physician to predict accurately the presence of certain genetic defects, amniocentesis significantly advances standard genetic counseling procedures that must rely on probabilities.[6]

If amniocentesis reveals a genetically defective fetus, the parents face the difficult choice of whether to abort the fetus. A couple informed of a genetically defective fetus may decide for religious, personal, or ethical reasons that they want to guarantee the birth of the life they created and therefore allow the pregnancy to continue. Such a choice raises the issue whether the child could bring a tort action against his parents for wrongful life. Under current law such a claim likely would fail.[7] However, " i[t]seems likely that Roe v. Wade, coupled with a shift in social attitudes concerning abortion and the sanctity of life will cause a reversal in the courts' negative attitude toward wrongful life suits."[8]

Whether a child may bring a wrongful life action against his parents involves a balancing of several interests. The parents' attitudes on the right to life and their declarations of parental love must be considered, although they may not justify a life of hardship and physical pain for a malformed child.[9] A strong enough social policy may exist to justify the imposition of a parental duty to prevent defective birth by contraception or abortion and to impose liability for the failure to do so.[10] A court might find alternatively that parents have a duty to avoid the affirmative act of birth that may further harm a defective fetus and subject it to a life of suffering outside the womb.[11] The societal interest in healthy offspring may exist not only to protect the prospective children but also to assure a more healthy society. Imposing a duty on the parents to abort a genetically malformed fetus after science identifies the malformation is consistent with the social ethic that seeks to minimize human suffering;[12] it would achieve the greatest good for society as a whole, for the prospective parents, and even for the fetus.[13]

Genetic Screening And Counseling Programs

Some of those concerned with negative eugenics currently have emphasized the need for the wide application of traditional screening procedures to identify the carriers of certain diseases.[14] Certain leaders of Jewish communities, for example, encourage citizens of their communities to participate in screening to identify carriers of the Tay Sachs recessive gene, which can cause a fatal debilitating illness.[15] Federal legislation permits the use of public funds to establish voluntary, genetic screening and counseling programs for carriers of sickle cell anemia;[16] some state legislatures have gone further to require genetic screening of school age children for that trait.[17] New York provides for premarital testing to identify carriers of the same defective gene.[18] Genetic screening programs also may include provisions for counseling.[19] Unfortunately, counseling efforts to date have been sporadic and ineffective.[20] If genetic screening programs are to have any significant impact, more effective counseling techniques must be devised and implemented.[21]

Public acceptance of mandatory genetic screening programs should not be impossible to achieve. Premarital genetic screening would be an easy addition to state statutes that presently require premarital testing for maternal rebella titre (although not itself considered to be a genetic defect), blood group, and Rh status.[22] One scholar asserts that statutes requiring genetic screening for the population at large would be a simple and readily acceptable extension of present laws requiring vaccinations and chest X-rays for school children.[23] Moreover, societal problems such as population control, the cost of supporting the handicapped, and the general welfare of the population favor the trend toward mandatory genetic screening.[24]

Some legal scholars maintain that compulsory genetic screening programs may be unconstitutional.[25] They assert that the taking of a child's blood sample would constitute a physical invasion of the body in violation of the fourth amendment and that a compulsory counseling program would interfere with the fundamental rights to marry and procreate.[26] These critics also contend that a less intrusive voluntary program, together with extensive dissemination of educational material, could accomplish the same objectives.[27] Although genetic screening involves a minor intrusion into an individual's body and may involve a "search" within the meaning of the fourth amendment, the search is not unreasonable and prohibited if executed in a proper manner and justified by a legitimate state interest.[28] Similarly, assuming arguendo that mere screening and counseling interfere with the right to procreate, such interference may be justified by a compelling state interest. The state's interest in improving the quality of a population's genetic pool in order to minimize suffering, to reduce the number of economically dependent persons, and possibly, to save mankind from extinction arguably justifies the infringement of individuals' civil liberties.[29]

Unfortunately, voluntary programs have little value in achieving the purposes for which they are structured. People are too preoccupied with the daily vicissitudes of life to be concerned with prospective occurrences of genetic possibilities. Therefore, although a voluntary program concededly is less intrusive, the only way to achieve positive, enduring results is to implement some form of mandatory genetic screening program.[30]

Restrictions On Marriage

An even more effective means of preventing the birth of genetically defective persons is prohibiting marriage between carriers of the same genetic defect. Both constitutional and social objections have raised to such a restriction on marriage.[31] Existing laws prohibiting marriage for eugenic reasons and proposals to restrict marriage between carriers of the same genetic defect are attacked as being excessively broad, and the suggestion is made that only procreation needs to be regulated to ensure both eugenic preservation and responsible parents.[32]

Since procreation traditionally is set within the marriage framework, however, establishing restrictions on marriage is the most practical mechanism for implementing a negative eugenics program. Moreover, married couples prohibited from procreation nonetheless might have children accidentally or intentionally.[33] Whether a state's pursuit of the public's health and welfare would justify an abridgment of the fundamental right of marriage between carriers of the same genetic defect is doubtful. Such restrictions also might well prove ineffective in the contemporary atmosphere that is increasingly tolerant of free love and common law marriage. Thus, it is unlikely that restrictions on marriage would prove to be an acceptable method of eugenic control.

Restrictions On Reproduction

Modern cases support the proposition that marital and procreative decisions fall within a constitutionally protected zone of privacy.[34] As long ago as 1941, the Supreme Court declared that man possesses the basic civil right to have offspring.[35] More recently, the Court has held that the choice of whether to give birth is within a constitutionally protected zone of privacy.[36] These broad pronouncements do not force the conclusion, however, that all restrictions on reproduction are per se unconstitutional. If a state may prevent a person from marrying more than one person at a time, should it not have the same power to prevent a person from having more than one or two children? The right to procreate may not include a right to breed without some restrictions.[38] Societal interests may be sufficiently powerful to justify at least some regulation for limitations on reproduction.[39]

Some legal precedents do uphold the constitutionality of eugenic sterilization. In Buck v. Bell[40] the Supreme Court upheld a Virginia statute providing for sterilization of inmates committed to state supported institutions who were found to have a hereditary form of insanity or imbecility.[41] Nearly half the states now have some form of compulsory sterilization legislation,[42] and courts typically uphold such statutes.[43]

The extension of Buck to sterilization of carriers of recessive defective genes cannot be accomplished without difficulty. Since its decision in that case, the Court increasingly has recognized the right to marry and have children as a fundamental right.[44] Existence of this right requires a state to show a compelling interest to justify the abridgment of this right.[45] Several factors seem to indicate that the state interest is not as compelling with regard to sterilization of carriers of defective genes as it is with regard to mental incompetents. A mental incompetent may well be unable to be an adequate parent, and the burden of care therefore would fall upon the state.[46] Moreover, the sterilization of mental incompetents in institutions can be said to benefit them directly in that it "enable[s] those who otherwise must be kept confined to be returned to the world. . . ."[47] The Court seemed to have assumed in Buck, however, that there is a strong likelihood that the child would in fact inherit the defect;[48] the child of two heterozygous individuals has only a one in four chance of exhibiting that defective trait.[49]

The distinguishing features of Buck v. Bell do not indicate that the state cannot offer compelling justifications to warrant mandatory restriction on reproduction. Such justification can be found in society's interest in the reduction of human suffering, in safeguarding the health and welfare of its citizens, in the allocation of economic resources, and in population control.[50] In Buck v. Bell Justice Holmes stressed that "it would be better for all the world . . . if society can prevent those who are manifestly unfit from continuing their kind."[51] Perhaps world conditions have become so complex and resources so valuable that society now has a compelling interest in restricting reproduction by those, who although not "manifestly unfit" themselves, perpetuate human suffering by giving birth to genetically defective offspring.

Notes--Chapter 2

1. Davis, "Ethical and Technical Aspects of Genetic Intervention," 285 NEW ENG. J. MED. 799 (1971).

2. Waltz & Thigpen, "Genetic Screening and Counseling: The Legal and Ethical Issues," 68 NW.U.L. REV. 696, 700 (1973).

3. Id.

4. Robinson, "Genetics and Society," 1971 UTAH L. REV. 487 at 488 n. 24.

5. Id. at 48

6. Id.

   The University of Colorado Medical Center permits amniocentesis only if there are signs of chromosomal abnormalities, bio-chemical diseases or "inborn errors of metabolism," or sex-linked recessive conditions such as muscular dystrophy and hemophilia. Id.

   Pregnant women over thirty-five years of age have a one and one-half per cent risk of having a child with abnormal chromosomes. Id. at 489. Down's syndrome, mongolism, is one of the most distressing examples of chromosomal abnormality. Professor Robinson reports:

   "The 10% of pregnant women who are thirty-five years of age or older are responsible for the birth of 50% of the children with Down's syndrome. Roughly 7,000 of these retarded babies are born each year in the United States and the majority of them eventually end their days in institutions for the retarded. The financial cost to our society has been estimated to be several billions of dollars per year." Id; cf. infra, note 51.

   Before amniocentesis will be adminstered, some physicians require that the risk that the fetus will be genetically defective must be higher than one or two per cent. Accordingly, a further decrease in the incidence of risk from amniocentesis compared with the incidence of genetic defects waiting to be discovered antenatally may be taken by reliance upon a maximum benefits ethic to be the way to determine what part of the population should be screened. It has been suggested that "incidence seems to be an odd point to light upon in balancing the two risks, since surely the depth of the genetic deformity and the depth of damage that might be caused by amniocentesis are at least equally important in a risks-costs-benefits analysis. Physicians should not avoid facing the quantity of the quality problem by concentrating on the quantifiable and comparable incidence, while neglecting the gravity of the afflictions to be treated or induced." Ramsey, "Screening: An Ethicists View," in ETHICAL ISSUES IN HUMAN GENETICS 147 at 154 (B. Hilton, D. Callahan, M. Harris, P. Condliffe, B. Berkley eds.1973).

7. Cf. Pinkney v. Pinkney, 198 So.2d 52, 54 (Fla. App. 1967) (court refuses to recognize tort of wrongful life for plaintiff, bastard, against father); Zepeda v. Zepeda, 41 Ill. App. 2d 240, 259, 190 N.E. 2d 849, 858 (1963), cert. denied, 379 U.S. 945 (1964) (same). Courts similarly have rejected a wrongful life action by a genetically defective child against doctors or hospitals for failure to inform the parents of genetic defects of which they knew or should have known. See Gleitman v. Cosgrove, 49 N.J. 22, 28, 227 A.2d 689, 692 (1967); Stewart v. Long Island College Hosp., 58 Misc. 2d 432, 296 N.Y.S. 2d

41, 46-47 (1968); modified, 35 App. Div. 2d 531, 313 N.Y.S.2d 502 (1970), aff'd, 30 N.Y.2d 695, 283 N.E.2d 616, 322 N.Y.S.2d 640 (1972). See generally Note, "A Cause of Action for 'Wrongful Life'," 55 MINN. L. REV. 58 (1970); Annot., 22 A.L.R.3d 1441 (1968).

8. Friedman, "Legal Implications of Amniocentesis," 123 U. PA. L. REV. 92, 155 (1974).

9. Some parents may procreate in order to gratify their own egos; the birth of a child gives them a "second chance" to catch glimpses of themselves through the child and to live again (through the child). See McGrath & McGrath, "Why Have a Baby?," N.Y. Times Mag., May 25, 1975, at 16, 26.

As Fletcher observes, dogmatic moralists maintain that a right is a right and--thus--if parents wish to reproduce it is their "God-given right." Humanistic or personistic moralists will, contrariwise, assert that a right depends upon human well-being. Therefore, if--for example--the parents are both carriers of a recessive gene causing lifelong pain and suffering for the child they would produce--the conception should not be undertaken. "The right is null and void." In other words, "the right to be parents ceases to run at the point of victimizing the offspring or society." J. FLETCHER, THE ETHICS OF GENETIC CONTROL 125, 126 (1974).

10. See generally W. PROSSER, HANDBOOK OF THE LAW OF TORTS § 55 (4th ed. 1971); Fried, "The Value of Life," 82 HARV. L. REV. 1415 (1969).

11. See W. PROSSER, supra, at § 56.

12. See Pauling, "Foreword, Symposium--Reflections on the New Biology," 15 U.C.L.A. L. Rev. 267, 270 (1968). For a discerning analysis of the moral, social and legal aspects of the problems of abortion in modern society, see D. GRANFIELD, THE ABORTION DECISION (1969).

13. Professor Robertson has suggested:

> "[T]he strongest claim for not treating the defective newborn is that treatment seriously harms the infant's own interests, whatever may be the effect on others. When maintaining his life involves great physical and psychological suffering for the patient, a reasonable person might conclude that such a life is not worth living. Presumably the patient, if fully informed and able to communicate, would agree."

Robertson, "Involuntary Euthanasia of Defective Newborns: A Legal Analysis," 27 STAN. L. REV. 213, 252 (1975). See generally Comment, "Proposed State Euthanasia Statutes: A Philosophical and Legal Analysis," 3 HOFSTRA L. REV. 115 (1975). One Catholic theologican has suggested that if a child with a birth defect would be born without any "potential for human relationships," the preservation of his life is futile. See Editorial Opinion and Comment, 61 A.B.A.J. 489, 490 (1975) (quoting Father Richard A. McCormick of Georgetown University). See also Howard, "Parents Tell How They Decided to Let Their Child Die," Wash. Star-News, Dec. 13, 1974, at B1, col. 1 (relating how parents concluded child dying from internal birth defects should not undergo life sustaining surgery).

14. See generally Rivers, "Grave New World," SATURDAY REV., Apr. 8, 1972, at 23, 26.

15. Screening efforts also are being conducted in several areas to identify male infants possessing the XYY chromosome pattern, the so-called "criminal disposition chromosome." Proponents of such screen-

ing believe that early identification and medical treatment of individuals born with the XYY chromosome could prevent the behavioral problems identified with the chromosomal pattern. Opponents maintain that the XYY syndrome is a dangerous myth and that the stigma of being identified as XYY creates a self-fulfillment prophecy. See Culliton, "Patients' Rights: Harvard is Site of Battle Over X and Y Chromosomes," 186 SCIENCE 715, 716 (1974); Smith, "Through A Test-Tube Darkly: Artificial Insemination and the Law," 67 MICH. L. REV. 127, 148, n. 114.

See also, Dershowitz, "Karyotype, Predictability and Culpability," in GENETICS AND THE LAW 63 (A. Milunsky, G. Annas eds. 1976); REPORT ON THE XYY CHROMOSOMAL ABNORMALITY, DHEW Pub. No. (HSM) 73-9109 (1970).

Tay Sachs is a clear example of a recessive disorder detectable through genetic screening. The disease, itself, is most common in Jewish persons of eastern European heritage and leads to the gradual debilitation and death of afflicted infants before the age of five. Thus, by application of Mendel's laws of heredity, one can predict that--on average--two Tay Sachs carriers who produce four children will have a family in which one child is afflicted with Tay Sachs disease, two children carry the recessive trait for Tay Sachs, and one child is completely free of the gene for Tay Sachs. Walters, "Introduction to Genetic Intervention and Reproduction Technologies" in CONTEMPORARY ISSUES IN BIOETHICS at 567 (T. Beauchamp, L. Walters eds. 1978).

16. National Sickle Cell Anemia Control Act, 42 U.S.C. § 3006 et seq. (Supp. III, 1973). The genetic etiology of sickle cell gene is most interesting. See D. PATERSON, APPLIED GENETICS 179 (1969).

Congressional concern has been registered over Tay Sachs and Cooley's anemia. A. ETZIONI, GENETIC FIX 132 (1973). See Culliton, "Cooley's Anemia: Special Treatment for Another Ethnic Disease," 178 SCIENCE 593 (1972); National Cooley's Anemia Control Act (Public Law 92-414) (1972). There has also been special congressional concern over the study and regulation of Huntington's chorea (89 Stat. 349) (1975) and hemophilia (90 Stat. 350) (1975).

Limited neonatal screening for phenylketonuria (PKU)--a single gene effect that produces severe mental retardation in children--was initiated in the United States and Britain during the 1950's. Today, some 43 states have PKU screening laws; another 14 test neonatally for a variety of screening problems other than PKU. Among such diseases may be listed: adenosine deaminase deficiency; galactosemia; homocystinuria; sickle cell anemia; tyrosinemia; histidinemia; branches chaisketonuria. Reilly, "State Supported Mass Genetic Screening Programs," in GENETICS AND THE LAW 159, 164 (A. Milunsky, G. Annas eds. 1976).

Multiple screening for genetic disease provides economies of scale for they require relatively minor expansion of facilities and manpower in order to add a disorder or two to a basic screening program. Reilly, supra at 168.

The Health Research and Health Service Amendments of 1976 (PL 94-278), provides Title IV which is designated The National Sickle Cell Anemia, Cooley's Anemia, Tay-Sachs and Genetic Diseases Act, 90 Stat. 407, 42 U.S.C. 300b (1976). The most interesting legislative history of this Act is to be found in U.S. Code Congressional & Ad. News 1071 passim, May 25, 1976.

The primary purpose of this new legislation is "to establish a national program to provide for basic and applied research, research training, testing, counseling and information and education programs with respect to genetic diseases...." The Secretary of the Department of Health Education and Welfare is allowed to expend grant monies through contracts with public and private entities for projects to establish and operate voluntary genetic testing and counseling programs--primarily in conjunction with other existing health programs assisted by Title V of the Social Security Act. (42 U.S.C. § 300 b-2).

Cf. Bazelon, "Medical Progress and the Legal Process," 32 PHAROS 34, 39 (April, 1969).

17. See, e.g., ILL. ANN. STAT. ch. 122 §27-8 (Smith-Hurd Supp. 1979) (exception for refusal of physical examination on constitutional grounds); MASS. GEN. LAWS ANN. ch. 76, § 15A (Supp. 1979) (mandatory only if child susceptible); N.Y. EDUC. § 904 (McKinney Supp. 1978-79) (exception for refusal based on religious beliefs). See also VA. CODE ANN. §§ 32-112.20 to 112.23 (Supp. 1979) (voluntary screening program).

Dr. Linus Pauling has suggested that sickle cell anemia carriers be identified by tattooing the forehead of every carrier. Other recessive genes, such as hemophilia and phenylketonuria, similarly could be identified. Dr. Pauling wistfully suggests that such identification would discourage carriers of the same defective gene "from falling in love with one another," and presumably, from procreating. See Pauling, supra note 12 at 269.

18. N.Y. DOM. REL. LAW § 13-aa (McKinney 1977). Other states provide for voluntary premarital testing for sickle cell anemia. See CAL. HEALTH & SAFETY CODE §§ 325-331 (West Supp. Pamph. 1978); GA. CODE ANN. § 53-216 (1974).

19. See VA. CODE ANN. § 32-122.22 (Supp. 1979).

The first step in the genetic counseling process is to establish an accurate diagnosis. After having accepted the validity of the diagnosis, the probability of recurrence of the disease is established. A number of procedures are employed here: family history, the diagnosis, the literature, Mendel's laws, the karyotype, and various special tests in order to establish the probability of recurrence. Then comes the critical process of assisting the family reach a decision. Here a variety of other considerations than the probability of recurrence and the severity of the disease are evaluated: religious, economic, cultural, familial, legal, etc. The best solution in terms of the immediate situation is sought. The end result is commonly drawn from--contraception, sterilization or abortion. Fraser, "Survey of Counseling Practices," in ETHICAL ISSUES IN HUMAN GENETICS 7 (B. Hilton, D. Callahan, M. Harris, P. Condliffe, B. Berkley eds. 1973). See also, Gustafson, "Genetic Counseling and the Uses of Genetic Knowledge--An Ethical Overview," in ETHICAL ISSUES IN HUMAN GENETICS, supra, at 101; and Kass, "Implications of Prenatal Diagnosis or the Human Right to Life," in ETHICAL ISSUES IN HUMAN GENETICS, supra, at 185. Dr. Kass posits four standards which are commonly used to judge whether a fetus with genetic abnormalities is unfit to live: societal goods; potential social worthiness; parental or familial good and natural selection.

It has been argued that a diagnosis of a genetic defect <u>before</u> the fact of an overt disease infringes not only upon the ethical values of security and well-being but even the parents' freedom and that of the patient. For many parents, knowledge of a potential problem disturbs their well being--without any other practical benefits. Knowing, then, infringes on the pleasure and well being of the parents--at least until such time as the first signs or symptoms of a particular condition appear. The genetic counselor has a duty after taking away the parents' well being by informing them of their child's defective genetic condition to enhance their well being by assuring them of their innocence in the causation of the condition. The fact most hereditary disorders now detected by screening have no effective therapy has been largely responsible for the emphasis on intrauterine diagnosis followed by therapeutic abortion. Here, therapeutic applies to the mother and father--not to the embryo which is to be considered the disease. Murray, "Screening: A Practitioner's View," in ETHICAL ISSUES IN HUMAN GENETICS, supra, 121 at 123.

Interestingly, approximately 90% of the people who rely on genetic counseling do so only _after_ a defective baby is born. The remaining ten per cent seek counseling because they are concerned about some known or "felt" weakness in the genetic line of the family. J. FLETCHER, THE ETHICS OF GENETIC CONTROL 59 (1974).

About one out of every six couples who are informed during a pregnancy that the fetus in question is genetically defective proceed with the pregnancy. FLETCHER, supra at 48, 49.

See also, Antley, "Variable in the Outcome of Genetic Counseling," 23 SOC. BIOLOGY 108 (1976).

A genetic counselor "has freedom to persuade, according to his personal convictions, but he does not have freedom to coerce, based upon his inherent power in the counseling milieu. He must accept the counselee as the ultimate decision maker. Different parents have a variety of motives for their ultimate decisions. Thus, the outcome of their deliberations will vary. And we will preserve our genetic heterogeneity!" Shaw, "Genetic Counseling" in HUMAN GENETICS: READINGS ON THE IMPLICATIONS OF GENETIC ENGINEERING 199 at 200 (T. Mertens ed. 1975).

20. Waltz & Thigpen, "Genetic Screening and Counseling: The Legal and Ethical Issues," 68 NW.U.L. REV.696 at 701-02 & n's. 28-29.

21. See id. at 701-02 & n's. 30-31.

Confusion as to the significance of possessing the defective gene not only renders screening programs less effective in discouraging reproduction, but the failure to differentiate between the disease and the trait also increases the stigmatization to which carriers are subjected. Id.

22. See Frankel, "The Specter of Eugenics," 57 COMMENTARY 25 at 29 (1974).

23. See id.

24. Id.

25. See Waltz & Thigpen, supra note 20 at 712.

26. Id. at 711-712.

27. Id. at 712.

28. Cf. Schmerber v. California, 384 U.S. 757, 772 (1966) (compulsory blood test to determine intoxication of automobile driver not unreasonable search).

29. See Vukowich, "The Dawning of the Brave New World: Legal,

Ethical and Social Issues of Eugenics," 1971 U. ILL. L.F. 189 at 208.

30. See Pauling supra note 12 at 270-271.

See generally, Note, "Legal Analysis and Population Control: The Problem of Coercion," 84 HARV. L. REV. 1856, 1865-75 (1971).

31. See Vukowich, supra note 29 at 215-216.

32. Id. at 216.

33. Id.

See Tsukahara, "Baby Making and the Public Interest," 6 THE HASTINGS CENTER REPORT 13 (Aug. 1976), for a discussion of the situation where a married couple participating in a group insurance plan from their employer knowingly increases insurance premiums for all those within their group plan by condoning the repeated use and application of fertility drugs--even though negative results (namely, the multiple birth of dead or malformed infants) occur repeatedly.

34. See, e.g., Eisenstadt v. Baird, 405 U.S. 438, 452-55 (1972) (forbidding on morality grounds, sale or gift of contraceptives to unmarried persons conflicts with fundamental constitutional rights); Loving v. Virginia, 388 U.S. 1, 12 (1967) (state may not infringe freedom to marry person of another race); Griswold v. Connecticut, 381 U.S. 479, 481-86 (1965) (statute forbidding use of contraceptives violates constitutionally protected right of marital privacy).

35. Skinner v. Oklahoma, 316 U.S. 535, 541 (1941). Concurring in Griswold v. Connecticut, Justice Goldberg commented that a compulsory birth control law unjustifiably would abridge the constitutional rights of marital privacy. 281 U.S. 479, 497 (1965) (Goldberg, J., with Warren, C.J., & Brennan, J., concurring).

36. See Roe v. Wade, 410 U.S. 113, 153 (1973). See generally Altman, "Doctor Guilty in Death of Fetus in Abortion," N.Y. Times, Feb. 16, 1975, at 1, col. 3; Kifner, "Doctor Disputes Abortion Method," N.Y. Times, Feb. 1, 1975, at C5, col. 6; Reinhold, "Abortion Trial's Crucial Issues: When Does Life Begin?," N.Y. Times, Jan. 12, 1975, at 34, col. 1.

A government report showed almost 7 out of 10 married couples (or 60%) of child bearing age in America use some form of contraception. This is an increase of nearly 20 percentage points in the 13 years between 1960 and 1973. The study also showed about 4.4 million American men and women became sterilized for contraceptive purposes or 12% women and 11% men. Auerbach, "Survey Shows Rise in the Use of Contraceptives," Wash. Post, Oct. 7, 1976, at A2, col. 4.

37. See Golding & Golding, "Ethical and Value Issues in Population Limitation and Distribution in the United States," 24 VAND. L. REV. 495, 511 (1971).

See generally, Pohlman & Callahan, "Food Incentives for Sterilization: Can They Be Just?" 3 THE HASTINGS CENTER REPORT 10 (Feb. 1973).

38. Id.

39. See id. at 512. Golding and Golding conclude, however, that the unrestricted freedom to procreate should be abridged only for a "good of momentous order." Id.

If there is a societal consensus that certain behavior is revolting or immoral and that consensus is crystalized into a law, the law survives because of the very consensus of its necessity to restrain the questioned behavior. See e.g., Poe v. Ullman, 367 U.S. 497, 545, 546 (1961) (Harlan, J., dissenting). Of course, the consensus to retain the law does not have to be universal. Ely, "The Wages of Crying Wolf: A Comment on Roe v. Wade," 82 YALE L. J. 920, 924 (1973).

India has 600 million people or 15% of the world's population on only 2.4% of its land. In order to meet its goal to bring the annual birth rate down to 30 per thousand from 35 per thousand by 1979 and to 25 per thousand in 1984, the government was offering vasectomies in sidewalk offices. (The annual birth rate in the U.S. is 14 per thousand.) Men who availed themself of this service received a bonus ranging from $8.00 to $16.00 and often a special gift such as a clock or cans of cooking oil. Since Mrs. Indira Ghandi's defeat as Prime Minister, the sterilization program is being de-emphasized. Borders, "India Reports Gains in Population Drive," N. Y. Times, Sept. 16, 1976, at 9, col. 1.

Only about a fifth of the 80 million women of fertile age in Latin America are believed to be employing contraception. deOnis, "Birth Control in Latin America Making Little Headway as Population Pressures Grow," N. Y. Times, June 20, 1977, at 126, col. 1.

40. 274 U.S. 200 (1927).

41. Id. at 207. Justice Holmes, speaking for the Court, stated: "We have seen more than once that the public welfare may call upon the best citizens for their lives. It would be strange if it could not call on those who already sap the strength of the State for these lesser sacrifices, often not felt to be such by those concerned, in order to prevent our being swamped with incompetence. It is better for all the world, if instead of waiting to execute degenerate offspring for crime, or to let them starve for their imbecility, society can prevent those who are manifestly unfit from continuing their kind." Id.

See also, In re Sterilization of Moore, 289 N.C. 95, 221 S.E. 2d 307 (1976).

42. Kindregan, "State Power over Human Fertility and Individual Liberty," 23 HASTINGS L. J. 1401, 1407 (1972); Paul, "State Eugenic Sterilization History: A Brief Overview," in EUGENIC STERILIZATION 25, 27 (J. Robitscher ed. 1973). Virginia's legislation in this area is typical:

> "Whenever the director of a hospital shall be of the opinion that a patient in such state hospital is afflicted with any form of hereditary mental illness or with mental deficiency and it is in the best interest of such patient and society that such patient should be sexually sterilized, the director is hereby authorized and directed to proceed. . . ."

VA. CODE ANN. § 37.1-171.1 (1976). It has been estimated that over 70,000 people have been sterilized under statutes similar to Virginia's. See STATISTICS FROM HUMAN BETTERMENT ASS'N OF AMERICA, SUMMARY OF U.S. STERILIZATION LAWS 2 (1958).

See generally, Greenawalt, "Criminal Law and Population Control," 24 VAND. L. REV. 465, 475 (1971); Vukowich, supra note 10 at 214-20; Gray, "Compulsory Sterilization in a Free Society; Choices and Dilemmas" in LIFE OR DEATH--WHO CONTROLS? at 169 (N. Ostheimer, J. Ostheimer eds. 1976).

The present eugenic sterilization statutes are: CAL. PENAL CODE § 645 (West 1970); CAL. WELF. & INST'NS CODE § 7254 (West Supp. 1979); DEL. CODE ANN. tit. 16, §§ 5701-5705 (1975); IDAHO CODE §§ 39-3901 to 3910 (1977); ME. REV. STAT. ANN. tit. 34, §§ 2461-2468 (1978); MINN. STAT. ANN. § 252A.13 (Supp. 1978); MISS. CODE ANN. §§ 41-45-1 to -19 (1972); MONT. REV. CODE §§ 69-6401 TO 6406 (1970); N. C. GEN. STAT. §§ 35-36 to -50 (1976); N. D. CENTURY CODE §§ 25-04.1-08 (1978); OKLA. STAT. ANN. tit. 43A, §§ 341-346 (1979); ORE. REV. STAT. § 436.010-.150 (1977); S. C. CODE ANN. §§ 44-47-10 to -100 (1977); UTAH CODE ANN. §§ 64-10-1 to -7 (1968); VT. STAT. ANN. tit. 18, §§ 8701-8704 (1968); VA. CODE ANN. § 37.1-171.1 (1976); W. VA. CODE ANN. §§ 27-16-1 to -5 (1976).

One should distinguish these eugenic sterilization statutes from those sterilization statutes which are wholly voluntary in nature. Among these type statutes are: GA. CODE ANN. §§ 84-931 et seq. (1979); ORE. REV. STAT. § 435.305 (Rpl 1977); N. M. STAT. ANN. §§ 24-1-14, 24-9-1 (1978); N. C. GEN. STAT. §§ 90-271 to -275 (1975); and VA. CODE ANN. §§ 32-423 et seq. (cum. Supp. 1978). These statutes are essentially contraceptive and therapeutic and not eugenic in nature.

For a detailed breakdown of the scope and application of the various state eugenic sterilization statutes, see Dunn, "Eugenic Sterilization Statutes: A Constitutional Re-evaluation," 14 J. FAM. L. 280, 304 passim (1975). To be noted is the fact that since this article was written, a growing number of states have repealed their eugenic sterilization statutes. The sixteen states listed above as having eugenic sterilization statutes is the most current listing.

43. See Oregon v. Cook, 9 Ore. App. 224, 230, 495 P. 2d 768, 771-72 (1972) (equal protection challenge based on indigency rejected); In re Cavitt, 182 Neb. 712, 721, 157 N. W. 2d 171, 178 (1968), appeal dismissed, 396 U. S. 996 (1970) (same).

44. See, notes 34-36 supra and accompanying text.

45. See, Shapiro v. Thompson, 394 U. S. 618, 638 (1969).

46. See, Oregon v. Cook, 9 Ore. App. 224, 230, 495 P. 2d 768, 771-72 (1972).

47. Buck v. Bell, 274 U. S. 200, 208 (1927). The Court's rationale acquires additional significance because it became the basis for distinguishing Buck v. Bell in Skinner v. Oklahoma, where the Supreme Court invalidated a statute providing for the sterilization of habitual criminals; the Court in Skinner concluded the statute violated the fourteenth amendment's equal protection clause. See 316 U. S. 535, 542 (1941).

48. The statute challenged in Buck required only that experience demonstrate heredity plays an important role in the transmission of the mental defect. See 274 U. S. at 206. The inmate involved, however, was the daughter of a feeble minded child. Id. at 205. See generally, Murray, "Marriage Contracts for the Mentally Retarded," 21 CATH. LAW. 182 (1975).

49. See Waltz & Thigpen, supra note 20 at 721 & n. 131.

50. See Vukowich, supra note 29 at 208.

A persuasive economic argument can be made for forced sterilization of mentally defectives. A 1971 study undertaken by the Federal government concerned one hundred and ninety public institutions for the mentally retarded and disclosed 15,370 patients were admitted for treatment during the 1971 calendar year. This is the equivalent of 7.5 patients per 100,000 people in the over-all population and represents an average daily resident patient population of 181,058. Even though this figure shows a slight decline from the peak year of 1968, during the same four year period, the annual cost of institutional care per patient rose from $3,472.00 to $5,537.00. Stated otherwise, the costs rose from $9.00 per day to $15.00 per day which is a 66% increase. UNITED STATES BUREAU OF THE CENSUS, STATISTICAL ABSTRACT OF THE UNITED STATES 82, 83 (95th ed. 1974).

See also, Landman, "The History of Human Sterilization in the United States: Theory, Statute and Adjudication," 23 ILL. L. REV. 463 (1929); Baron, "Voluntary Sterilization of the Mentally Retarded," in GENETICS AND THE LAW 267 (A. Milunsky, G. Annas eds. 1976).

See generally, N. MORRIS, H. ARTHUR, STERILIZATION AS A MEANS OF BIRTH CONTROL IN MEN AND WOMEN (1976); J.P. BOYLE, THE STERILIZATION CONTROVERSY: A NEW CRISIS FOR THE CATHOLIC HOSPITAL? (1976).

One respected authority has observed that ethics cannot be viewed or understood as an independent of economics and utilitarian or distributive justice. "Economics deals with preferences among competing choices, and utility aims at spreading expectable benefits." J. FLETCHER, THE ETHICS OF GENETIC CONTROL 160 (1974).

51. 274 U.S. at 207. Unrestricted genetic transmission forces a heavy burden upon society. The Juke and Kallikak family histories reveal clearly this point. Max Juke resided in Ulster County, New York. Max had two sons who married two of six sisters of a local feeble-minded family. One other sister left the area; the other three married mental defectives. From these five sisters, 2,094 direct descendants and 726 consortium descendants were traced by 1915 into fourteen states. All of them were feeble-minded and the cost to society from their welfare payments, illicit enterprises, jail terms, and prostitution brothels had reached $2,516,685.00. J. WALLIN, MENTAL DEFICIENCY 43-44 (1956).

Martin Kallikak, Sr., fostered a son--Martin Jr.--by a feeble-minded girl during the Revolutionary War. Martin Jr. married a feeble-minded girl and they, in turn, had seven children: five of whom were similarly afflicted. From these progeny sprung 480 descendants, 143 feeble-minded, 46 normals, and 291 of unknown mental stature. When Martin Sr. returned from the War, he married a normal woman and started a line culminating in 496 descendants--all of whom were normal. WALLIN, supra, at 44-45. Environmental deprivation has been recognized by some as an important--if not the determining--factor in the Kallikak "saga."

One can but speculate regarding the number of similar unreported case histories and of the cost to society of these tragic occurrences.

Various estimates have been made relative to the lifetime costs of various genetic diseases--often with rather astonishing results. For example, it has been calculated that the lifetime costs of maintaining a seriously defective individual is $250,000.00; this assumes, of course, institutionalization. Conservative estimates place the number of new cases of Down's syndrome in this country at 5,000 each year--or, one in every 700 live births. Using the $250,000.00 figure for the cost of maintenance, the lifetime committed expenditure for new cases of Down's syndrome standing alone comes to at least $1.25 billion yearly which is, admittedly, a staggering figure for but one disease entity. With available technology, however--

including greater use of prenatal diagnosis--it is possible to prevent the birth of a high proportion of such affected infants. Screening high-risk (older) pregnant women for Down's syndrome in utero might well prevent 1,000 cases yearly, which would otherwise cost approcimately $250 million in lifetime care. To detect 1,000 cases, 96,000 pregnancies would have to be monitored at a cost of between $15 million and $25 million, or less than a tenth of the cost of lifetime care.

Another way of calculating the toll of genetic disease is to estimate the future life years cost. One widely cited estimate indicates that some 36 million future life years are lost in this country by birth defects--putting the figure for recognized genetic disease (80% of birth defects being genetic in whole or in part) at 29 million future life years lost, or several times as much as from heart disease, cancer, and stroke.

Consider the costs of Tay-Sachs disease. Intensive care of the child over its brief lifespan runs about $35,000.00. Since there are about 50 cases of this disease every year in the United States, the annual cost for this rare genetic disease amounts to $1,750,000.00. Recent research advances permit the identification of carriers, intra-uterine diagnosis, and elective abortion of affected fetuses. Aside from the economic saving, families can be spared the heartache associated with this distressing genetic anomaly.

Phenylketonuria, or PKU, is another interesting case in point. This mentally retarding genetic disease occurs approximately once in every 14,000 births. Screening newborns for the disease costs $1.25 per test; thus, approximately $17,000.00 is spent to detect each case. An additional $8,000.00 to $16,000.00 must then be spent for dietary treatment over a five to ten year period, in order to prevent the retarding effects of the disease. This brings the total cost of prevention to about $33,000.00 per child. Untreated, severe mental retardation care for, say, 50 years in an institution at a cost of $20.00 a day, would run to $365,000.00--or more than ten times the cost of prevention. WHAT ARE THE FACTS ABOUT GENETIC DISEASE? at 27, 29, U.S. Dept. of H.E.W., Public Health Service, N.I.H., DHEW Public. No. (NIH) 75-370. See also, M. FRANKEL, GENETIC TECHNOLOGY: PROMISES AND PROBLEMS 46-77 (1973); R. VEATCH, DEATH, DYING AND THE BIOLOGICAL REVOLUTION (1976).

CHAPTER 3

THE VAGARIES OF INFORMED CONSENT

Free and informed consent is necessary to justify an intentional invasion of one's bodily integrity. 1 It is the pivotal issue on which hangs the majority of not only the ethical problems of human experimentation, but the growing number of cases dealing with malpractice--and particularly those concerned with surgical misfeasance and nonfeasance. 2 The principal reasons why the consent process fails here is that pressures and constraints often act to prevent free choice and subjects often lack sufficient information which prevent them from making an informed decision. Thus, a lack of basic information often acts to inhibit one's freedom to refuse. 3

In order that a person's consent to medical treatment be valid and, further in order that the attending physician not be subjected to liability for assault, 4 the physician must carefully and cautiously structure his disclosures to the patient. This need for complete and accurate information includes a full description of the nature of the patient's condition, what treatment the doctor proposed, the risks and alternatives of the proposed treatment and the risks of alternate treatment or non treatment, as the case may be. Only material facts need be disclosed. Thus, the failure to disclose an exceedingly remote risk would be consistent with obtaining a valid consent and effectively bar a patient's suit against a doctor. 5 It reamins the ultimate responsibility of the surgeon to provide the patient with enough information to allow the patient's decision regarding the questioned treatment to be termed "informed." There are other requirements which must be met before the consent is understood as effective.

Consent must be given by one who has the capacity to consent, or by one empowered to consent for the individual who has the incapacity The consent must be to the particular conduct proposed and subsequently undertaken or substantially the same conduct. 6 The material presented to the patient must be given in such a manner as to remove any and all taint of coercion. This can be readily accomplished through the use of visual aids, a description of the procedure along with its attached risks and alternatives and a statement of expectation of improvement. 7

While normally, in the absence of consent, the medical practitioner subjects himself to a tort claim of battery, the emerging basis for liability sounds in negligence. In order to establish negligence, it must be shown that under the particular facts--the doctor in question owed a duty to his patient, which thereby defined the standard of conduct to which he was held, 8 that his conduct violated that duty and that there was a causal connection between the action or inaction taken by the doctor and the resulting injury. The rationale of this action is not well accepted by those in the medical profession and its development has been one of only recent emergence in the law of tort. 9

The Historical Perspective

The earliest case dealing with informed consent, though only impliedly, was Slater v. Baker & Stapleton, C.B. decided in 1767. 10 There, it was held that it was improper in the treatment of a broken leg to disunite the callous without consent and that "indeed it is reasonable that a patient should be told what is about to done to him, that he may take courage and put himself in such a situation as to enable him to undergo the operation." 11 The doctrine received little notice and actual use until the 1960's--with but brief, scattered reference in the two hundred year interim. 12 The early cases were such that a battery theory of "unlawful touching" was used to subject the physician to liability; however, since Salgo v. Leland Stanford Jr. University Bd. of Trustees,[13] the courts have been emphasizing and developing a negligence or "failure to use due care" theory of liability.[14] This theory in turn places the physician in default for failing adequately to educate his patient to the collateral risks involved in the treatment.

The groundwork for the explosion of cases dealing with informed consent was laid in Salgo. There, the problem confronting the doctor was made abundantly clear: he must balance the patient's need to know the risks and alternative to treatment in order to give an informed consent with that individual patient's mental and emotional condition to accept and understand the medical information.[15] The case held that a physician owes a duty to his patient to disclose facts necessary to the formation of an intelligent consent and, further, subjects himself to liability for violation of that duty.[16] The standard enunciated in Salgo, though, is at best confusing. Plaintiff's counsel may use it to argue that there must be full disclosure since there is an established duty to disclose while, on the other hand, a court may cite the case as justification for holding for the defendant on the issue of the adequacy of disclosure.[17] This discretion allowed to physicians has tended to subject them to liability in cases involving high risk and to exonerate them when the court considered the risk to be slight.[18]

The highly sensitive fact situation in Natanson v. Kline[19] gave informed consent the firm impetus necessary to establish as a pillar of malpractice law. The possibility of a radical mastectomy and the fear of the unknown inherent in treatment with radioactive cobalt strike fear into the minds of those women who must undergo such treatment. Such was the situation in Natanson. The patient, Mrs. Natanson, received injuries as the alleged result of an excessive dose of radioactive cobalt during radiation therapy. The Kansas Supreme Court found that where the physician failed to warn or explain the hazards involved in the therapy, he failed to meet his duty to make a reasonable disclosure as a matter of law.[20] The decision recognized that there is ". . . probably a privilege, on therapeutic grounds, to withhold the specific diagnosis where the disclosure of cancer or some other dread disease would seriously jeopardize the recovery of an unstable, temperamental or severely depressed patient."[21] This is not, though, the situation normally; for, in the standard medical case, there would be no compulsion to supress facts or not to make a substantial disclosure. The instant case is important for the reason that it sounds in negligence and not battery and, thus, affords physicians some protection.[22] In order to recover for a tortious battery, all that must be proved is the consent was invalid for some reason or another (immaturity, incapacity, fraud, duress, etc.). The physician thereby becomes liable for every touch--both harmful and benign.[23] With negligence as the cause of action, the physician will only be held liable upon breach of a duty to disclose and the establishment of some causal connection between that breach and the resulting harm.[24]

The 1972 case of Canterbury v. Spence[25] presents a modern, comprehensive or focal paradigm of the legal concept of informed consent in application. There, a young boy complaining of back pain submitted to a myelogram which revealed a filling defect. The boy's mother was contacted after the test and an operation was recommended by Dr. William T. Spence--the attending physician--stating that such an operation was "not anymore [serious] than any other operation."[26] The boy submitted to the operation without being informed that the risk carried with it was one of paralysis. Mrs. Canterbury arrived at the hospital <u>after</u> the operation and signed a consent form. The boy fell from his bed a day after the operation while, without assistance, he attempted to void. He thereupon became paralyzed and required to undergo surgery yet another time. This time, Mrs. Canterbury signed a consent form before the operation. Years later, the youth hobbled about on crutches, ". . . a victim of paralysis of the bowels and urinary incontinence."[27] At trial, Dr. Spence testified that paralysis is to be expected in the neighborhood of one percent of the cases of operations of the type performed here.[28] The central issue of the case was the scope and application of the doctrine of informed consent. The court's treatment of this doctrine is properly considered at this point in greater elaboration.

It is, admittedly, a difficult task for a physician to explain to a layman/patient what medical therapy is being proposed. Most patients understand little of medicine and rely almost exclusively on their doctor for explanations of conditions and treatments. Therefore, the duty of the practitioner does not usually end with proficient diagnosis and therapy--but rather extends to subsequent communication and counsel with his patient. Since the patient will be the one to determine what will happen to him, it is mainly through the efforts of explanation on the part of his doctor he becomes familiar with the alternatives and corresponding risks of his decisions.[29] Every person is master of his own body.[30] There is, consequently, a great dependence on practitioners by patients, that a disclosure of all foreseeable potential dangers be required.

Because there is a duty to disclose, the scope of that duty should be known. Any standard set in terms of what is done in the profession will be at odds with the patient's prerogative to decide on prospective therapy. This right of self-decision shapes the boundary of the duty to reveal.[31] In order that the patient's interest in achieving his own determination of treatment is fulfilled, it is the law which must set the standard for adequate disclosure.[32] The test enunciated in Canterbury, then, is ". . . [a] risk is thus material when a reasonable person, in what the physician knows or should know to be the patient's position, would be likely to attach significance to the risk or clusters of risks in deciding whether or not to forego the proposed therapy."[33] This includes a discussion of the inherent and potential dangers of the proposed treatment, the alternatives to that treatment and the results likely if the patient remains untreated.

The courts have noted two exceptions to the general rule of disclosure. The first is where the person is unconscious or otherwise incapable on consenting and there is imminent harm which would result from failure to treat which in turn outweighs any harm threatened by the proposed treatment.[34] If possible, consent of relatives should then be obtained. The second exception arises when the disclosure threatens the patient so as to become infeasible from a medical point of view. The critical inquiry, then, would be whether the physician was guided by sound medical judgment.[35] This privilege does not carry with it the paternalistic notion that the physician may remain silent simply because diligence might prompt the patient to forego therapy the physician maintains the patient needs.[36]

There is always a danger in this area of consideration that a subjective, hindsight test might be employed. Canterbury speaks to this concern and resolves it by requiring a determination of whether a prudent person in the patient's position would have decided to undergo treatment if informed of all perils bearing significance.[37] This affords opportunity for medical testimony regarding the relevance of certain risks as will other testimony by anyone having sufficient knowledge and capacity to testify. The courts thus assume a determinative role in assessing liability.[38]

Consent denotes both awareness and assent. This involves a communication that is understandable--the failure of doing so being a risk the physician assumes. To determine the nature of the disclosure which will satisfy the law, a standard of reasonableness is employed. The standard, itself, will be structured differently among jurisdictions. Either the reasonable patient standard is used--which tests what the reasonable patient would, under the particular facts of the case, want to know about the operation in question or, the reasonable medical practitioner standard is applied. Of course, under the second standard, the practice of physicians in the medical profession relative to the need or nature of the questioned disclosure is followed.[39] In order for a physician to meet his duty of disclosure, he must disclose information pertinent to collateral risks and not proceed without consent to the risks which were or should have been disclosed.[40] The physician owes to the patient an obligation not to mislead him.[41] In turn, out of this obligation arise the requirements of informed consent to treatment.

The consent must be knowing.[42] The patient must be made to understand that to which he is consenting. The physician cannot confuse the patient with professional terminology. The language used, hence, cannot be vague or ambiguous.[43] The disclosure must be "full"--but not necessarily "total."[44] And, of course, the consent must be given by one having the capacity to do so.[45] If these standards are met, the practitioner will encounter little difficulty with medical malpractice litigation in this area and--at the same time--the very real functions of informed consent will be met.

The functions of informed consent are much more than merely requiring that a patient understand and assent to medical treatment. Our jurisprudential premise which permeates the entire concept is that of individual autonomy and self-determination. Requiring informed consent promotes this individualism in three ways: it grants a measure of freedom--one may choose what is to be done to one's body; it protects the person's status as a human being by guaranteeing the total use of the mind; and it safeguards against the use of force and duress in medical treatments.[46] Yet another function is the encouragement of rational decision-making through the promotion of knowledge and, thus, aiding the decision-making process.[47] By informing the patient, the medical practitioner is educating the general public, as well as protecting the experimental process from the natural fears harbored of the unknown or highly technical sciences.

Implied Consent and Eugenic Sterilization

With racial progress rather than racial deterioration as an avowed goal of eugenics,[48] both voluntary and involuntary sterilization programs play a significant role in the achievement of this goal. Reference has already been made to the state's power to enforce involuntary sterilization laws. It is beneficial at this point to go beyond discussion of the need for showing reliance upon a rational basis and a compelling state interest by the state to justify legislation which approves involuntary sterilization programs.[49] A central concern to this whole area is the validation and protection of the standard of substantive due process and, to be certain, its enforcement by the courts.

Whether the principle of substantive due process is an "oxymoronic concept" which has distorted history and logic,[50] or whether it is a "nightmare from which the Supreme Court ought to awake,"[51] the fact remains the Court has not yet chosen to awake from its disturbed slumber and abandon the principle.[52]

Substantive due process requires once a law is challenged, the law must be shown to serve or rationally relate to an aspect of the public welfare. If the public welfare is not in fact served by a law, it is because the law fails to achieve its legitimate objective or--while achieving its stated objective--the objective is outside the scope of the public welfare. So it is, then, that judicial inquiry under the principle of substantive due process involves two primary questions: 1.) Whether the law achieves its objectives and 2.) Whether the objective, or perhaps another objective not specifically intended, is within the scope of the public welfare. In order to resolve the first question, empirical and analytical inquiry must be undertaken. The second question--perhaps more difficult--necessitates an investigation into so-called conventional social attitudes, which may be regarded as the ultimate determinants of that very scope of the public welfare.[53]

Those items which are believed to constitute the public welfare in any given society at any given period are the elements forming the "public welfare." Thus, the public welfare is neither an objective entity nor an a priori concept. Giving content to and defining a limit to the public welfare is largely a problem of giving content to its public moral element.[54] The searching question, "Does this law serve the public welfare?" is--in the main--a question which probes the extent and application of particular social conventions. The effort to discern the content of the questioned social conventions, then, may repose with the judiciary or with the legislature--depending on which has the greater degree of expertise under any given situation.[55] Generally, as of late, the courts have assumed this task more efficiently than the legislatures. What does remain clear, in spite of the confusing search for an arbiter of social conventions, is that the police power is limited "inherently, by and to the pursuit of the public welfare."[56]

Since the basic determinants of the public welfare are properly viewed as "conventional attitudes of the socio-political culture," when the Supreme Court applies the substantive imperative of the due process clause--or, in other words, the public welfare limit--it is effectively called upon to discover the ramifications of the questioned social conventions.[57] In performing this task, it "exercises the ethical function of judicial review" by bringing "culturally shared ideals, sensibilities, and norms to bear on the political process."[58] It is asserted that the technical function of judicial review implicit in the principle of substantive due process should not be renounced by the Court.[59]

Over the past fifty years, familial and procreative rights have gained the status of being fundamental rights.[60] The leading case, from an historical perspective, is Skinner v. Oklahoma.[61] Although Skinner recognized "marriage and procreation are fundamental to the very existence and survival of the race,"[62] the right to procreate certainly does not include a right to breed without restriction of any nature.[63]

The state must meet certain requirements in order to justify a compelling interest in legislating for sterilization. First, it must be shown that the particular defect of the individual in conjunction with the ability to procreate is a present danger to the health, safety and welfare of the public. Secondly, the state must show the only way to negate the harmful effect of such conduct is by sterilization. Finally, the statute drawn to accomplish the sterilization must be drawn narrowly. Exactly what defects and under what circumstances the procedure is allowed and the class of individuals to which applicable must be detailed.[64]

Under the inherent equity powers of a court to supervise the affairs of incompetents when questions arise as to the advisability of their sterilization, the court will normally appoint a guardian, conservator or committee to protect the rights of the mental defectives during the time it is being detetmined whether sterilization should be performed. If necessary, independent counsel may be appointed to represent the defective.[65]

Although by statute a guardian ad litem, a conservator or committee may be empowered to give consent to sterilization of an incompetent, the court must make the determination that such acts are in the "best interests" of the incompetent. Given the possibility that an operation of this nature may produce adverse psychological reactions in the patient, and the general uncertainty of the result of such operation, there has been reluctance to move expeditiously here.[66]

Perhaps less hesitancy would be encountered if the courts were to be convinced that the incompetent--if fully aware of his own mental state--would not want to burden society with other defective individuals such as himself. Thus, a societal standard of the greatest benefit or good would be made more significant than the fundamental right of individual procreation and the general public welfare served.

The issue of involuntary sterilization is fraught with complexities. Perhaps a more practical way to achieve the end of positive eugenics and guarantee preservation of so-called "individual liberties" of all members of society to procreate is to be found within voluntary sterilization programs.

Voluntary Sterilization

Opinion is divided on the issue of whether programs of voluntary sterilization should be allowed and promoted.[67] In the United States, at least, the right to subject one's self to sterilization is well established.[68] For example, it has been held that a hospital policy barring the use of its facilities for consensual sterilizations while allowing other operations of equal risk prevents the exercise of the fundamental right to control pregnancy and thus violates the Equal Protection clause of the Fourteenth Amendment.[69] As previously discussed, in order that such treatment be considered voluntary, the patient's consent to it must be informed. Since there must be informed consent for sterilization operations, the sterilization could--arguably--be considered part of a comprehensive eugenics program. That is, there should be both disclosure to the patient of the risks, benefits, results and effects of such an operation. As part of the program, proper counseling should be required. Every effort should be made to relieve the feelings of "guilt" and/or "worthlessness" which oftentimes result from such an operation. In addition to full disclosure, the test of the nature of the voluntary consent is that the individual giving it must have the physical and mental capacity to render the actual consent given as a result of proper disclosure made by the doctor.

The question of voluntariness is one which provides most of the difficulty in administering a program of voluntary sterilization. The sterilization cannot be used as a condition for something beneficial for the patient. It should not be a pre-requisite to release from an institution.[70] However, the case of In re Cavitt[71] allowed that where a woman of thirty-five with an I.Q. of 71 was given an opportunity to be released from a state home on condition that she be sterilized, the choice was still the woman's and the procedure was upheld. One would have to question whether, in a situation of this nature, the choice was truly a free one. Nevertheless, the court considered not only what would be in Mrs. Cavitt's best interests, but also the interest of society; for, if she were to have other defective children, they, too, would obviously require public assistance.[72]

Voluntary sterilization should not be used upon threat of cut-off from welfare benefits. Relf v. Weinberger,[73] a challenge to the voluntary sterilization programs funded by family planning services of the Public Health Service[74] and also the Social and Rehabilitation

Service,75 discussed the nature of voluntary and knowing consent to
sterilization procedures. The court held there that the regulations
needed amendment to include a requirement that individuals seeking
sterilization be orally informed at the very outset that no federal
benefits can be withdrawn because of a failure to accept steriliza-
tion.76 There should, however, be no objection to informing the
patient/client that additional welfare benefits can be withheld.
Dandridge v. Williams,77 upheld a Maryland maximum grant regulation
which placed an upper limit on the amount of Aid to Families with Dependent Children78 payments a family unit could receive. The court
indicated there that in the area of social welfare and economics, a
law need only satisfy a "reasonable basis" test to meet the Equal Protection Clause. That is, the means must be a reasonable manner to
achieve the stated ends of the program. Thus, it appears that while
there can be no threat to withhold benefits already being enjoyed, as
part of a voluntary program, the patient can and should be told that
additional benefits may not accrue should the decision be made to have
more children.

Part of the attacks on the statutes involved in Relf 79 brought
out the charge that certain provisions of the acts in question rendered
the sterilization involuntary. This was based upon the assertion that
the sterilization of minors or incompetents is, by nature, involuntary.
The Guidelines of Sterilization Procedures under HEW supported programs80
allowed that a state court and a Review Committee pass on the steriliz-
ation--with the mandate that for it to be approved, the sterilization
must be determined in the best interests of the patient. Since the
family planning programs of HEW are designed to be voluntary and not a
pre-requisite to eligibility for receipt of any other service or assistance, the court held provisions allowing what amounts to involuntary
sterilization to be arbitrary and unreasonable. The court concluded by
advocating a balancing test: "We should not drift into a policy which
has infathomed implication and which permanently deprives unwilling or
immature citizens of their ability to procreate without adequate legal
safeguards and a legislative determination of the appropriate standard
in light of the general welfare and of individual rights." 81

As noted, consent is not voluntary if it is not informed. Patients
should be made to understand the nature of the operation, the fact that
they will no longer be able to procreate, as well as the attendant
dangers inherent in such a surgical procedure. Perhaps the greatest
inadequacy at this time of such programs is the lack of education
offered to the general public on this subject. Making society aware
of the concept and desirability of a superior gene pool might alleviate
to a measurable extent the guilt and shame associated with sterilization. 82 Yet, developing a proper medium through which an educative process required here may be undertaken presents serious difficulties.

The question of who can consent to an operation for sterilization
is no less intriguing. Consent to a surgical procedure where the
patient is incapable of giving his own consent, has been upheld where
it was given by direction of a court (exercising its parens patri
power), by members of the patient's family or by one empowered by the
court to give consent. 83 The first consent given by the court under
parens patri powers can be modeled by the case of In re Willoughby. 84
This would allow the court to use its "substituted judgment" doctrine
for that of the patient. In other words, the court must be absolutely
convinced that the incompetent person, fully informed, would grant the
sterilization if he were fully competent. Yet, in order to exercise
this power, the courts must find some benefit to the incompetent, which
has oftentimes been difficult. 85

Consent--as has been observed--by one other than the person undergoing the sterilization can be a treacherous avenue to follow. The person giving the consent may have interests which are at odds with those of the person being sterilized.[86] Authorizing such an operation within the scheme of an appointed guardian may not always guarantee that the patient's physical condition as well as the possible adverse reactions he may suffer as a consequence of the surgery will be considered. A surrogate's consent might well be motivated by a sense of over-protectiveness which is so often characteristic of parents of retarded children.[87]

Since allegedly some ninety per cent of the mentally retarded can appreciate or even understand in a limited way the responsibilities of parenthood, as well as the implications of sterilization, an argument could presumably be made that the consent by another to a sterilization violates the Equal Protection Clause of the Constitution.[88] Since there would be no determination of the mentally deficient's fitness for parenthood before deprivation of the choice of whether or not to be a parent, these actions might raise an Equal Protection argument because an invidious discrimination of a class (or, ninety per cent of the mental defectives) was perpetrated. This group of defectives would not be given the same protection (i.e., necessity of informed consent of the person undergoing the procedure) as anyone else capable of giving informed consent. Therefore, the argument would conclude that since ninety per cent of the mentally defectives may well understand and appreciate the responsibilities of parenthood together with the implications of sterilization, they should not be treated differently from normal people who are given the right to informed consent under similar conditions.

Control

Which level of government should have the control over a eugenics program of this nature? Perhaps the strongest argument lies in favor of the national government having this power exclusively. Using Gibbons v. Ogden[89] as a lynchpin, it is possible to justify the Commerce Clause as a source for validating the use of national authority in this area. This idea has been extended and fortified by NLRB v. Jones and Laughlin Steel Corp.[90] There, the test for such a use was determined bo be a close and substantial relationship to interstate commerce. Certainly, maintaining a superior gene pool is a national interest; its effects are far reaching and not contained within any one state's borders. Another source of national power is the General Welfare clause of the Constitution. By defining the interest of the government in promoting the health, safety and welfare of its citizens, the power to regulate and maintain a voluntary sterilization program is, thus, substantiated.

Much of the private, non-government funded research being conducted in this area falls within the scope or ambit of academic research and thus raises the correlative issue of academic freedom. The court has recognized this and reduced the government interest/academic freedom conflict issue down to a balancing of interests test.[91] Thus structured, the test devolves into a consideration of the need for governmental control in the specified area and the interest being furthered and compares it with the necessity and advisability of unhindered academic freedom in the area.[92] With a program of voluntary sterilization, the governmental interest in providing protection for the individual's right to procreate might conceivably outweigh the interest of the academic world in unrestricted experimentation with gene pool manipulation.

Consent in Human Experimentation

There are two forms of human experimentation: therapeutic and

and nontherapeutic. When experimentation is undertaken on a person solely in order to obtain information which will be useful to others--thus in no way treating illnesses the experimental subject may have--it is considered as nontherapeutic in nature. Contrariwise, when a therapy is used with the primary view of ascertaining the best form of treatment for a particular patient, it is therapeutic. 93

In nontherapeutic research, the doctor confronts the subject of his research as a scientist. The doctor in such cases has no patient--only a subject of investigation. It can be argued, therefore, that the usual privileges accorded to a doctor's work and those doctrines to which the liabilities of doctors are judged should not be applicable here. 94 This is the situation simply because those privileges and liability doctrines "proceed from the premise that the doctor must be given considerable latitude as he works in the personal interests of his patient." 95

A strict duty of disclosure is nevertheless imposed by the law wherever one is exposed to a risk or asked to abridge certain fundamental rights by someone who possesses greater expertise. 96 Without too much difficulty, an argument could be made that under the developing doctrines of strict liability, liability should be imposed without fault--irrespective of disclosures for harm occasioned during the course of nontherapeutic experimentation. 97

All hypothetical or experimental remedies being conducted by various researchers do not come within the ambit of a physician's obligation to advise a patient of alternative therapies. 98 Yet, where the particular therapy which is being used is, itself, or an experimental nature, this fact and the very existence of either alternatives of professional doubts become material facts. Consequently, as such, these facts should be disclosed. 99

The most vexatious type of medical experimentation to the lawyer, ethician and physician is commonly referred to as mixed therapeutic and nontherapeutic experimentation. Although a patient may be undergoing valid treatment for a particular illness, the treatment may not have been chosen with the sole view of curing that particular patient of his own illness. Treatment, then, is administered as part of an experiment or as a research program to either test new procedures or--as the case may be--to compare the efficacy of previously established procedures. Patients are assigned to various treatment categories not as a consequence of a careful consideration of the patient's needs, but in measure to the specific needs of the research design. 100

Randomized clinical trial (RCT) is a procedure under which patients are placed in categories of treatment by a randomizing device; hence, experimenter bias can be statistically eliminated. 101 There is mixed reaction in cases of this nature to the need for disclosure to the patient that his treatment will not be determined by an independent judgment made by the physician but, rather, by a random procedure. 102 There are no decided cases which resolve the conundrum of whether it is necessary to disclose to the patient that an experiment is being conducted and the very nature of the experiment itself. 103

The legal status of mixed therapeutic and nontherapeutic experimentation is, in part, unquestioned and predictable. The general obligation to obtain patient consent to a therapy which will be used on the patient adheres. With this also goes the need by the physician to make a full disclosure of expected benefits and possible hazards. 104

Fetal Consent

The issue of fetal consent is fraught with more ambiguities and uncertainties than the general area of consent.[105] Here, a determination must be made whether consent should be required before fetal research may be commenced and, if so, from whom such consent must be obtained. It has been maintained that since meaningful life outside the uterus may be sustained by a viable fetus, it should therefore be regarded--for purposes of consent--as a premature infant.[106] Although consistent with Roe v. Wade in the determination that the fetus is not to be regarded as a person, it is proper "to procure consent for experimentation on a viable fetus in a manner similar to that in which it is procured in the case of any other premature infant."[107]

Usually, with cases of premature infants, the parents exercise their judgment with the best interests of the fetus in mind. When--because of danger to her health--a mother seeks an abortion of a viable fetus it would be both humane and equitable for the court to appoint a guardian for the fetus to the abortion. Such an appointment would hopefully allow procedures to be structured to save the life of the fetus and prevent experimentation which would endanger its health.[108]

The extent of the state's interest in a nonviable fetus is unclear. If, for purposes of consent, a nonviable aborted fetus is treated as but a collection of the mother's tissue, then, experimentation on a nonviable fetus in utero would likely be considered an experimentation on the mother.[108] Although medically sound, this is regarded by some as both morally and psychologically repugnant. Nonviable fetuses should, it is argued, be distinguished from other human tissue. Yet the mother, herself, should have no particular concern with experimentations performed on disorganized fetal tissue and, hence, should not be requested to give her consent to such acts.[109] An exception must, of course, be allowed and maternal consent sought when a woman desiring an abortion is, for moral reasons, opposed to experimentation on an intact or disorganized aborted fetus.[110]

So it is seen, then, in the final analysis, that the wishes and directives of parents make abortion moral. It is also the parents who determine whether the particular fetal research is moral--so long as the experimenters respect the range of the parental decision or, as in many cases in modern society, a single woman's decision.[111] Particularly in matters of fetal research and generally as to human experimentation, the contemporary medical practitioner needs not only to do that which is considered professionally right--but, in addition, he must be viewed by others as doing that which is correct.[112]

Structuring A Right Of Refusal--A Synthesis

The very essence of informed consent is recognition of a right to refuse treatment, experimentation, efforts at behavior modification or any act interferring with one's autonomy. Indeed, as has been seen, the right to protect one's physical integrity or to maintain his selfhood is a fundamental right fully recognized by law. Among the enumerated rights of the First Amendment to the United States Constitution is the right to freely exercise religion. The Eighth Amendment protects against the cruel and unusual infliction of punishments; and the Civil Rights Act underscores the right of all citizens to be protected against any deprivation of "any rights, privileges or immunities secured by the Constitution and laws. . . ."[113]

The right to refuse medical treatment because of religious belief or in spite of criminal or civil imprisonment are two areas of central concern. There has been previous discussion of proxy consent for children, under age, to participate in scientific or medical experimentations.[114]

In 1944, the United States Supreme Court held that the state could protect children from burdensome and exploitative labor even though such was condoned and, indeed, encouraged by either religious motives on the part of parents or guardians.[115] Having decreed this, it was relatively easy for the court to decide summarily that a parent could not withhold a blood transfusion from its child who required it in order to sustain life even though religious beliefs of the parents precluded such an act being administered.[116] Many lower courts have followed this posture and found little, if any, difficulty in concluding that parents cannot prevent their children--on whatever religious grounds--from receiving treatment without which the health of their children would be called into serious question.[117] In fact, some courts have even gone so far as to hold that parents who have minor children dependent upon them "are not even free to risk their own lives in a religious cause--not out of paternalistic concern for the parents but out of concern for the welfare of the children."[118]

Perhaps the leading state court decision enunciating the rule that there is no right to refuse medical treatment and thereby die is, John F. Kennedy Memorial Hospital v. Heston.[119] Here, the New Jersey Supreme Court was faced with an application of a hospital to order a patient to accept a blood transfusion considered essential to a lifesaving operation. The patient, a twenty-two year old adult, was not lucid--but her mother opposed the transfusion because of her daughter's membership in the Jehovah's Witness sect.[120] The court, nonetheless, granted the relief sought and observed, "It seems correct to say there is no constitutional right to choose to die. . ."[121] Chief Justice Weintraub observed that:

> "Appellant suggests there is a difference between passively submitting to death and actively seeking it. The distinction may be merely verbal, as it would be if an adult sought death by starvation instead of a drug. If the State may interrupt one mode of self-destruction, it may with equal authority interfere with the other. It is arguably different when an individual, overtaken by illness decides to let it run a fatal course. But unless the medical option itself is laden with the risk of death or of serious infirmity, the State's interest in sustaining life in such circumstances is hardly distinguishable from its interest in the case of suicide."[122]

The recognition of the right to refuse medical treatment based upon a religious belief or otherwise has nonetheless been upheld--and with growing regularity.[123] In 1976, the New York Supreme Court held that members of the Jehovah's Witness religious sect could, indeed, refuse to submit to blood transfusions.[124] An application by a hospital was made for an order authorizing a qualified physician from the hospital to perform a blood transfusion on a twenty-three year old married patient--fully competent, not pregnant and with no children. A dilation and curettage had been performed on her December 3, 1976. Some twenty-four hours after the operation, she developed a uterine hemorrhage which resulted in lowering her hemoglobin count dramatically. She, in turn, developed anemia. The patient and her husband, as members of the Jehovah's Witness sect, had both signed documents which refused to permit blood transfusions. The application sought permission to perform--given a favorable opinion for such action by a qualified physician--a blood transfusion or undertake any other surgery considered necessary to save the life or protect the patient's health.

The court denied the application stating in part:

"As a general rule, every human being of adult
years and sound mind has a right to determine what
shall be done with his own body and cannot be subjected to medical treatment without his consent. . .
specially where there is no compelling interest which
justifies overriding an adult patient's decision not
to receive blood transfusions because of religious
beliefs, such transfusions should not be ordered. . .
Such an order would constitute a violation of the
First Amendment's freedom of exercise clause. . .
However, judicial power to order compulsory medical
treatment over an adult patient's objection exists in
some situations. . . It may be the duty of the court
to assume the responsibility of guardianship for a
patient who is not compos mentis to the extent of
authorizing treatment necessary to save his life even
though the medical treatment authorized may be contrary to the patient's religious belief. . . Here the
patient is fully competent, is not pregnant and has no
children. Her refusal must be upheld, even if the procedure is necessary to save her life."[125]

Applying a basic substituted judgment principle previously analyzed in connection with the Saikewicz case, the New Jersey Supreme
Court in the now famous case of Matter of Quinlan[126] determined that
as to the condition of Karen Quinlan--a twenty-two year old woman surviving at a public hospital in a debilitated and moribund state--"no
external compelling interest of the State could compel Karen to endure
the unendurable, only to beget a few measurable months with no realistic possibility of returning to any semblance of cognitive or sapient
life."[127] The court continued by observing that the "only practical
way to prevent destruction of [her] right "to resist further bodily
invasion and thus to die in dignity is to permit the guardian and
family of Karen to render their best judgment. . . as to whether she
would exercise [her right] in these circumstances."[128] It is inter-
resting to note the change in attitude of the New Jersey Supreme Court
from 1971 and its decision in Heston recognizing no right to die, and
five years later in the Quinlan case. The latter view is certainly
the more humane and enlightened one and the one that should be
ascribed to by the courts.

The United States Supreme Court departed from an enlightened and
compassionate view of life--and of the circumstances under which could
be ended--in a decision rendered on June 18, 1979. The court, in construing provisions of the Federal Food, Drug and Cosmetic Act relevant
to the interstate distribution of the drug laetrile, determined that
the Commissioner of the Food and Drug Administration had correctly determined that the drug was a "new drug" within the meaning of the Act
and, furthermore, that: no uniform definition of Laetrile exists;
that no adequate well-controlled scientific studies of its safety or
effectiveness exist; and that the drug, when administered orally,
shows itself to be toxic and--therefore--it could not be marketed in
interstate commerce. The Act was found to make no express exception
for drugs taken by the terminally ill.[129] Thus, by this decision,
dying persons suffering from cancer are prevented from receiving a
course of treatment which they not only want, but believe in--regard-
less of the questioned usefulness of alleged fraudulent nature of such
treatment. The court is also holding, in effect, that one has only a
conditional right to die, conditioned by those circumstances prescribed
solely by the government.

Although the major thrust of the legal determination in this case
is tied to a basic exercise in statutory interpretation, it presents a
far more complex study in political and social values. Of principal

concern here is the extent to which a citizen may have his constitutionally protected right of privacy infringed upon. Are not both the court and the federal government implicitly using a type of substituted judgment which in effect states to the citizenry that in certain matters of health and safety it is not sufficiently possessed of rational powers of thought and analysis, nor has the vast information resources available to it that the government does and, thus, is not capable of making definitive judgments in such matters?

Just as a clear pattern begins to emerge in the more current cases of a right to refuse treatment when the decision is made rationally, the Supreme Court implies, admittedly, more by its silence than by its direct action, that a more informed or superior rationality imposes in the federal government and that its exercise is superior to individual autonomy. By refusing to show a spirit of judicial activism here, the high court has allowed critical policy areas to go uncharted and, as such, bend to the winds of indecisiveness and capriciousness. If this attitude of passivity is subscribed to by the lower courts, a disappointing retrenchment of personal liberty will be recorded and the heretofore waning principle of parens patri resuscitated.

Finally, in state imposed situations of confinement such as normally seen in prisons, the right of the prisoner to refuse rehabilitative therapy or to participate in an experimental drug program is a right which at best is qualified and at worse, subject to coercion. There are severe difficulties in showing a voluntary, knowing capacity to giving an informed consent under such conditions.[130]

In Clonce v. Richardson, a Missouri federal district court implied that the full protection of due process must always be provided whenever a prisoner's status within an institution is sought to be altered to his detriment.[131] And, of course, in the landmark Kaimowitz case, in 1973, a lower state court in Michigan held that one who is involuntarily committed to an institution is, from a legal standpoint, incapable of consenting to a surgical intervention (psychosurgery).[132] A number of other cases has put a large segment of the judiciary on record in support of providing basic constitutional protections to institutionalized persons.[133]

Today, research undertaken which does not have a significant likelihood of providing a benefit for the patient is at an end in public institutions. In fact, it has been predicted that large scale treatment of mental retards, prisoners or other patient populations without psychotic disorder for purpose of achieving behavior control or modification will yield a good number of cases in malpractice litigation.[134]

Notes--Chapter 3

1. C. FRIED, MEDICAL EXPERIMENTATION: PERSONAL INTEGRITY AND SOCIAL POLICY 18 (1974).

See also, EXPERIMENTATION WITH HUMAN BEINGS, Chs. 8, 9, 10 (J. Katz ed. 1972); A. ETZIONI, GENETIC FIX, Ch. 5 (1973).

See Appendix A for AMA Model Consent Form.

The canon of loyalty joining men together in medical practice as well as investigation is the principle of informed consent. The consensual relation in medical experimentation is to be found in the common cause of the advancement of medicine to benefit others. In therapy--be it in the nature of diagnostic or therapeutic investigations--the common cause is some benefit to the patient himself; "but this is still a joint venture in which the patient and physician can say and ideally should both say, 'I cure.'" P. RAMSEY, THE PATIENT AS PERSON 5, 6 (1970).

See generally, Comment, "Informed Consent to Medical Treatment: The Oregon Standard of Disclosure," 57 ORE.L. REV. 322 (1978).

"No right is held more sacred, or is more carefully guarded, by the Common Law, than the right of every individual to the possession and control of his own person, free from all restraint or interference of others, unless by clear and unquestionable authority of law." Union Pacific Railway Co. v. Botsford, 141 U.S. 250, 251 (1891).

John Stuart Mills' concept of freedom of choice has been identified by many as the premise upon which the decision maker has the ability to choose in matters which have a direct adverse effect on no one but himself. Winters v. Miller, 446 F.2d 65, 73-74 (2d Cir. 1971), Moore, J., concurring and dissenting; cert. denied, 404 U.S. 985 (1971).

The functions of informed consent are to: promote individual autonomy; protect the patient-subject's status as a human being; encourage self-scrutiny by the physician-investigator; encourage rational decision making; and involve the public. J. KATZ, A. CAPRON, CATASTROPHIC DISEASES: WHO DECIDES WHAT? 82 passim (1975).

It has been suggested that the descriptive word, "informed," should not be used to qualify consent here. Rather, what should be done is to strive to set a standard of conduct not for the citizen, but for the authority. In order to effectively safeguard a patient's right of self decision, the authorities should provide him with a real opportunity (not with an obligation) to possess what information he and a reasonable person might in fact require in order to exercise a choice. Hence, the process of informing for decision rather than informed consent should be developed. Goldstein, "For Harold Laswell: Some Reflections on Human Dignity, Entrapment, Informed Consent and the Plea Bargain," 84 YALE L. J. 683, 692 passim (1975).

See also, Note, "Restructuring Informed Consent: Legal Therapy for the Doctor-Patient Relationship," 79 YALE L. J. 1533 (1970).

2. B. GRAY, HUMAN SUBJECTS IN MEDICAL EXPERIMENTATION 202 (1975); G. ANNAS, L. GLANTZ, B. KATZ, INFORMED CONSENT TO HUMAN EXPERIMENTATION: THE SUBJECT DILEMMA, Ch. 1 (1977).

3. GRAY, supra at 204.

See also, Annas, "Problems of Informed Consent and Confidentiality in Genetic Counseling," in GENETICS AND THE LAW (A. Milunsky, G. J. Annas ed. 1976)

Dr. Jay Katz terms the present law of informed consent as "substantially mythic and fairy tale-like as far as advancing patients' rights to self-decision-making is concerned." While acknowledging human capacities for intelligent choice, the law nonetheless conveys at the same time a distinct tone of permission regarding the extent of human capacity to be active participants as "choice-makers". Katz, "Informed Consent--A Fairy Tale? Law's Vision," 39 U.PITT.L. REV. 137, 1974 (1977).

4. Schloendorff v. New York Hospital, 211 N.Y. 127, 105 N.E. 92 (1914).

"Every human being of adult years and sound mind has a right to determine what shall be done with his own body; and a surgeon who performs an operation without his patient's consent, commits an assault for which he is liable in damages." 211 N.Y. at 129, 105 N.E. at 93. Schloendorff was overruled by Bing v. Thunig, 2 N.Y. 2d 656, 143 N.E. 2d 3 (1957), on the grounds that charitable immunity was no longer to be extended to hospitals in New York. Bing, however, did not affect the necessity of consent from the patient to treatment by the doctor.

5. FRIED, supra, note 1 at 20.

Some courts have nonetheless required that patients be told about the hazards and advantages of alternative forms of treatment. Canterbury v. Spence, 464 F.2d 722 (D.C. Cir. 1972; Durham v. Wright, 423 F.2d 940 (3rd Cir. 1970); Campbell v. Oliva, 424 F.2d 1244 (6th Cir. 1970)

6. Restatement (Second) of Torts, § 892A, (Tentative Draft No. 18, April 26, 1972). Comment "e" states that: "The consent must be to the actor's conduct, or to substantially the same conduct, rather than to the invasion which results from it. Consent to an invasion by particular conduct is not consent to the same invasion by entirely different conduct." FRIED, supra at 24.

7. See generally, N. HERSHEY, INFORMED CONSENT STUDY: THE SURGEON'S RESPONSIBILITY FOR DISCLOSURE TO PATIENTS. (Aspen Systems Corporation 1969). This study was followed to determine what standards are used by physicians in disclosures to patients. From the collected responses a model consent form was drawn. FRIED, supra note 1 at 21.

8. There are generally two views of what information should be disclosed to the patient:

1. Aiken v. Clary, 396 S.W. 2d 668 (Mo. 1965), lists the reasonable medical practitioner standard--which is regarded as the majority view. Here, the controlling point is tied *not* to the degree to which the juror would relate to the patient under similar circumstances, but rather, what the physician would disclose taking into account the state of the patient's health, the condition of his heart and nervous system, his mental state, and furthermore, whether the risks involved were remote possibilities and not something which happened with any frequency or regularity. See also, DeFillippo v. Preston, 3 Storey 539, 53 Del. 539, 173 A.2d 333 (1961); Haggarty v. McCarthy, 344 Mass. 136, 181 N.E. 2d 562 (1966); Roberts v. Young, 369 Mich. 133, 119 N.W. 2d 627 (1963); W PROSSER, TORTS § 32 at 65 (4th ed. 1971).

2. Other states follow the minority view and require what the

reasonable patient would want to know under the circumstances. See Cooper v. Roberts, 220 Pa. Super. 260, 286 A. 2d 647 (1971); Gray v. Grunnagle, 423 Pa. 144, 223 A.2d 663 (1966). For a criticism of the majority standard see, Note, "Informed Consent in Malpractice," 55 CAL.L. REV. 1396 (1967).

One could structure a third viewpoint here--that being the subjective patient's right to refuse treatment for any reason (irrational, religious or otherwise). The so-called Jehovah Witness cases where members of this sect have commonly refused medical treatment (i.e., blood transfusions) because of their religious convictions forbidding such acts are illustrative. These and other cases will be analyzed in depth at notes 115-124, infra.

See Canterbury v. Spence, 464 F.2d 772 (D.C. Cir. 1972), where what must be disclosed is measured against an objective standard: "[a] risk is thus material when a reasonable person, in what the physician knows or should know to the patient's position, would be likely to attach significance to the risk or cluster of risks in deciding whether or not to forego the proposed therapy." 464 F.2d at 787. See generally, Laforet, "The Fiction of Implied Consent," 235 J.A.M.A. 1579 (April 12, 1976); Fellner & Marshall, "The Myth of Informed Consent," 126 AM. J. PSYCHIATRY 1245-1250 (1970).

10. 2 Wils. K.B. 359, 95 Eng. Rep. 860 (1967).

11. 2 Wils. K.B. 359, 362, 95 Eng. Rep. 860, 862 (1967).

12. See, Wall v. Brim, 138, F.2d 478 (5th Cir. 1943), consent issue tried on remand, verdict for plaintiff aff'd, 145 F.2d (5th Cir. 1944), cert. denied 324 U.S. 857 (1945); Mohr v. Williams, 95 Minn. 261, 104 N.W. 12 (1905); Pratt v. Davis, 118 Ill. App. 161 (1905); Carpenter v. Blake, 60 Barb. 488 (N.Y. Sup. Ct. 1871), reversed on other grounds, 50 N.Y. 696 (1872). In Williams and Davis, plaintiff alleged a total lack of consent to the surgery which was performed on them and recovered.

The principal reason why the doctrine of informed consent was not recognized sooner is due to the general ignorance of the public to basic medical practices and the very jargon or terminology used by the profession. Doctors also feared that disclosing to the patient too much information would unnecessarily worry those patients in therapy.

A patient's right not to know has been upheld. Putsenson v. Clay Adams, Inc., 91 Cal. Rptr. 319 (1970).

13. 154 Cal. App. 560, 578, 317 P. 2d 170, 181 (1957).

14. See Note, 75 HARV.L. REV. 1445, 1446 (1962). If the battery theory were used, the physician may be held liable for acts which the patient consented to. However, using the negligent nondisclosure theory will only subject the practitioner to risks to which the plaintiff did not consent. See also, Oppenheim, "Informed Consent to Medical Treatment," 11 CLEV-MAR.L. REV. 249 (1962); A. MORITZ, R. MORRIS, C. HIRSCH, HANDBOOK OF LEGAL MEDICINE, Ch. 19 (4th ed. 1975).

Battery actions have largely in recent years--although not exclusively--involved cases in which an improper operation has been performed, due usually to a confusion or "mix up" in hospital charts. Byron v. Stone, 190 N.W. 2d 336, (Mich. 1971); Ebaugh v. Rankin, 99 Cal. Rptr. 706 (1972). No physical damage need be proven in battery cases of this nature simply because the interest which is legally protected here is of a dignitary nature. Yet, interestingly, the courts appear to bar claims grounded on a theory of battery unless the patient gave no consent at all to the surgical procedure. Dow v. Kaiser Foundation Hospitals, 90 Cal. Rptr. 747 (1970); Trogun v. Fruchtman, 207 N.W. 2nd 297 (Wisc. 1973). The Trogun decision provides an excellent discussion of the conflicts between the theories of assault and battery and negligence. As early as 1940, it was held in Bednarik v. Bednarik, 18 N.J. Misc. 633, 16 A.2d 80 that a physician who treats a patient in spite of an informed refusal of treatment by the patient commits a battery unless there is an emergency and the patient is unconscious.

To recover under a theory of negligence—as opposed to battery in informed consent cases—the patient must meet a complicated and sometimes onerous burden of evidentiary proof. The requirements under negligence of establishing a duty of care and causation, as well as the more limited measure of damages, make an action grounded in a negligence theory less attractive than battery. Yet, the prospect of recovering an award of considerable amount in negligence acts as a strong compensating factor. To be remembered is the fact that awards for dignitary injuries under a theory of battery normally are but nominal. The extent to which a plaintiff encounters difficulty working with a negligence theory, however, will depend in very large part upon a particular court's views on the scope of the physician's duty under the facts of the case and its corresponding view of causation. Expert testimony in charting the parameters of both will be of crucial significance. See Riskin, "Informed Consent: Looking for the Action," 1975 U. ILL. L.F. 580, 585.

One commentator has observed, further, that the very ambivalence of judges toward patients' self-determination was evidenced in the early competition between battery and negligence doctrines for deciding the issues created by claims of lack of informed consent. "In virtually every jurisdiction, they resolved the conflict in favor of negligence law, disguising a basic policy choice between patients' self-determination and doctors' paternalism as a choice between battery and negligence doctrines." KATZ, supra note 3 at 165.

See generally, Meisel, "The 'Exceptions' to the Informed Consent Doctrine: Striking a Balance Between Compelling Values in Medical Decisionmaking," 1979 WISC.L. REV. 413.

15. Supra, note 13.

16. Id.

Salgo sets out with clarity the parameters of the physician's duty in cases where consent is an issue:

"A physician violates his duty to his patient and subjects himself to liability if he withholds any facts which are necessary to form the basis of an intelligent consent by the patient to the proposed treatment. Likewise the physician may not minimize the known dangers of a procedure or operation in order to induce his patient's consent. At the same time, the physician must place the welfare of his patient above all else and this very fact places him in a position in which he sometimes must choose between two alternative courses of action. One is to explain to the patient every risk attendant upon any surgical procedure or operation, no matter how remote; this may well result in alarming a patient who is already unduly apprehensive and who may as a result refuse to undertake surgery in which there is in fact minimal risk; it may also result in actually increasing the risks by reason of the psychological results of the apprehension itself. The other is to recognize that each patient presents a separate problem, that the patient's mental and emotional condition is important and in certain cases may be crucial, and that in discussing the element of risk a certain amount of discretion must be employed consistent with the full disclosure of facts necessary to an informed consent."

17. See Comment, "Informed Consent in Medical Malpractice," 55 CAL.L. REV. 1936, 1399 (1967).

Plaintiff's counsel, using the Salgo precedent, would assert that had plaintiff been forewarned of the very risk which occurred, he would not have asserted to the procedure in question.

Contrariwise, a court holding for the defendant could use Salgo for sustaining the proposition that the amount of information the physician gave the patient was sufficient for the patient to understand the risks and alternatives of the proposed treatment.

18. Id., at n's 34-45 and accompanying text, pp. 1402-1403.

19. 186 Kan. 393, 350 P.2d 1093 (1960); motion for rehearing denied, 187 Kan. 186, 354, P.2d 670 (1960).

20. Id.

Two days after Natanson was decided, Mitchell v. Robinson, 334 S.W. 2d 11 (Mo. 1960) was announced. There it was held, as well, that a physician has a duty to inform his patient of the hazards of the treatment being proposed. The treatment offered in this case was electro-shock and insulin subcoma therapy and there are many attendant risks, i.e., fractured bones, serious paralysis of limbs, irreversible coma and even death. The patient went into convulsions and fractured several vertebrae, sued and won claiming he was not warned of the risks.

21. Supra, note 19 at 1103.

Under the doctrine of therapeutic privilege, the general requirement for full and informed consent is qualified. A doctor is, thus, justified in withholding information from his patient if he has a reasonable belief that a disclosure of the information would not be in the patient's best interests in that receipt of it would interfere with the patient's effective treatment.

To be carefully noted is the fact that this therapeutic privilege is regarded as a modification of a doctor's duty to obtain consent "only insofar as it may be assumed that the patient would, if he were in a position to judge, ratify the doctor's decision." When the questioned information is withheld in an attempt to "subvert what the patient would choose if he had the information, then the privilege itself is abused, and liability imposed, FRIED, supra, note 1 at 21, 22 (1974).

See generally, Plotkin, "Limiting the Therapeutic Orgy: Mental Patients' Right to Refuse Treatment," 72 N.W.U.L. REV. 461 (1977).

See also, Patrick v. Sedwick, 391 P.2d 453 (Alaska 1964).

22. See Note, 74 HARV.L. REV. 1445, 1446 (1962).

See also, Plante, "An Analysis of Informed Consent," 36 FORD.L. REV. 639, 645 (1968).

23. Id.

24. Id.

24. 464 F.2d 772 (D.C. Cir. 1972); cert. denied, 409 U.S. 1064 (1972). Accord, Wilkinson v. Vesey, 295 A.2d 676 (R.I. 1972).

26. Id. at 777.

27. Id. at 776.

28. The procedure used is referred to as a laminectomy--the excision of the posterior arch of the vertebrae to correct a ruptured disc.

It has been argued--although not too persuasively--that Canterbury was renounced (sub silentio), by Haven v. Randolph, 161 U.S. App. D.C. 150, 494 F.2d 1069 (D.C. Cir, 1974). Laskey, "Canterbury v. Spence--Informed Consent Revisited," 9 THE FORUM 713 (1976). The per curiam decision in Haven affirmed the lower court consideration by Judge Flannery, 342 F. Supp. (D.C., D.C., 1972). Judge Flannery held there could be no recovery against a surgeon on the theory his failure to mention the risk of paraplegia in connection with a retrograde femoral arteriorgram performed on a minor plaintiff rendered the parents unable to have an informed consent to the procedure followed. The court found no evidence that such an omission was a deviation from the normal standard of care. This was the case particularly when such a risk had never before been encountered in a child of the minor plaintiff's age. Furthermore, there was no evidence that the parents would not have given their consent if they had been told of a possible paralysis. Id., at 543, 544.

In order for a plaintiff to recover in a case of this nature, the court determined it "must produce some evidence as to each of the following five elements:"

Plaintiff must show: 1. By expert medical evidence that which a reasonable medical practitioner practicing in the same or similar communities, under the same or similar circumstances, would have disclosed to the patient's parents about the reasonably foreseeable risks incident to the proposed diagnosis or treatment; and 2. That the physician departed from this standard by not disclosing this information to the patient's parents; and 3. That had all the reasonably foreseeable risks been disclosed to the patient's parents, as well as the risks of failing to proceed with the proposed operation, that the patient's parents would not have consented to the proposed diagnosis or treatment; and 4. That because the diagnosis or treatment proceeded without the informed consent of the patient's parents, that the patient was injured thereby; and 5. That the injury to the patient is measurably greater than the injury that would have occurred if the patient's parents had been fully informed and had chosen to prevent the treatment or diagnosis by refusing to consent.

Id. at 544.

29. Canterbury v. Spence, 464 F.2d 772 (D.C. Cir. 1972).

Interestingly, the Hippocratic Oath makes no mention of any duty of physicians to either inform, or, for that matter, converse with their patients. Hippocrates admonished physicians in his day, as deans of medical schools do for their graduating seniors today, to "Perform (these duties) calmly and adroitly, concealing most things from the patient while you are attending to him. Give necessary orders with cheerfulness and sincerity, turning his attention away from what is being done to him; sometimes reprove sharply and emphatically, and sometimes comfort with solicitude and attention, revealing nothing of the patient's future or present condition. . ."

The Pennsylvania Legislature has chosen to codify the parameters of informed consent in its Health Care Services Malpractice Act (Pa. Stat. Ann. Tit. 40, § 1301.101-1006 [Supp. 1979] at § 1301.103) thusly:

"Informed consent means for the purposes of this act and of any proceedings arising under the provisions of this act, the consent of a patient to the performance of health care services by a physician or podiatrist: Provided, That prior to the consent having been given, the physician or podiatrist has informed the patient of the nature of the proposed procedure or treatment and of those risks, and alternative to treatment or diagnosis that a reasonable patient would consider material to the decision whether or not to undergo treatment or diagnosis. No physician or podiatrist shall be liable for a failure to obtain an informed consent in the event of an emergency which prevents consulting the patient. No physician or podiatrist shall be liable for failure to obtain an informed consent if it is established by a preponderance of the evidence that furnishing the information in question to the patient would have resulted in a seriously adverse effect on the patient or on the therapeutic process to the material detriment of the patient's health."

30. Supra, note 13.

31. Supra, note 29 at 786.

32. Supra, note 17 at 1407-10.

33. Supra, note 29 at 787.

It is difficult to predict the full tide of Canterbury in judicial circles during the years ahead. Many state courts have renewed their adherence to standards of informed consent--albeit differently from those posited in Canterbury. More and more legislatures are becoming involved in drafting statutes, as those presently on the books in Georgia, which require only a writing setting forth in general terms the treatment proposed and making that writing a conclusively presumed valid consent in the absence of fraudulent misrepresentation of material facts in obtaining it.

The controversy over informed consent has had a salutary effect on the integrity and honesty required in a doctor-patient relationship in that a growing number of doctors are assuming an affirmative duty to disclose methods of treatment being used on their patients. Murphy, "Canterbury v. Spence, The Case and a Few Comments," 9 THE FORUM 716, 724, 725 (1976).

Professor Guido Calabresi, Sterling Professor of Law at Yale, has suggested the doctrine of informed consent is but a subterfuge or incentive for improved medical services. Lecture, "Risk Sharing in Medical Practice," Institute of Human Relations, Yale Medical School, Nov. 10, 1976, New Haven, Conn.

34. See, Dunham v. Wright, 423 F.2d 940, 941-942 (3rd Cir. 1970).

35. See, Roberts v. Wood, 206 F. Supp. 579, 583 (S.D. Ala. 1962).

As to the issues of autonomy and informed consent, see a thoughtful analysis in T. BEAUCHAMP, J. CHILDRESS, PRINCIPLES OF BIOMEDICAL ETHICS, Ch. 3 (1979).

36. Supra, note 32 at 1409-10; Note, 74 HARV.L. REV. 1445, 1448 (1962).

37. Supra, note 29 at 791.

38. Examples of how courts have treated the application of informed consent include: Corn v. French, 71 Nev. 280, 289 P.2d 173 (1955) (physician held liable where mastectomy was performed with a signed consent form, but patient had told physician that she did not want anything removed); Di Rosse v. Wein, 261 N.Y.S. 2d 623, 24 App. Div. 2d 510 (1965) (failure to tell of the danger of exfolliative dermatitis from "gold" treatment for rheumatoid arthritis, resulting in exfolliative dermatitis, imposed liability on the physician); Darrah v. Kite, 301 N.Y.S. 2d 286, 32 App. Div. 2d 208 (1969) (failure to give adequate and timely explanation of the risks of ventricolograms, imposed liability on a neurologist).

39. Supra, note 8.

As discussed previously, there are actually two reasonable man (or person) standards which can be applied when attempting to discern what should be disclosed. With the majority of states holding doctors to a reasonable medical practitioner standard, some degree of uniformity re standard setting may be achieved. This would be validated or established through testimony of other practitioners. Using the reasonable patient standard will but add to the present confusion in the field. For, by adherence to it, it will be possible for a good number of courts hearing the same case to arrive at differing conclusions as to what information must be disclosed--since the choice is left largely to the jury and a reliance upon their particular sympathies in each case. See also, Isele, "After Canterbury: The Need for Medical Experts in the Informed Consent Suit" 4 J. LEGAL MED. 17 (May 1976).

40. See, Waltz & Scheuneman, "Informed Consent to Therapy," 64 N.W.U. L. REV. 628 (1969); see also, Comment, "Informed Consent in Medical Malpractice," 55 CAL.L. REV. 1396 (1967).

41. Plante, "Analysis of Informed Consent," 36 FORD.L. REV. 639, 651 (1968), citing Woods v. Brumlop, 71 N.M. 221, 377 P.2d 520 (1962)

42. See, Gray v. Grunnagle 423 Pa. 144, 223 A.2d 663 (1966), where the patient understood "exploratory surgery" to mean diagnosis and not correction. To the surgeons, this meant if they found something, they would make every effort to correct it. The court found that there was no consent to the treatment and therefore, the doctor was found to be liable.

43. See Paulson v. Gonderson, 218 Wis. 578, 260 N.W. 448 (1935), holding that there was ambiguity between the "simple" mastoid operation described and the "radical" mastoid operation performed.

44. See, Wall v. Brim, 138 F.2d 478 (5th Cir. 1943), aff'd, 145 F.2d 492 (5th Cir. 1944), where the surgeon had previously described the operation as simple, and it turned out to be complicated. The patient was conscious, being under a local anesthetic, and the physician therefore had a duty to advise him, failure to do subjecting the physician to liability as such for the unpermitted operation.

45. See, Moss v. Rishworth, 222 S.W. 225 (Ct. of Appeals, Tex. 1920), where the sister of a child gave consent for removal of the child's tonsils. There was no emergency situation, hence the consent was not given by one able to do so and vitiated.

In litigation concerning informed consent, nearly all of the cases where patients have received awards clearly deficient treatment--if not legally determined negligence--was given by the physician and established by the facts of each case. Thus, inadequate medical treatment, not a genuinely unpreventable and untoward occurrence has been and continues to be the standard of judicial concern. Other than being a highly topical matter in the professional literature, and one which has tended to make the average physician unduly concerned and anxious, the practical effects of informed consent litigative "successes" appear, in reality, negligible. See A. HOLDER, MEDICAL MALPRACTICE LAW, Ch. 8 (1975); Holder, "Failure to Inform Patient of Nature and Hazards of Surgery," 8 PROOF OF FACTS 2d 145 (1976).

46. John Sutart Mill has observed that, "In the part which merely concerns himself, his (the individual's) independence is, of right, absolute. Over himself, over his own body and mind, the individual is sovereign." Introduction to Essay, "Of the Liberty of Thought and Discussion," in ESSAYS ON POLITICS AND SOCIETY by J.S. MILL 217 at 223, 224 (J. Robson ed. 1977).

See also, F. HARPER & F. JAMES, THE LAW OF TORTS (1968 Supp.) § 171.1 at 61: (writing of informed consent) "Individual freedom here is guaranteed only if people are given the right to make choices which would generally be regarded as foolish." See generally, Ingelfinger, "Informed (But Uneducated) Consent," 287 N. ENG. J. MED. 465 (1972).

47. See, EXPERIMENTATION WITH HUMAN BEINGS, 569-587 (J. Katz ed. 1972).

48. H. LAUGHLIN, THE LEGAL STATUS OF EUGENIC STERILIZATION 60 (1930).

49. In other words, the questioned legislation must have a factually rational basis or purpose. The classification set forth must be reasonable in light of the stated purpose of the legislation. There must also be a demonstration--upon attack, that there is no rational basis--of the absence of any alternative which would affect the statutory purpose while infringing on the fundamental right to a lesser degree. Lindsley v. Natural Carbonic Gas Co., 220 U.S. 61 (1911).

The argument is made by opponents of eugenic sterilization that there are two viable alternatives to sterilization which should be utilized: 1.) Genetic Screening and Counseling and 2.) Euthenics-- which, as previously noted, is the alternation of the environment in order to allow aberrant individuals to develop "normally" and lead a so-called normal life consistent with their defects. See Comment, "Eugenic Sterilization Statutes: A Constitutional Re-Evaluation," 14 J. FAM. L. 280, 301 passim (1975). The problem with the first alternative is that, because of the voluntary nature of such activities, they will never within the foreseeable future be able to accomplish their objectives.

Normalization--which is the center focus of the proposal for euthenics--while of admirable design, also can never really be an effective alternative to sterilization so long as society, rightly or wrongly, looks with fear and intrepidation upon those afflicted with a mental disease. A mentally retarded person cannot be sufficently normalized within the general social fabric. Within a closed institutional setting, however, efforts should be made whenever possible to provide opportunities for normalization. See Nirje, "The Normalization Principle and Its Human Management Implications," in CHANGING PATTERNS IN RESIDENTIAL SERVICES FOR THE MENTALLY RETARDED 179, 181 (R. Kugel, N.W. Wolfenberger eds.1969)

In O'Connor v. Donaldson, 422 U.S. 563 (1975), the court held that a non-dangerous mental patient who was not undergoing medical treatment and showed an ability to live safely in the community by himself, or with friends or relatives, could not be held in simple custodial confinement. With this decision, the court appears to be telling society that--under specified circumstances, normalization will be expected and, indeed, imposed.

50. Perry, "Abortion: The Public Morals, and the Police Power: The Ethical Function of Substantive Due Process," 23 U.C.L.A.L. REV. 689, 702 (1976).

51. Id., at 689.

52. Id., at 704.

53. Id., at 703.

   In 1887, the formal parameters of substantive due process were structured by Mr. Justice Harlan in the landmark case of Mugler v. Kansas, 123 U.S. 623, 661 (1887).

   "It does not at all follow that every statute enacted ostensibly for the promotion of [the public morals, the public health, or the public safety], is to be accepted as a legitimate exertion of the police powers of the State. There are of necessity, limits beyond which legislation cannot rightfully go. While every possible presumption is to be indulged in favor of the validity of a statute, the courts must obey the Constitution rather than the law-making department of government, and must, upon their own responsibility, determine whether, in any particular case, these limits have been passed. . . The courts are not bound by mere forms, nor are they to be misled by mere pretenses. They are at liberty--indeed, are under a solemn duty-- to look at the substance of things, whenever they enter upon the inquiry whether the legislature has transcended the limits of its authority. If, therefore, a statute purporting to have been enacted to protect the public health, the public morals or the public safety, has no real or substantial relation to those objects, or is a palpable invasion of rights secured by the fundamental law, it is the duty of courts to so adjudge, and thereby give effect to the Constitution."

54. Supra, note 50 at 726.

55. Id., at 723

56. Id., at 706, n. 83.

57. Id., at 735.

58. Id.

59. Id., at 736.

60. Meyer v. Nebraska, 262 U.S. 390, (1923) (parents have right that children study German language); Pierce v. Society of Sisters, 268 U.S. 510 (1925) (parents have right to direct education of children); Skinner v. Oklahoma, 316 U.S. 535 (1942) (person has right not to be compulsorily sterilized); Griswold v. Connecticut, 381 U.S. 479 (1965) (married persons have right to use contraceptives); Loving v. Virginia, 388 U.S. (1967) (black and white persons have right to marry each other); Eisenstadt v. Baird, 405 U.S. 438 (1972) (unmarried persons have right to use contraceptives); Stanley v. Illinois, 405 U.S. 645 (1972) (unmarried father has custodial rights to his child); Roe v. Wade, 410 U.S. 113, and Doe v. Bolton, 410 U.S. 179(1973) (person has right to procure abortion at certain times and under certain circumstances).

   In Roe, the court set three stages where state interest in the mother's health or that of potential human life becomes significant and, thus, the mother's privacy must be considered in relation to other issues and weighed against the validity of the state interest. Id., at 163. The first stage is to be understood as being prior to approximately the end of the first trimester. Here an abortion decision and its effectuation are left to a medical decision made by the attending physican to the pregnant woman, with no state interference. Id., at 164. After the first stage, (at approximately the end of the first trimester) the state--if it elects--may reasonably regulate abortion procedures in order to preserve and protect maternal health. Id. At the point subsequent to viability, which is a flexible point based upon a professional determination of developing medical skill and technical ability the state may act to regulate abortion to protect the fetus' life and, where appropriate medically, for the mother's health as well. Id., at 163-165.

See generally, Hellegers, "Fetal Development" in CONTEMPORARY ISSUES IN BIOETHICS 184 (T. Beauchamp, L. Walters eds. 1978); THE TEACHINGS OF THE SECOND VATICAN COUNCIL, Ch. 1, "The Dignity of the Human Person" (1966).

In Harris v. McRae, et al, decided in June 1980, in a 5-4 decision by the United States Supreme Court, it was held that government funding of medically necessary abortions is not required by either the Constitution or Title XIX of the Social Security Act. The decision thus upheld the Hyde Amendment and similar state statutes excluding most abortions from Medicaid coverage. 48 U.S. LAW WEEK 4941 (June 24, 1980). Mr. Justice White, in a concurring opinion sought to distinguish the instant case from Roe v. Wade by observing Roe did not "purport to adjudicate a right to have abortions funded by the government, but only to be free from unreasonable official interference with private choice. . ." The government does not in the instant case seek to interfere with or to impose any coercive restraint on the choice of any woman to have an abortion. The woman's choice remains unfettered." Id., at 4949.

61. 316 U.S. 535 (1942).

62. Id., at 541.

63. See Golding & Golding, "Ethical and Value Issues in Population Limitation and Distribution in the United States," 24 VAND.L. REV. 495 (1971).

64. See Comment, "Sterilization of Mental Defectives: Compulsion and Consent," 27 BAYLOR L. REV. 174 (1975).

See also, G. ANNAS, L. GLANTZ, B. KATZ, INFORMED CONSENT TO HUMAN EXPERIMENTATION: THE SUBJECT'S DILEMMA, Ch. 5 (1977).

In Jackson v. Indiana, 406 U.S. 715, 738 (1972), the United States Supreme Court held that basic concepts of due process require both the nature and the duration of commitment by one to an institution have some reasonable relation to the very purpose for which that individual is in fact committed. The right of protection from harm and the right to treatment are complementary and powerful tools in the arsenal of the institutionalized. See e.g., the Willowbrook Consent Order case in New York Assn. for Retarded Citizens, Inc. v. Carey, 393 F. Supp. 715 (E.D.N.Y. 1975) which covers some 23 categories for separate relief.

See generally, Note, "Implied Executive Authority to Bring Suit to Enforce the Rights of Institutionalized Citizens," 26 CATH.U.L. REV. 794 (1977).

To be remembered is the fact that "just as man is a sacredness in the social and political order, so he is a sacredness in the natural, biological order. He is a sacredness in bodily life." P. RAMSEY, THE

PATIENT AS PERSON, xiii (1970).

See Appendix I, The United Nations Declaration on the Rights of Mentally Retarded Persons.

See also, Plotkin, "Limiting the Therapeutic Orgy: Mental Patients' Right to Refuse Treatment," 72 NW.U.L. REV. 461 (1977); A. STONE, MENTAL HEALTH AND THE LAW: A SYSTEM IN TRANSITION (1976); Schapiro, "Legislating the Control of Behavior Control: Autonomy and the Coercive Use of Organic Therapies," 47 SO. CAL.L. REV. 237 (1974).

65. Comment, BAYLOR L. REV., supra at 190.

66. Id., passim.

See Note, "Procedural Safeguards for Periodic Review: A New Commitment to Mental Patients' Rights," 88 YALE L. J. 850 (1979); READINGS IN BIOETHICS, Pt. 7, (T. Shannon ed. 1976); Lewin, "Disposition of the Irresponsible: Protection Following Commitment," 66 MICH.L. REV. 721 (1968); Wolff, "Legal and Psychiatric Aspects of Voluntary Sterilization," 3 J. FAMILY LAW 103, 115-122 (1963).

67. Pope Pius XII said, in the "Moral Limits of Medical Research and Treatment," 44 ACTS APOSTOLICAE SEDIS 779 (Rome 1952), 3 PROCEEDINGS OF THE FIRST INTERNATIONAL CONGRESS OF NEUROTHERAPY 137 (1952):

"Moreover, in exercising his right to dispose of himself, his faculties and his organs, the individual must observe the hierarchy of the orders of values--or within a single order of values, the hierarchy of particular rights--insofar as the rules of morality demand. Thus, for example, a man cannot perform on himself or allow doctors to perform acts of a physical or somatic nature which doubtless relieve heavy physical or psychic burdens or infirmities, but which bring about the same time permanent abolition or considerable and durable diminution of his freedom, that is, of his human personality in its typical and characteristic function. Such an act degrades a man to the level of a being reacting only to acquired reflexes or to a living automation. The moral law does not allow such a reversal of values. Here it sets up its limits to the 'medical interests of the patient.'"

See also, Bravery v. Bravery, 3 All. E.R. 59 (Ct. of Appeals 1954):

". . . Take a case where a sterilization operation is done so as to enable a man to have the pleasure of sexual intercourse without shouldering the responsibilities attaching to it. The operation then is plainly injurious to the public interest. It is degrading to the man himself. It is injurious to his wife and to any woman whom he may marry, to say nothing of the way it opens to licentiousness; and, unlike contraceptives, it allows no room for a change of mind on either side. It is illegal, even though the man consents to it. . ." Id., at 68.

68. The decision whether to have children is, as has been noted, a fundamental right. Since voluntary sterilization is a decision whether to have children, it, too, should be regarded as such a fundamental right. See, Note, "Sexual Sterilization--Constitutional Validity of Involuntary Sterilization and Consent Determinative of Voluntariness," 40 MO.L. REV. 509, 520 (1975). See also, Note, 20 WAYNE L. REV. 1309 (1974). See Appendix B for a model voluntary sterilization statute.

69. See, Hathaway v. Worcester City Hospital, 475 F.2d 701 (1st Cir. 1973).

70. See, Murdock, "Sterilization of the Retarded: A Problem or a Solution," 62 CAL.L. REV. 917, 923 (1974).

See also, J. KATZ, A. CAPRON, CATASTROPHIC DISEASES: WHO DECIDES WHAT? 108 passim (1975).

71. 182 Neb. 712, 157 N.W. 2d 171 (1968), rehearing denied, 183 Neb. 243, 159 N.W. 2d 566 (1968), appeal dismissed, 396 U.S. 996 (1970).

72. 182 Neb. 712 at 721, 157 N.W. 2d 171 at 178 (1968).

73. 372 F. Supp. 1196 (D.D.C. 1974). See generally, Michelman, "In Pursuit of Constitutional Welfare Rights: One View of Rawl's Theory of Justice," 121 U. PA.L. REV. 962 (1973).

74. 42 U.S.C.§§ 300 et seq., 708(a), July 1, 1944, c. 373, Title X, § 1007, as added Dec. 24, 1970, Pub. L. 91-572, § 6(c), 84 Stat. 1508.

75. 42 U.S.C. §§ 601 et seq., Jan. 2, 1968, Pub. L. 90-248, Title II, § 241(b)(1), 81 Stat. 916.

76. Supra, note 73 at 1203.

In 1973, Minnie and May Alice Relf, aged 14 and 12, were sterilized under the authority of the previously cited federally funded family planning program. A million dollar damage suit was filed, alleging that the operations had been performed without the knowledgeable consent of the girls' parents. The agency contended that the operations were performed with the written consent of the girls' illiterate mother. The mother, though, asserted that she only gave consent for "shots" to be administered to her daughters. See 2 FAMILY PLANNING POPULATION REP. 77 (1973).

As can be seen, at issue was not only the question of who can consent for one incompetent to do so, but also, the qualifications of one giving consent for such an operation to be performed on another person. The result here was that the family planning sections of the Social Security Act (42 U.S.C. § 703(a) (1970); 42 U.S.C. § 602(a)(15), 1396 d(a)(vi) (Supp. III 1973), and the Public Health Service Act, 42 U.S.C. § 300a-5 (1970) were found deficient as to their statutory authority to fund the sterilization of any person incompetent under state law to consent to such an operation--either because of minority or mental deficiency.

See Walters, "Sterilizing the Retarded Child," 6 THE HASTINGS CENTER REPORT 13 (April 1976).

77. 397 U.S. 471 (1970).

The doctrine of informed consent is applied through the mechanism of proxy or substituted consent given by a parent or guardian to only two groups: children and those adjudicated individually as legally incompetent. For all other categories--such as, for example, mental competence (without a declaration of legal incompetence)--the subject or individual patient's capacity to consent is controlling in and of itself. For minors, their personal capacity to understand is in many cases largely irrelevant. So long as there is no discernible risk of harm, there is considerable professional support that parents should be allowed to consent to non-therapeutic research on a child. The law remains unsettled and without a definite point of direction here, however. Obviously, where research is undertaken for the unquestioned benefit of a particular child, there can be no question that the parent may consent to standard therapy--and arguably should--particularly in a life threatening situation where no standard therapy exists.

See McCormick, "Experimentation in Children: Sharing in Sociality," 6 THE HASTINGS CENTER REPORT 39 (Dec. 1976); McCormick, "Experimental Subjects--Who Should They Be?," 235 J.A.M.A. 2197 (1976); Shaw, Dilemmas of Informed Consent in Children, 289 N. ENG. J. MED. 885 (1976); Note, "The Minor's Right to Consent to Medical Treatment: A Corollary to the Constitutional Right of Privacy," 48 SO. CAL.L. REV. 1417 (1975); G. ANNAS, L. GLANTZ, B. KATZ, INFORMED CONSENT TO HUMAN EXPERIMENTATION: THE SUBJECT'S DILEMMA, Ch. 3 (1977).

In a Superior Court case decided in San Francisco in 1973, it was held that a child has certain constitutional rights and a parent cannot waive those rights, in a proxy consent situation. To be more specific, the case can be construed as holding that a parent should not be able to consent to non-beneficial research on his or her child. The obvious question that follows is why should the parent be allowed to consent to beneficial research which entails greater pain, risk or stigma than that produced by available treatment methods. Nielsen v. The Regents of the Univ. of Calif. Civ. No. 665-049 (S. F. Super. Ct., filed Sept. 11, 1973). See Komkle, "Nielsen v. The Regents: Children as Pawns or Persons?" 2 HASTINGS CONST. L. Q. 1151 (1975). It has been suggested that a parent should be allowed to waive its child's constitutional rights and admit it--even though under the age of twelve--to a mental health facility or enlist a child under twelve in a research project unless it is demonstrated that there is probable cause to believe that the conditions of the hospital or the research are of such a condition as to constitute neglect, abuse or abandonment. Stone, "The History of Future Litigation in Psychopharmacologic Research and Treatment," in LEGAL AND ETHICAL ISSUES IN HUMAN RESEARCH AND TREATMENT 19 at 27-28, 32 (D. Gallant, R. Force eds. 1978).

78. 42 U.S.C. § 601 et seq. (1976 ed. & Supp. IV).

A recent survey regarding attitudes by gynecologists on the issue of sterilization that 95% of them favored compulsory sterilization for welfare mothers with three or more illegitimate children, Dr. Curtis Wood, the President of the Association for Voluntary Sterilization stated: "People pollute, and too many people crowded too close together cause many of our social and economic problems. . . As physicians, we have obligations to our individual patients, but we also have obligations to the society. . . The welfare mess, as it has been called, cries out for solutions, one of which is fertility control." Herman, "Fighting Sterilization Abuse," 1 INSIDE 17, 18 (Jan.-Feb. 1977).

79. Supra, note 73.

See Hyer, "U.S. Bishops Bar Tubal Ligations--Catholics Tighten Sterilization Ban," Wash. Post, July 10, 1980, at 1, col. 2.

See generally, J. BOYLE, THE STERILIZATION CONTROVERSY: A NEW CRISIS FOR THE CATHOLIC HOSPITAL (1977).

80. 38 Fed. Reg. 20930 (1973).

81. Supra, note 73 at 1204.

82. Interview with Dr. M. W. Steele, Children's Hospital, Pittsburgh, Pa. (June 7, 1976, conducted by Mr. Stephen J. Stabler, the author's former research assistant).

83. Practicing Law Institute, HOSPITAL LIABILITY: CURRENT PROBLEMS 9-10 (1975).

84. 11 Page 257 (N.Y. 1844). Cf. Comment, "Sterilization of Mental

Defectives: Compulsion and Consent," 27 BAYLOR L. REV. 174, 190 (1975).

85. See, Strunk v. Strunk, 455 S.W. 2d 145 (Ky. 1969), where the Kentucky Court of Appeals upheld the power of a county court to order a kidney transplant from an incompetent adult to his dying brother. The justification for such use of the power was found in the benefit to the incompetent adult in having his brother alive and the doctrine of substituted judgment--if the incompetent were in full control of his mental powers he would have elected to donate the kidney.

However, in Ohio, while the case of In Re Simpson, 180 N.E. 2d 206 (Ohio Prob. Ct. 1962), used the power to authorize a sterilization of a sexually promiscuous, mentally retarded eighteen year old girl, Wader v. Bethesda Naval Hospital, 337 F. Supp. 671 (S.D. Ohio 1971) held that without express authority to authorize sterilizations there was no implied powers in equity to allow such an order.

86. See, In the Interest of M.K.R., 515 S.W. 2d 467, (Mo. 1974), where the court would not exclude sterilization from the operations which parents could give consent for their children. This case neglected to consider two important facts: that the operation, itself, was irreversible and arguably was a violation of fundamental right. But see, Prince v. Massachusetts, 321 U.S. 158, 170 (1944) where the court stated generally that: "Parents may be free to become martyrs themselves. But it does not follow that they are free, in identical circumstances, to make martyrs of their children before they have reached the age of full and legal discretion when they can make that choice themselves."

The general rule is that parental proxy consent is sufficient to authorize treatment of a medical nature for a minor child primarily because such treatment is presumed to be of a potential beneficial nature to the child. This rule does not hold in application, however, to a normal adult child. There is a split of authority regarding the rule's application to parental consent for a mentally retarded child. No case law has been found which challenged a parent's informed consent to allow the sterilization of a retarded minor child, however. See, Holmes v. Powers, 439 S.W. 2d 579 (Ky. 1968); Neuwrith, Heisler & Goldrich, "Capacity Competence, Consent: Voluntary Sterilization of the Mentally Retarded," 6 COLUM. HUMAN RIGHTS L. REV. 447, 453, 454 (1974-1975). See also, Ellis, "Volunteering Children: Parental Commitment of Minors to Mental Institutions," 62 CAL.L. REV. 840 (1974).

At Common Law, all minors were held to be incapable of consenting to surgical operations. No age below which minors were automatically incapable of giving consent to medical procedures on them was set by the Common Law. The degree of comprehension by the minor to the procedure involved would be the crucial factual test here. A number of children under the age of 10 today are probably capable of consenting to relatively minor medical procedures. The current position would be that a parent in giving a proxy consent for a minor has wide latitude to act since the courts assume a reasonable parent would not put his child's interests in jeoparty--either for the benefit of another individual or for the public in general. Skegg, "Consent to Medical Procedures on Minors," 36 MOD.L. REV. 370 (1973).

The American Medical Association's Ethical Guidelines for Clinical Investigation provide that, "minors or mentally incompetent subjects may be used as subjects only if the nature of the investigation is such that mentally competent adults would not be suitable subjects." A.M.A. OPINIONS & REPORTS OF THE JUDICIAL COUNCIL at 11 (1969).

Dr. Alan A. Stone of the Harvard Law School Faculty, has suggested that so long as the child is not abused or neglected, the parents should be allowed to give a proxy consent. Lecture, "Psychiatry for Lawyers," July 23, 1979, Harvard Law School, Cambridge, Mass.

87. See Murdock, supra, note 70 at 933.

Roe v. Wade, and its companion case, Doe v. Bolton (supra, note 60) did not decide the constitutionality of statutes requiring parental consent in addition to the minor's consent before the minor could have an abortion. The court failed to conclude in either of these cases that--on demand--a woman could have an abortion. Parental consent statutes have been justified by a rather strained argument that the compelling state interest which directs state intervention and proscription here is the state's desire to maintain the family as a social unit and to promote parental control over children. Comment, "Abortion: An Unresolved Issue--Are Parental Consent Statutes Unconstitutional?" 55 NEB.L. REV. 256, 262 (1976). See Swan, "Abortion on Maternal Demand: Parental Support Liability Implications," 9 VALP. UNIV.L. REV. 243 (1975); Roupas, "The Value of Life," 7 PHILOSOPHY & PUB. AFFAIRS 154 (1978) where the author posits the argument that abortion is not morally wrong. But see, THE TEACHINGS OF THE SECOND VATICAN COUNCIL, Ch. 1, "The Dignity of the Human Person" (1966).

See generally, S. NICHOLSON, ABORTION AND THE ROMAN CATHOLIC CHURCH (1978).

Statistics for 1978 show nearly 30% of all pregnancies in the United States uninterrupted by natural causes were terminated by legal abortions. The study concerned estimated that there were 1,374,000 legal abortions in 1978, up about 54,000 over 1977. Rich, "Legal Abortions Seen in 30 Pct. of Pregnancies," Wash. Post, Jan. 9, 1980 at A10, col. 1.

Interestingly, The British Medical Association recently published a precedent setting code of ethics allowing doctors to perform abortions on girls under 16 years of age without parental permission. Wash. Post, Jan. 25, 1980, at A19, col. 1.

Mr. Justice William J. Brennan, Jr., barred officials in Massachusetts from enforcing a state law passed in 1974 requiring unmarried women under the age of 18 to obtain parental consent or a court order before having an abortion. This action grows from the High Court's decision in Bellotti et al v. Baird et al, 428 U.S. 132 (1976) which required--irrespective of a decision by a three judge federal court that the statute in question was unconstitutional because it imposed a type of parental veto over abortions performed on minors--that the Massachusetts Supreme Judicial Court should be given the first opportunity to construe the questioned statute. N.Y. Times, Aug. 1, 1976, at 51, col. 6. On remand, after the Massachusetts Supreme Judicial Court construed the challenged statute, and the District Court again declared the statute unconstitutional and enjoined its enforcement, its judgment was affirmed by the U.S. Supreme Court. Bellotti v. Baird, 443 U.S. 622 (1979).

In Planned Parenthood of Central Missouri et al v. Danforth et al, 428 U.S. 52 (1976), which is regarded as a "logical and anticipated corollary to Roe v. Wade," by Mr. Justice Blackmun who delivered the opinion, it was decided that a Missouri statute which required parental consent for an unmarried woman under 18 years of age and required (during the first 12 weeks of pregnancy) the written consent of the spouse of the woman seeking an abortion unless a licensed physician certified that the abortion is necessary to preserve the mother's life, was unconstitutional. These two specific consent issues were held invalid because the state had no significant interest in safeguarding the family unit or otherwise in conditioning an abortion on the consent of the parents of an unmarried woman, under-18-year-old-minor, during the first twelve weeks of her pregnancy. The spousal consent was regarded as being unconstitutional for the reason that since the state cannot interfere with a woman's decision to have an abortion, it cannot delegate a veto power to the spouse. For an enlightened philosophical discussion of Danforth, see, P. RAMSEY, ETHICS AT THE EDGE OF LIFE, Ch. 1 (1978).

In a related case, Singleton v. Wulff, et al, 428 U.S. 106 (1976), the court found a Missouri statute in violation of the Equal Protection Clause of the Federal Constitution. The statute in question excluded abortions that were not "medically indicated" for the purposes for which Medicaid benefits are available to needy persons. The court, again through Mr. Justice Blackmun, observed that impecunious women cannot secure an abortion without their physician being paid by the state and that such women have a right to obtain abortions if they so wish and should not be hindered in the exercise of this right by this particular Missouri statute.

On March 28, 1978, the United States Supreme Court decided that when an Indiana Circuit Court judge, acting on the petition of the mother of a "somewhat retarded" 15 year old daughter, approved the petition for sterilization--within a broad grant of general jurisdiction conferred by statute upon the court--the judge was immune from suit under the doctrine of judicial immunity. The daughter had been told initially that she was to have her appendix removed rather than be sterilized. Two years later, she got married. Her inability to become pregnant led her to the discovery that she had been sterilized-- whereupon she and her husband filed suit in Federal District Court under the Federal Civil Rights Act, 42 U.S.C. § 1983. Under this particular section, one may have an action at law or in equity for deprivations of rights, privileges or immunities secured as such by the Constitution. Stump et al v. Sparkman, 435 U.S. 349 (1978). See also, Rosenberg, "Stump v. Sparkman: The Doctrine of Judicial Impunity," 64 VA.L. REV. 833 (1978).

See generally, Comment, "The Minor's Right to Abortion and the Requirement of Parental Consent," 60 VA.L. REV. 304 (1974); Lee & Paxman, "Pregnancy and Abortion in Adolescence: A Comparative Legal Survey and Proposals for Reform," 6 COLUM. HUMAN RIGHTS L. REV. 307 (1974-1975); Rich, "Teen Births Cost $8.3 Billion a Year," Wash. Post, May 13, 1979, at A10, col. 4.

88. Nat. Assoc. for Retarded Children, FACTS ON MENTAL RETARDATION 4 (1971).

In an interview with Dr. M. W. Steele, supra, note 82, he stated that when he is advising patients, he does not encourage or advise sterilization, but rather if the patient/client brings up the option, he will then explain it to them. It appears, however, that--as part of a comprehensive genetic counseling program--the option of sterilization be discussed as objectively as possible. Many patients/clients may not even know of this option.

See generally, J. KATZ, A. CAPRON, CATASTROPHIC DISEASES: WHO DECIDES WHAT? 108-112 (1975).

89. 22 U.S. (9 Wheat) 1 (1824).

90. 301 U.S. 1 (1937).

Using the power extended by the Jones & Laughlin case, which was originally espoused by Chief Justice Marshall in Gibbons, Congress can be easily recognized as having the power to control (all) research in eugenics. "Since research activities in this field have not only national, but international consequences, the Jones & Laughlin test and the need for uniformity of research controls in this field should make the federal action appropriate." Comment, "Government Control of Research in Positive Eugenics," 7 MICH. J. L. REFORM 615, 631 (1974).

91. See, Keyishian v. Board of Regents, 385 U.S. 589 (1967). There the court stated that legitimate governmental interests have to be tailored as narrowly as possible in order to avoid encroachment upon academic freedom. See generally, Comment, supra.

Academe is no longer able to support and sustain scientific research out of operating budgets. Today, almost unavoidably, a major part of the bill for research is paid by the federal government. See Metzer, "Academic Freedom and Scientific Freedom" 107 DAEDALUS 93, 107 (1978).

In balancing the government interest in regulation with the fundamental right of academic freedom--recognized as such in Sweezy v. New Hampshire, 354 U.S. 234 (1957)--the choice will be between the individual rights of the researcher and student and the protection of society against the consequences of potentially harmful scientific advances. Since this is an infringement of a fundamental right, there must be shown a compelling state interest in the regulation; and in the area of eugenics programs, it is more to society's advantage that the courts choose now, rather than be forced to accept later. Once the decisions have been made, the program implemented, the effects of such programs are put into operation. See, Bates v. City of Little Rock, 361 U.S. 516 (1960).

Mr. Justice Brandeis, in an eloquent dissenting opinion in New State Ice Co. v. Liebmann, 285 U.S. 262, 311 (1932) observed that, "To stay experimentation in things social and economic is a grave responsibility. Denial of the right to experiment may be fraught with serious consequences fo the Nation. It is one of the happy incidents of the federal system that a single courageous state may, if its citizens choose, serve as a laboratory, and try novel social and economic experiments without risk to the country." See also, Traux v. Corrigan, 257 U.S. 312, 344 (1921) where Mr. Chief Justice Holmes in a dissenting opinion cautioned that the Fourteenth Amendment to the Constitution should not be used in order to prevent social experiments in "the insulated chambers afforded by the several states."

See Robertson, "The Scientist's Right to Research: A Constitutional Analysis," SO. CAL.L. REV. 1203 (1978); Neville, "Philosophic Perspectives on Freedom of Inquiry," 51 SO. CAL.L. REV. 1115 (1978).

Dr. Richard C. Atkinson, the Director of the National Science Foundation, has wisely noted that efforts to censor or to suppress ideas--even if considered absurd--violate a fundamental freedom. Freedom of inquiry and research should not be taken as freedom from social responsibility, however. Atkinson, "Rights and Responsibilities in Scientific Research," THE BULLETIN OF ATOMIC SCIENTISTS 10 (Dec. 1978).

92. In Sweezy, supra, the court held that, "To impose a strait jacket upon the intellectual leaders in our colleges and universities would imperil the future of our Nation. No field of education is so thoroughly comprehended by man that new discoveries cannot yet be made. Particularly is that true in the social sciences, where few, if any, principles are accepted as absolutes. Scholarship cannot flourish in an atmosphere of sucpicion and distrust. Teachers and students must

always remain free to inquire, to study and to evaluate, to gain new maturity and understanding, otherwise our society will stagnate and die." Id., at 250.

It is rather safe to assume that anything possible in the scientific field will eventually be attempted--particularly as to private, non-government supported research projects. This fact is particularly disturbing to some when DNA research and experimentation is considered. The Environmental Defense Fund and Natural Resources Council petitioned The Department of Health, Education and Welfare in Washington, D.C., charging its safety regulations concerning DNA did not cover private industry and were, thus, inadequate. It was submitted that HEW has legal powers to impose regulations on private, non-federally funded laboratories through the Public Health Services Act which empowers the Secretary to enforce whatever regulations "as in his judgment may be necessary" to prevent the introduction and spread of communicable diseases from foreign countries into the United States or from one state to another. It is generally believed, however, that HEW will not seek to control private experimentation here for, if they did act in this area, great and justifiable alarm would be registered that such actions would be commonplace and extended to other areas of vital concern; and of course the problems of enforcement would be significant. Rensberger, "Lawyers Seek Broader Curbs on DNA Study," N.Y. Times, Nov. 11, 1976 at 18, col. 5.

See also, Callahan, "Recombinant DNA: Science and the Public," 7 THE HASTINGS CENTER REPORT 20 (April 1977); Dismukes, "Recombinant DNA: A Proposal for Regulation," 7 THE HASTINGS CENTER REPORT 25 (April 1977); Cohn, "'Products' of Genetic Engineering Seen Less Than Five Years Away," Wash. Post, Nov. 11, 1977, at A2, col. 3.

See generally, supra, note 9, Introduction, and Appendix G, infra; Hutt, "Research on Recombinant DNA Molecules: The Regulatory Issue," 51 SO. CAL.L. REV. 1435 (1978).

93. C. FRIED, MEDICAL EXPERIMENTATION: PERSONAL INTEGRITY AND SOCIAL POLICY 25 (1974).

94. Id., at 26.

See Whalen v. Roe, 429 U.S. 589, 604 (1977), where the right of a physician-scientist to administer new or experimental medical treatments to a willing patient-subject was validated as a right derivative of the patients.

95. Id.

96. Id., at 27.

See, R. Bonnie, P. Hoffman, "Regulation of Human Experimentation: A Re-appraisal of Informed Consent," in HUMAN EXPERIMENTATION 52-79 (R. Bogomolny ed. 1976); J. KATZ, A. CAPRON, CATASTROPHIC DISEASES: WHO DECIDES WHAT?, Ch. 8 (1975); R. M. VEATCH, CASE STUDIES IN MEDICAL ETHICS (1977); Gauvey, Leviton, Shuger & Sykes, "Informed and Substitute Consent to Health Care Procedures: A Proposal for State Legislation," 15 HARV. J. LEG. 431 (1978); DuVal, "The Human Subjects Protection Committee: An Experiment in Decentralized Federal Regulation," 1979 AM. BAR. FOUND. J. 571.

97. Supra, note 93 at 27.

98. Id., at 29.

99. Id.

100. Id., at 30.

101. Id., at 31.

102. Id., at 32.

103. Supra, note 101.

104. Id.

See, N. HERSHEY, R. MILLER, HUMAN EXPERIMENTATION AND THE LAW at 111 passim (1976) for a survey of state laws and regulations involving experimentation with human subjects and their informed consent to acts thereof.

The Federal Government's guide for the protection of human subjects--structured before the recent work of the National Commission for the Protection of Human Subjects of Biomedical and Behavorial Research--set forth the following elements required for a valid informed consent as:

1. A fair explanation of the procedures to be followed, and their purposes, including identification of any procedures which are experimental;

2. A description of any attendant discomforts and risks reasonably to be expected;

3. A description of any benefits reasonably to be expected;

4. A disclosure of any appropriate alternative procedures that might be advantageous for the subject;

5. An offer to answer any inquiries concerning the procedures; and

6. An instruction that the person is free to withdraw his consent and to discontinue participation in the project or activity at any time without prejudice to the subject.

No such informed consent, oral or written. . . shall include any exculpatory language through which the subject is made to waive, or to appear to waive, any of his legal rights, including any release of the organization or its agents from liability for negligence.

U.S. DEPT. OF HEALTH, EDUCATION & WELFARE, THE INSTITUTIONAL GUIDE TO DHEW POLICY ON PROTECTION OF HUMAN SUBJECTS 8 (1971), amended by 39 FED. REG. 18914 (1974). Cf. 21 C.F.R. § 130.37 (1972)--FDA policy on informed consent.

See also, Koskoff, "The Kaimowitz Case: A Short Term Legal Restraint Contrary to the Long Term Public Good," 13 DUQUESNE L. REV. 879 (1975). The Kaimowitz case is the leading authority regarding the scope of patients rights re compulsory psychotherapy. See Kaimowitz v. Dept. of Mental Health, Civil No. 73-19434-AW (Cir. Ct. Wayne Co., Mich., July 10, 1973) summarized at 42 U.S.L.W. 2063 (July 31, 1973), holding that an involuntarily institutionalized patient was legally incapable of consenting to psychosurgery; Ayres & Holbrook, "Psychotherapy and the Duty to Warm: A Tragic Trilogy?" 27 BAYLOR L. REV. 677 (1975); G. ANNAS, L. GLANTZ, B. KATZ, INFORMED CONSENT TO HUMAN EXPERIMENTATION: THE SUBJECT'S DILEMMA, Ch. 7 (1977).

The Kaimowitz court accepted a rather unique argument, drawn from a thesis propounded in Shapiro, "Legislating the Control of Behavior

Control: Autonomy and Coercive Use of Organic Therapies," 47 SO. CAL.L. REV. 237 (1974). The thesis was that a person's mental processes, communication of ideas and the generation of ideas, come within the ambit of the First Amendment to the United States Constitution. Since the brain is reponsible for generating ideas, brain surgery raises First Amendment questions. Therefore, to allow experimental psychosurgery under these circumstances is "to condone state action in violation of basic First Amendment rights of such patients, because impairing the power to generate ideas inhibits the full dissemination of ideas." Thus, Kaimowitz can be read as suggesting prisoners cannot give their informed consent to anything. Stone, "The History and Future of Litigation in Psychopharmacologic Research and Treatment" in LEGAL AND ETHICAL ISSUES IN HUMAN RESEARCH AND TREATMENT 19, 31-32 (D. Gallant, R. Force eds. 1978).

See CAL. PENAL CODE §§ 2670-2680 (West 1979 Supp.) for a strong law on organic therapy and informed consent to psychosurgery and shock treatment.

See also, Schmeck, "Panel Urges that Psychosurgery be Continued on Research Basis," N.Y. Times, Sept. 11, 1976, at 8, col. 1, reporting the recommendation of the National Commission for the Protection of Human Subjects of Biomedical and Behavorial Research.

See Appendix J, infra, The Belmont Report of The National Commission for the Protection of Human Subjects of Biomedical and Behavioral Research for a review of various ethical considerations to be undertaken in experimentation with humans.

105. Reback, "Fetal Experimentation: Moral, Legal and Medical Implications," 26 STAN.L. REV. 1191, 1201 (1974).

106. Id.

107. Id., at 1202.

108. Id.

109. Id.

110. About 20 to 30% of conceptuses, embryos and fetuses "die" by spontaneous or natural abortion. Almost all such aborts are defection--genetically or congenitally. "Nature takes the same way medicine does: it closes the book on failures." J. FLETCHER, THE ETHICS OF GENETIC CONTROL 51 (1974).

111. P. RAMSEY, THE ETHICS OF FETAL RESEARCH 39 (1975).

See generally, Comment, "Abortion: An Unresolved Issue--Are Parental Consent Statutes Unconstitutional?," 55 NEB.L. REV. 256 (1976) and the discussion of current Supreme Court decisions related thereto at supra, note 87.

See also, Camenisch, "Abortion: For the Fetus's Own Sake?" 6 THE HASTINGS CENTER REPORT 38 (April, 1976); D. GRANFIELD, THE ABORTION DECISION (1969).

112. RAMSEY, supra at xix.

It has been suggested that the more certain we are that the benefits of experimentation with human fetuses will accrue only to society or to other fetuses and the more the still-living abortus alone will be damaged or suffer pain or injury, the more we should approve the research. The closer socially beneficial experimentation comes to bestowing some benefit also on the research subject, the more it deserves moral condemnation. RAMSEY, supra at 39.

See also, Tiefel, "The Costs of Fetal Research: Ethical Considerations," 294 N. ENG. J. MED. 85 (1976); Friedman, "The Federal Fetal Experimentation Regulations: An Establishment Analysis," 61 MINN.L. REV. 961 (1977).

113. "Every person who, under color of any statute, ordinance, regulation, custom or usage, of any state or territory, subjects, or causes to be subjected any citizen of the United States or other person within the jurisdiction thereof to the deprivation of any rights, privileges, or immunities secured by the Constitution and laws, shall be liable to the party injured in an action at law, suit in equity or other proper proceeding for redress, 42 U.S.C. § 1983 (1977)."

114. Supra, notes 77, 86, 114.

115. Prince v. Massachusetts, 321 U.S. 158 (1944). Supra, note 86.

116. Jehovah's Witness v. King County Hospital, 390 U.S. 598 (1968). (per curiam), aff'g 278 F. Supp. 488 (W.D. Wash. 1967). See L. TRIBE, AMERICAN CONSTITUTIONAL LAW 859 (1978).

117. TRIBE, supra.

See generally, Woolley, "Informed Consent to Immunization: The Risks and Benefits of Individual Autonomy," 65 CAL.L. REV. 1286 (1977).

118. Id.

The leading case here, which relied upon a theory of parens patriae to sustain government intervention was Application of President and Directors of Georgetown College, Inc., 331 F.2d 1000 (D.C. Cir.), rehearing en banc den., 331 F.2d 1010 (D.C. Cir.), cert. den. 337 U.S. 978 (1964). Here, the order of a trial judge authorizing a hospital to administer blood transfusions was upheld. The patient and her husband had refused to consent to the transfusions for religious reasons. Their objections were found insufficient to override the parens patriae interest of the sovereign.

See also, United States v. George, 239 F. Supp. 752 (D. Conn. 1965); Powell v. Columbia Presbyterian Medical Center, 49 Misc. 2d 215, 267 N.Y.S. 2d 450 (Sup. Ct. 1968). But see, In re Brooks Estate, 32 Ill. 2d 361, 205 N.E. 2nd 435 (1965) (in the case of a critically ill mother of adult children, court does not have the same right to order lifesaving measures against her religious beliefs as it would if minor children were involved.

See generally, Annot., Power of Courts or Other Public Agencies, In the Absence of Statutory Authority, To Order Compulsory Medical Care for Adult, 9 A.L.R. 1391 (1966); TRIBE, supra.

119. 58 N.J. 576, 279 A.2d 670 (1971).

120. The Jehovah's Witnesses have opposed blood transfusions primarily because of an interpretation of a biblical passage from Leviticus 17:10 (King James): "And whatsoever man be in the house of Israel, or of the strangers that sojourn among you, that eateth any manner of blood; I will even set My face against that soul that eateth blood, and will cut him off from among his people."

121. 58 N.J. at 580.

122. Id., at 581-582.

See also, Raleigh Fitkin-Paul Morgan Memorial Hospital v. Anderson, 42 N.J. 421, cert. den., 377 U.S. 985 (1964) (pregnant Jehovah's Witness ordered to submit to life sustaining blood transfusions); Hoener v. Bertinato, 67 N.J. Super. 517 (J.D.R.C. 1961) (blood transfusions ordered for infant over objections of Jehovah's Witness parents); United States v. George, supra, note 118; People v. Labrenz, 411 Ill. 618, 104 N.E. 2d 769 (Sup. Ct.), cert. den., 344 U.S. 824 (1952).

123. In re Osborne, 294 A.2d 372 (D.C. Ct. App. 1972); Erickson v. Dilgard, 44 Misc. 2d 27, 252 N.Y.S. 2d 705 (Sup. Ct. 1962); In re Maida Yetter, 62 D.C. 2nd 619 (Northampton County Orphans Ct., No. 1973-533, 1973).

In the Yetter case, the questioned mental competency of the patient was in issue over the appointment of a guardian whose sole purpose of appointment was to give consent to the performance of diagnostic and corrective surgery. Judge Alfred T. Williams, Jr. acknowledged that the constitutional right of privacy, "includes the right of a mature competent adult to refuse to accept medical recommendations that may prolong one's life and which, to a third person at least, appear to be in his best interest; in short, the right of privacy includes a right to die with which the state should not interfere where there are no minor or unborn children and no clear and present danger to public health, welfare or morals. If the person was competent while being presented with the decision and in making the decision which she did, the court should not interfere even though her decision might be considered unwise, foolish or ridiculous." Id.

124. Matter of Melideo, 390 N.Y.S. 2d 523 (Sup. Ct. 1976).

125. Id., at 524.

See Holmes v. Silver Cross Hospital of Joliet, 340 F. Supp. 125 (N.D. ILL. 1972). There the court upheld an action for damages against doctors and a hospital for violating the civil rights of a patient, who subsequently died, by medically treating him in a manner that was considered inconsistent with his religious beliefs. The facts do not reveal the patient's religious preference; presumably he was either a Jehovah's Witness or a Christian Scientist. They do disclose the fact that while fully conscious and competent, the patient advised the doctors that religious convictions prevented his acceptance of blood transfusions. Four hours after his admission and declaration, he lapsed into a state of unconsciousness. Thereupon the hospital petitioned the Circuit Court of Will County for the appointment of a conservator who would authorize the transfusion--owing to the patient's alleged incompetence. The authorization was given and the transfusion given.

See generally, Hegland, "Unauthorized Rendition of Life Saving Medical Treatment" in CONTEMPORARY ISSUES IN BIOETHICS at 162 (T. Beauchamp, L. Walters eds. 1978); T. BEAUCHAMP, J. CHILDRESS, PRINCIPLES OF BIOMEDICAL ETHICS 82 passim, (1979).

126. 70 N.J. 10, 355 A.2d 647 (1976).

See generally, Annas, "Reconciling Quinlan and Saikewicz: Decision Making for the Terminally Ill Incompetent," 4 AM. J. L. & MED. 367 (1979).

127. 335 A.2d at 663.

128. Id., at 664.

See TRIBE, supra note 116 at 936, 938.

In Eichner v. Dillon, 425 N.Y.S. ]d 517 (App. Div. 1980), it was held that the circumstances which would support the removal of life support systems were: if doctors certified the patient as being in an irreversible and permanent vegetative coma; or that the prospects of retaining brain function were extremely remote. A hospital committee, in turn, must agree with the diagnosis. Thereupon, a court hearing on the matter must be scheduled to precede the final decision at which time the Attorney General is to be notified and a guardian appointed to protect the interests of the patient. In Eichner, the patient had previously expressed the fact that he would not want "extraordinary business" done for him if he were to become terminally ill.

The court observed that the Constitutional Right of Privacy was the basis for sustaining a competent patient's right to refuse medical treatment and "to be let alone" to die with dignity. Finally, the standard of proof used to determine or test the medical criteria for activation of terminally ill comatose patients' right to refuse medical treatment must be established by "clear and convincing evidence."

Karen Quinlan still lives after four and one-half years since her parents asked that her respirator be turned off. She is not sustained by artificial life supports and weighs about 70 pounds. See Cohen, "We'll Go On--Quinlan Visit Comatose Daughter Daily," Wash. Post, Dec. 18, 1979, at 1, col. 1.

129. United States et al v. Rutherford, et al, 442 U.S. 544 (1979).

See also, Greenhouse, "Cases Force Value Conflicts on High Court," N.Y. Times, June 25, 1979, at A15, col. 4.

On December 21, 1979, the Food and Drug Administration announced a moderation of its strong opposition to laetril and observed preliminary tests had begun in order to see if the drug has any anti-cancer effect on humans. See, Wash. Post, Jan. 4, 1980, at A5, col. 1; Sheiring, "The Laetril Movement: A Challenge to the FDA's Regulating Authority," 1 J. LEGAL MED. 103 (1980).

Concerning the freedom to commit suicide, see generally, H. RESNIK, suicidal behaviors, Ch. 3 (1968).

130. See Szasz, "The Theology of Therapy: The Breach of the First Amendment Through the Medicalization of Morals," 5 N.Y. UNIV. REV. L. & SOC. CHANGE 127 (1975); Appendix f, infra (guidelines for experimentation with prisoners); E. PELLEGRINO, HUMANISM AND THE PHYSICIAN, Ch. 8, (1979).

131. 379 F. Supp. 338 (W.D. Mo. 1974).

See Gobert, "Psychosurgery, Conditioning and the Prisoner's Right to Refuse Rehabilitation," 61 VA. L. REV. 155 (1975).

See also, 3 ENCYCLOPEDIA OF BIOETHICS 1349-1354(W. Reich, ed. (1978).

132. Supra, note 104.

See also, A. STONE, MENTAL HEALTH AND LAW: A SYSTEM IN TRANSITION, Ch. 6 (1975).

133. Wyatt v. Stickney, 344 F. Supp. 373, 344 F. Supp. 387 (M.D. Ala. 1972) enforcing 325 F. Supp. 781, 334 F. Supp. 1341 (M.D. Ala. 1971), aff'd in part, decision reversed in part sub. nom; Wyatt v. Aderholt, 503 F.2d 1305 (5th Cir. 1974); Donaldson v. O'Connor, 493 F.2d 507 (5th Cir. 1974).

See Note, "The Wyatt Case: Implementation of a Judicial Decree Ordering Institutional Change," 84 YALE L. J. 1338 (1975).

134. Stone, "The History of Future Litigation in Psychopharmacologic Research and Treatment," in LEGAL AND ETHICAL ISSUES IN HUMAN RESEARCH AND TREATMENT, 19 at 24, 36 (D. Gallant, R. Force eds. 1978).

Rogers v. Okin, 478 F. Supp. 1342 (D. Mass. 1979), rev'd 634 F. 2d 650 (1980), structures a judicial approach to determining the parameters of the right of the mentally ill to refuse psychotrophic drug treatment. The Circuit Court rejected a strict emergency standard devised by the District Court and substituted its own flexible standard of balancing of competing interests based on a case-by-case basis and determined by a qualified physician. While recognizing the right of a psychiatric patient--found in the Due Process Clause of the Fourteenth Amendment to the United States Constitution--to be free to decide for himself whether to be subjected to treatment with potentially harmful drugs, the court also recognized the state's responsibility to care for the mentally ill who are at times dangerous to either themselves or others. When, owing to their illness, such patients are unable to make treatment choices, the state may assert its police and parens patriae powers as justification for the forced administration of anti-psychotic drugs. The right to refuse medication during an emergency is a medical decision which must be left to the judgment of physicians in each case. See also, Rennie v. Klein, 462 F. Supp. 1131 (D. N.J. 1978), 478 F. Supp. (1979).

See generally, Symonds,"Mental Patients' Rights to Refuse Drugs: Involuntary Medication as Cruel and Unusual Punishment," 7 HASTINGS CONST. L. Q. 701 (1980); Comment, "Madness and Medicine: The Forcible Administration of Psychotropic Drugs," 1980 WISC. L. REV. 497.

CHAPTER 4

WRONGFUL LIFE v. WRONGFUL BIRTH--AN IN DEPTH CONSIDERATION

In 1966, a Rhode Island court held that simple justice required the principle be recognized "that a child has a legal right to begin life with a sound mind and body."[1] The law of torts has been marked by a capacity to develop new theories of recovery when presented with new situations.[2] Entrance into new fields of recovery may at times be a slow process and this is true of the tort of wrongful life. Wrongful life claims, in theory, are made by an infant asserting that recovery should be allowed for injuries which are caused by the defendant's negligence which led to the plaintiff's wrongful existence.[3] Although recognized as a proper legal theory for recovery for approximately fourteen years,[4] it was recently validated and sustained in a lower court New York case, Park v. Chessin,[5] for but a brief period of time until "invalidated" by the highest tribunal in the state, the New York Court of Appeals.

An action for wrongful birth is often mistaken for a suit for wrongful life. The two are not the same cause of action. A suit for wrongful birth is maintained by the parents of a child which is usually unplanned. It is brought predominantly against a doctor for negligent performance of a sterilization operation,[6] but suits have in fact been allowed under this theory against, for example, a pharmacist for negligent filling of a prescription.[7] As will become apparent, recoveries have been allowed for wrongful birth, and analogies can be drawn which indicate that actions for wrongful life should be given the same treatment by the courts. While discussion of torts of negligence should focus on duty, breach of that duty and resulting injury that is a foreseeable consequence of such a breach, the courts have issued opinions with pronounced degrees of circularity--citing as reasons for not allowing the suit the uncertainty of calculating damages, the harmful effect on the family structure and the public policy against abortion.[8] These avenues of reasoning will also be examined and drawing an analogy from recovery for prenatal injuries, the logic of allowing wrongful life to take its place as a theory of recovery in tort will be demonstrated.

Elements Of A Cause Of Action

An action for wrongful life breaks down into the elements common to all negligence actions. At the base of the action there must exist a duty on the part of someone, either to insure that the infant is not born, or that the proper disclosure to be made to its parents, so that a decision can be made whether or not to continue the pregnancy to term. Two cases dealing with this proposition have held that there is no duty on the part of a doctor. In Gleitman v. Cosgrove,[9] a mother was informed by her physician that there was no possibility of a defective child being born even though she had contracted rubella during pregnancy. This advice was sought during the first trimester of pregnancy, which would have allowed ample time for a therapeutic abortion had one been determined advisable. The child was subsequently born deaf, blind and mentally retarded. The theory of the suit brought on behalf of the infant was that the doctor's failure to inform the parents of the possibility of birth defects was the proximate cause of the child's injuries. The reason for this allegation was that had the parents been so informed they would have sought an eugenic abortion. The action was dismissed for failure to show proximate cause.[10] The reasons given for the holding are two: 1) it was impossible to ascertain damages as a comparison would have to be made between life with defects and nonexistence,[11] and 2) the "preciousness" of human life outweighed the need for recovery.[12]

In Stewart v. Long Island College Hospital,[13] suit was brought by an infant whose mother had contracted rubella during the first trimester of pregnancy. The child was born with congenital defects and the trial court awarded damages to the plaintiff. However, the Supreme Court of King's County, New York, dismissed the cause of action after setting aside the verdict,[14] citing Gleitman for the reason that there exists no remedy for being born with a defect when the only alternative is nonexistence. As will be seen, both cases are suspect for the reasons given for their decisions.

Difficulty in ascertaining damages should provide no bar to recovery. The decisions in Gleitman and Stewart both ignore the directive issued by the Supreme Court in Story Parchment Co. v. Paterson Parchment Paper Co.[15] which stated unequivocally that it would be a perversion of justice to deny recovery on a specious point of this nature. The second reason--that of not wanting to encourage abortions--has been seriously challenged if, indeed, not eroded by recent Supreme Court decisions.[16] Birth control has been deemed within a constitutionally protected "zone of privacy" surrounding the marital union, which the State cannot infringe on the right to use contraceptives to limit family size.[17] More recently, the Court has held that during the first trimester of pregnancy, a woman has right to have the pregnancy terminated.[18] Thus, the fear of encouraging abortions fails as a valid reason for disallowing suit by a child against one who negligently causes injury to him by promoting or allowing his birth.

A more recent Wisconsin case, Dumer v. St. Michael's Hospital,[19] addressed itself to a situation similar to that found in Gleitman and Stewart. The plaintiff, Mrs. Dumer, entered defendant hospital with an upper body rash she suspected to be rubella. She was misdiagnosed as having an allergic reaction and released. A child was thereafter born suffering from mental and physical retardation, cataracts and heart malfunctions. These conditions were diagnosed as rubella syndrome, indicating that Mrs. Dumer had, in fact, contracted rubella during the first trimester of pregnancy. Faced with malpractice suits on behalf of both the parents and the child, the trial court dismissed the action.[20] On appeal, though, the Supreme Court of Wisconsin held that while the damages to the child for wrongful life may be immeasurable, those suffered by the parents by way of additional medical expenses and supportive expenses could be recovered. The rationale for this holding was that the physician did have a duty to inform the mother of her condition and its possible consequences.[21]

The Dumer Court cited and distinguished a previous Wisconsin case which denied recovery for an unwanted child--Reick v. Medical Protective Co.[22] There, the child involved was born healthy and its parents were suing for the expenses associated with its unwanted birth. This was not the situation in Dumer where the parents had a defective child and were seeking only the additional expenses resulting from congenital birth defects. Dumer, then simply stands for the proposition that if a physician is negligent in diagnosing or recognizing a condition which might indicate an abnormal fetus, and fails to inform the parents of the risks, he stands to be liable to the parents if they can show that an abortion was legally available to them and also, that they would have sought it.[23]

The tort of wrongful life was given recognition in Zepeda v. Zepeda,[24] and Williams v. State.[25] The Zepeda case was an action brought by a child against her father for damages for being born an adulterine bastard. The suit alleged that the defendant induced the plaintiff's mother to have sexual relations upon promise of marriage (which was never kept, nor could have been because the defendant was at the time married) in willful disregard of possible injury to the plaintiff.[26] While the court recognized the father's conduct as tortious,[27] it did not allow plaintiff recovery. Three reasons were given for this action: first, the court indicated that the growing number of illegitimate births would give rise to a great number of claims being presented for the courts to handle--the typical "floodgate" argument; second, that a change in the law should be initiated by the legislature, and should not be ground for denying relief unless a claim is found to be invalid under common law; finally, the court foresaw difficulties in distinguishing other claims for being born disadvantaged.[28] It is important to note that the court did recognize the tortious conduct toward the child, even though it occurred before conception. The court reasoned:

> "But what if the wrongful conduct takes place before conception? Can the defendant be held accountable if his act was completed before the plaintiff was conceived? Yes, for it is possible to incur, as Justice Holmes phrased it in the Dietrich case, 'a conditional prospective liability in tort to one not yet in being.' It makes no difference how much time elapses between a wrongful act and a resulting injury if there is a causal relation between them. . . . If a child is born malformed or an imbecile because of the genetic effect on his father and mother of a negligently or intentionally caused atomic explosion, will he be denied recovery because he was not in being at the time of the explosion?"[29]

In Williams, suit was brought on behalf of a child born to a mentally deficient mother as the result of a rape by another patient in a state mental institution. The New York Court of Appeals affirmed the dismissal of the claim holding:

> "Impossibility of entertaining this suit comes not so much from the difficulty in measuring the alleged 'damages' as from the absence from our legal concepts of any such idea as a 'wrong' to a later born child caused by permitting a woman to be violated and to bear an out-of-wedlock infant. . . . Being born under one set of circumstances rather than another or to one pair of parents rather than another is not a suable wrong that is cognizable in court."[30]

While the court in Williams recognized that one born illegitimate suffered injuries that were substantial and merited compensation, it ruled that since the law knew of no cure or compensation for it, that it should originate from the legislature.[31]

Park v. Chessin: A Study In Conflicts

In June of 1969 Hetty B. Park gave birth to a child. The infant lived for only a few hours. Its cause of death was determined to be a hereditary genetic disorder known as polycystic kidneys. Mrs. Park was at the time a patient of Doctor Chessin, a physician who practices in the field of obstetrics and gynecology. She continued as his patient thereafter and in July of 1970 he and his associate, Doctor Allan Gibstein, delivered her second child. This child, Lara, too, was afflicted with the genetic disorder. She lived for only two and one half years. It is the birth of the second child, Lara, which is the basis of the current litigation.

In April of 1972 Mrs. Park, along with her husband, Steven, commenced an action against the two physicians individually and in behalf of their deceased daughter. The complaint, as amended, stated a number of causes of action against the defendants.[32] In the first, Hetty Park sought to recover for her own personal injuries, pain, and mental anguish which she suffered allegedly as a result of the defendant physicians' medical malpractice. In the second, the father, Steven, also alleged malpractice and sought to recover the expenses which he incurred for the care and treatment of Lara during her lifetime. In addition, he sought compensation for his own mental anguish. The third, alleging fraud, was an attempt by Hetty Park to recover for her own emotional distress arising from the care and raising of "a chronically and severely ill child."[33] In the fourth, the father, Steven, sought to recover for emotional distress and anguish and for the expenses incurred by him for the care and treatment of Lara after her birth. The fifth cause of action, alleging malpractice and fraud, was brought individually by Steven for the loss of services and companionship of his wife. The sixth cause will be discussed shortly. In the seventh, Steven sought to recover expenses incurred for Lara's care and treatment after her birth and also for "loss of society and companionship" of his daughter.[34]

In the face of the defendants' motion to dismiss for failure to state causes of action,[35] Judge Harold Hyman of the New York Supreme Court took the following action. On the basis of the New York Court of Appeals decision in the case of Endresz v. Friedberg[36] which had held that the mother of stillborn infants could recover for pain and emotional upset resulting from defendants' negligence, Judge Hyman denied defendants' motion to dismiss the first cause of action. Defendants' motion was granted, however, as to the second cause of action, citing Palsgraf v. Long Island Railroad.[37] The court held that a father is not within the "orbit of danger" so as to yield a right of recovery for the negligent infliction of emotional distress. Likewise, the court dismissed the third cause of action, that was brought by Hetty Park sounding in fraud. The court characterized it as "essentially the tort action for medical malpractice and for emotional distress as set forth in the first cause of action."[38] Defendants' motion was also granted as to fourth cause of action on the rationale that no such action would lie for negligent infliction of mental distress on the father. While stating that Steven Park's attempt to recover for expenses incurred due to Lara's polycystic kidney condition could properly be demanded in the fourth cause of action, Judge Hyman did not allow the complaint to be repleaded since the plaintiff similarly requested such compensation in the seventh cause of action. Holding that Steven Park's claim for the loss of services, society and companionship of his wife due to the alleged malpractice by Doctor Chessin stated a viable cause of action, Judge Hyman denied defendants' motion to dismiss. Defendants' motion as to the seventh cause of action was granted in part and denied in part. It was held that when repleaded, plaintiff's claim for the expenses incurred in caring for Lara during her lifetime stated a viable cause of action. Steven Park's claim on the basis of loss of companionship of his daughter, Lara, however, was not allowed to stand.

The Sixth Cause Of Action

By far the most significant of the claims presented in the Parks' complaint was set forth in the sixth cause of action. Judge Hyman recognized the possible import of this aspect of the case when he stated at the outset of his opinion: "The present claim brings directly to the forefront the highly controversial issue of whether. . . there exists after birth a legal right to make a claim for pain and suffering resulting from a tort committed prior to conception. . . ."[39] In the sixth cause of action characterized by the court as an "admixture of alleged malpractice and fraud"[40] the deceased child, Lara, through her estate sought to recover for her own pain and suffering which resulted from the violation of her alleged "right not to be conceived and therefore not to be born."[41]

The gravaman of this claim was that following the birth of the Parks' first child with the kidney condition the defendants knew or should have known that the fatal disorder was hereditary and that there was a high probability that it would recur in subsequent children born to them. It was alleged that the couple specifically inquired of the physicians about this matter.[42] Further, it was alleged that the defendants, without performing any available genetic tests or working histories, advised the Parks that the chances of future pregnancies resulting in congenitally defective children was slight. Relying on such advice the Parks decided to continue their efforts towards having a family, the result of which was the birth of Lara, who shared her late brother's disorder.

Judge Hyman approached the Park case with care. He took pains to point out that as the case was postured before him, the issue turned on a matter of pleading only and that his decision would not determine the merits of the case.[43] His rationale for denying defendant's motion to dismiss was grounded on the basis that since "life begins at the moment of conception"[44] Lara, in her fetal stage, enjoyed "the rights of a human person"[45] among which is "the legal right of every human being to begin life unimpaired by physical or mental defects resulting from the negligence of another."[46] The court rejected the argument that to allow a cause of action for wrongful life to stand would thrust upon the medical professional an unreasonably heavy burden. Generally, the profession being charged with the "very lives" of those entrusting themselves to its care, could not be allowed to escape liability for negligence because of "the worn out rejected cliche of public policy."[47] To do so would "single out and grant preferential treatment to the medical profession over all other professions and enterprises where malpractice would result in payment of ensuing resultant damages."[48] The damage question which has often proved fatal in other wrongful life cases at the initial motion stage,[49] was felt by the court to be without the scope of discussion at this stage in Park. The court refused, however, to hold the possible difficulty in determining an aware sufficient ground for dismissal of the complaint. The trial court likewise declined to be bound by previous New York cases which had dealt unfavorably with complaints postured in terms of wrongful life. It distinguished the Park case regarding it as the first "definitive tort action brought by a 'child' after its birth for 'conscious pain and suffering' based upon the tort committed prior to its conception."[50] Judge Hyman refused "to absolve (the defendants) of their tort liability merely because no legal precedent exists in common law" because to do so would "stultify legal progress."[51]

## The Appeal

Discussion in the opinion rendered by the Supreme Court Appellate Division was devoted, as in the court below, chiefly to the aspect of the complaint dealing with wrongful life. The plaintiffs did not appeal from the trial court's dismissal of four of their causes of action. The defendants did appeal the court's denial of their motion to dismiss the remaining aspects of the complaint. The Appellate Court affirmed the trial court's denial of the defendants' motion to dismiss as to the first cause of action brought by the mother for pain and suffering and the seventh by the father for costs incurred in the care of Lara with the modification that no recovery could be had in either case for emotional distress.[52] In addition, the portion of the suit in which Steven Park sought to recover for the loss of his wife's services was validated.[53]

Focusing on the wrongful life aspects of the case, the opinion[54] of Justice Damiani, in which Justice Margett and Rabin concurred, viewed the question of the existence of such a cause of action in New York as theretofore undecided by the State Court of Appeals. Finding a "right within certain statutory and case law limits not to have a child"[55] in the legislature's abolition of a ban on abortions in New York,[56] the opinion held that this right encompassed "instances in which it can be determined with reasonable medical certainty that the child would be born deformed."[57] Given the circumstances presented by the fact pattern in Park, the court surmised that this right had been breached by defendants' negligent conduct and that they should, therefore, be held liable in tort. Noting that wrongful life cases had not met with favor in the courts of New York or those of other jurisdictions, the opinion reviewed the various grounds upon which such cases had failed in the past[58] and rejected them stating "cases are not decided in a vacuum; rather, decisional law must keep pace with expanding technological, economic, and social change."[59]

In a vigorous dissent,[60] Justice Titone expressed the belief that plaintiff's remaining causes of action should be dismissed. He reasoned that the court's recognition of a cause of action for wrongful life ignored a "plethora of judicial precedent to the contrary not only from other jurisdictions (citations omitted) but likewise from this state as well."[61] Rejecting the majority's reasoning that the law must keep pace with technological change, he characterized the court's action as "rushing into the adoption of a radical, social concept having no basis in law."[62] The dissent focused upon a detailed analysis of the legislative basis[63] for the cause of action for wrongful death and analogized the development of that legal action to the situation presented by the facts in Park calling for judicial restraint until such time as a cause of action for wrongful life is created by the legislature.

A reading of the two Park opinions discloses the unclear condition of the actual case law on wrongful life in the New York courts. In affirming the validity of Lara Park's complaint, the majority of the Appellate Division held that previous case precedents constitute no bar to the recognition of a cause of action for wrongful life in the state of New York.[64] It has been more than a decade since the court was presented with a case raising the issue of wrongful life. A final resolution of the problem which would provide direction for the state's inferior courts and guidance to New York physicians as to the standards of care expected of them and the extent of liability to be imposed upon them should be made.

A Suggested Resolution

As will be seen in due course, the basic question confronting the New York Court of Appeals in wrongful life cases of this nature--is the very title of cause, itself. Captions of wrongful birth, wrongful pregnancy, wrongful conception, wrongful diagnosis--to name the most important ones--have all appeared at one time or other in cases as well as legal literature. While commendable efforts have been undertaken to distinguish the various titles from one another, they have generally been regarded as confusing and essentially meaningless.[65] The lower court's perception of the term, wrongful life, as it is thus postured in Park should be accepted as a final resolution to the controversy. Likewise, irrelevant to a conclusive judicial determination of the problem area are past cases maintained by healthy illegitimate children against their parents for being born bastards. The Williams decision regarding cases of this nature has of course been conclusively settled--at least in New York. Implicit in the court's post Williams' statement that the wrongful life question was still open in New York is the fact, however, that suits brought on the basis of illegitimacy, are without the scope of the cause of action of the state. Cases wherein parents of healthy children have sought recovery from physicians or health-care institutions for unplanned births are also of no importance to reaching a final resolution of the wrongful life conundrum. Eschewing consideration of such extraneous matters would enable the judiciary to focus sharply on the salient issues demanding resolution.

A truly definitive decision to the problem area would speak to the questions of duty, causation and damages as well as considerations of public policy as they impact on the determination of the matter. Such an evaluation should not be held as a bar to the subsequent validation of a cause of action for wrongful life. Rather, courts should take judicial notice of the marked advances made in recent years by medical science in the fields of genetic research[66] and prenatal care. Technological breakthroughs now enable physicians, as has been noted previously, to predict the possibility of genetic defects in offspring from prospective parents. Such individuals can thus be advised of the approximate level of risk attendant to the birth of a child prior to that child's conception. After conception, the fetus can be tested and monitored during the early stages of pregnancy in such a way that any existing defects of deformities can be discovered easily.[67] Given the wide latitude of choice during the first trimester of pregnancy,[68] the prospective mother is able to make an informed decision whether or not to carry the fetus to term.

Given the existence of such tools for gathering information, physicians should be held liable for negligently disregarding a patient's request for guidance when planning a family. With the refinement of contraceptive technologies, a birth in contemporary society often results from a choice rather than chance. If in the course of such informed decision making, individuals are misadvised by their physicians after affirmatively seeking out genetic counseling information so that a child with deformities or defects is born, the physician should not be held immune from liability. The pain and suffering experienced by issue--such as Lara Park--who are born as a result of a physician's failure to respond to reasonable parental inquiries relative to genetic inheritance, should not go uncompensated.

Wrongful Conception

The Supreme Court of Minnesota recently added a new legal concept to the wrongful life/wrongful birth lexicon: "wrongful conception."[69] Here, the parents of seven children consulted a physician upon the birth of the last child in order to learn what steps could be taken in order to prevent even more children. The husband subsequently had a vasectomy and--on the initial advice of his physician regarding the success of surgery--resumed sexual relations with his wife. Thereafter, an eighth child was born to the couple and they thereupon sued the physician for negligent post operative care of the father, for medical expenses and damages incident to the birth, for pain and suffering caused to the mother by virtue of her pregnancy and the delivery, for the husband's loss of consortium and--perhaps most significantly--for the costs of supporting and educating the child until the age of majority. A jury returned a verdict of $19,500.00 and this was challenged by the defendants on the grounds that the evidence was both insufficient to support the verdict and that the verdict, itself, was contrary to law.

Although reversing and remanding the case solely on the issue of damages--inasmuch as the jury was not specifically instructed to endeavor to offset the value of the child's aid, comfort, and society against the projected rearing costs and--furthermore--because the damage award was made upon general instructions of negligence and without the aid of a special verdict form,[70] the Supreme Court made a significant determination. Specifically, that "the time-honored command to 'be fruitful and multiply' has not only lost contemporary significance to a growing number of potential parents but is contrary to public policies embodied in the statutes encouraging family planning."[71] The court--albeit rather feebly--endeavored to distinguish the differences between wrongful birth, wrongful life and wrongful conception by observing wrongful birth actions are brought by parents for what is more commonly regarded as negligent sterilization cases of ordinary medical malpractice,[72] wrongful life actions would be brought by a child to recover damages for wrongfully being born[73] and wrongful conception actions (as was this) are maintained by parents for the harm they have suffered at the very point of conception--harm, physical and financial --that they, not the unplanned child, have sustained as a consequence of a physician's negligence.[74] Allowing damages for acts of wrongful conception, the Minnesota court maintained, was wholly compatible with elementary principles of compensatory damages "which seeks to place injured plaintiffs in the position that they would have been in had no wrong occurred."[75]

Wrongful Birth And Allowable Recovery

Actions for damages resulting from pregnancy following the failure of a sterilization operation or negligent filling of prescriptions have met with greater success than their counterparts for wrongful life. Plaintiffs, Berdella Custodia and Bravlio Custodia, in an action for damages as the result of pregnancy following negligent performance of a sterilization operation, were allowed recovery of more than nominal damages even if the child was born normal and healthy.[76] Custodia v. Bauer[77] is a case cited often for its discussion of what damages are compensable in a wrongful birth claim. Underlying the whole notion of recovery in this area is the principle known in tort as the benefits rule:

> "Where the defendant's tortious conduct has caused harm to the plaintiff or to his property and in so doing has conferred upon the plaintiff a special benefit to the interest which was harmed, the value of the benefit conferred is considered in mitigation of damages, where this is equitable."[78]

This is regarded as applicable here because of the benefit to the parents of a "blessed" event such as having a child.[79]

The Custodia court, in addressing itself to the issue of recoverable injuries, held that there could be recovery for: 1) the mental suffering attendant to the unexpected pregnancy because of complications which may or may not result, the complications that do result, and the delivery of a child--all forseeable consequences of the failure of the operation;[80] 2) in the event of the mother's death, the surviving children and husband would be compensated for the value of her society, comfort, care, protection and right to receive support which they lost;[81] 3) if the change in the family status can be measured economically, it, too, should be compensated, i.e., if she can spread her society, care, confort, protection and support over a larger group;[82] 4) the expenses of bearing a child.[83] It is important to remember that the plaintiff has the burden of introducing evidence to show that the defendant was negligent and to establish a causal connection between the defendant's negligence and the plaintiff's injury, since recovery for any of the above expenses depends upon sustaining these burdens. Custodia was cited with approval three years later by the Florida court in Jackson v. Anderson.[84] There, the court allowed recovery in an action for breach of express warranty and negligence by a surgeon in the performance of a sterilization operation. The action for the resultant pregnancy was allowed not withstanding that the pregnancy was normal and resulted in a healthy child being born.

Suits for wrongful birth have been attached on public policy grounds. These have generally been largely of a moral character-- involving the so-called sanctity of life argument, as well as the opinion that abortions should not be encouraged. Courts in recent years have been turning away from using these public policy considerations as reason for denying recovery in these cases primarily because of the recent and widening shift in a major segment of the public opinion itself relative to the approved use of abortion as a personal method of family planning and population control.

In Coleman v. Garrison,[85] a husband, his wife and four children brought suit against a surgeon to recover for damages following the wife's unexpected pregnancy after a sterilization operation. The court recognized that married couples have a right not to have children[86] and read this as an indicator that public policy did not disfavor abortions. The decision indicated that the jury should be allowed to weigh the benefits with the burdens of the unwanted (unplanned) child,[87] and allowed recovery for pain and suffering, medical expenses and loss of consortium.[88] While this action held that there could be recovery for these injuries, and such recovery would not violate public policy, the subsequent action did not meet with success.[89] In the four year interim, Mrs. Coleman had conceived yet another child (her sixth), and the fifth child (also unplanned and the subject of the earlier suit) joined the husband, wife and four previous children in the suit. The action resulted in summary judgment for the defendant on two grounds. Using the comparative benefits rule,[90] the court held that the value of human life outweighs any damage resulting from birth; and secondly, that to compel a surgeon to assume financial responsibility for raising and educating a child born subsequent to a sterilization operation would violate public policy.[91]

The sentiment expressed in the final Coleman decision was an echo of that of a previous Texas opinion, Terrell v. Garcia.[92] This case involved a suit by parents of an unwanted child following an unsuccessful sterilization operation. In disallowing recovery, the court there commented on the difficulty of ascertaining damages, but countered that argument with the fact that juries have been determining similar costs in wrongful death suits.[93] In those types of suit, the jury is asked to put a value on the life of a child in order that there can be recovery for its termination. It is no more difficult to determine the care and maintenance costs of a child, even though such variables as standard of living and extent of education would require some speculation and hypothesis. The court, however, retreated to the benefits rule in order to disallow recovery.[94] The dissent, written by Justice Cadena, preferred to attach to the benefits rule a provision that a negligent person is liable for all foreseeable consequences of his negligence. While there may be mitigating factors, as recognized in the original benefits rule, they do not result in vitiating the cause of action. Justice Cadena concludes:

> "There is no justification for holding as a matter of law, that the birth of an 'unwanted' child is a 'blessing'. The birth of such a child may be a catastrophe not only for the parents and the child itself, but also for previously born siblings. The doctor whose negligence brings about an undesired birth should not be allowed to say 'I did you a favor,' secure in the knowledge that the courts will give to this claim the effect of an irrebuttable presumption."[95]

The dissent is a far more realistic approach to the problem--for it detaches itself from emotional entanglements. If there has been a wrong, the courts should not succumb to the difficulties of ascertaining damages.[96]

More recent cases have given support to the allowance of recovery for wrongful life as the result of negligent surgical procedures. The New Jersey case, Betancourt v. Gaylor,[97] allowed an action for a negligent sterilization. While applying the benefit rule, by allowing mitigation for the joy and happiness of a healthy child, the court there allowed recovery for the cost, emotional upset and physical inconvenience of rearing a child. The damages allowed there are in accord with the Custodia case,[98] as with a California case, Stills v. Gratton.[99] That case involved an action for damages for negligent performance of an abortion. The court allowed the suit by the mother, with damages being guided by Custodia. The suit on behalf of the child though was not allowed. The reasons given for such a decision are not precise. The mother, Hannah Stills, was allowed recovery on tort principles of negligence. The child was not accorded such treatment. The denial of recovery to the child was based on a form of circular reasoning. The court stated that the purpose of tort damages is to restore the plaintiff to a position he would have been in were it not for the occurrence of the tort. To the court, this meant nonexistence; and therefore, made the computing of damages impossible because of the difficulty in comparing non-existence with existence. In so doing, this court chose to ignore the fact that this was an unwanted birth--one that the woman, once she knew she was pregnant, had taken every step to terminate. By non-suiting the child's case and limiting damages to the mother so as not to include the costs of raising and educating a child, the court, in effect, insulated the physician from a gross error.

In an action much the same as in Stills, the New York Supreme Court in Ziemba v. Sternberg[100] allowed a medical malpractice suit when the plaintiff became pregnant and was not advised in time to terminate it. The plaintiff-wife had sought medication to prevent pregnancy, but became pregnant and delivered a healthy baby. The woman alleged that the doctor was negligent in not detecting and advising her of her pregnant condition within a time which would have allowed a legal abortion to be sought. The court distinguished Stewart v. Long Island College Hospital[101] by indicating that at the time of that decision the abortion would have been illegal. This was no longer true in 1974 when Ziemba was decided because of the new legal environment characterized by Roe v. Wade.[102]

A Texas court one year later used the same Roe precedent to find a duty on the part of a physician to inform a pregnant woman of the risk to her child when she contracted rubella during the first trimester of pregnancy. Jacobs v. Theimer was a suit for medical expenses for pain and suffering associated with the birth of a child with defective organs. The court found that the doctor had failed to give proper disclosure by not discovering the rubella and advising of the attendant risks resulting therefrom to a continuation of the pregnancy. While finding damages for the parents' emotional suffering too speculative and not allowable, it did allow that public policy considerations did not bar a suit for a recovery of those expenses considered reasonably necessary for the care and the treatment of the child's subsequent physical impairment.[103] The trend toward allowing recovery for defective or unwanted birth by the parents is clearly evidenced by these various wrongful birth actions. It is obvious, then, that in light of the new abortion decisions of the Supreme Court,[104] there is no longer a public policy that would prohibit suits by parents for wrongful birth.

One further case merits discussion in order to complete the treatment by the courts of the concept of wrongful birth. While the shift toward holding doctors responsible for malpractice for negligent sterilization operations or negligent failure to disclose in time to procure an abortion is a relatively new trend, it was heralded by a suit against a pharmacist. In Troppi v. Scarf,[105] the court used tort theory to establish the liability of a pharmacist who negligently filled a prescription for birth control pills. Instead of filling the prescription with birth control pills, the pharmacist gave Mrs. Troppi a mild tranquilizer. This mistake resulted in the addition to the Troppi family of a healthy, but unwanted child. The Court of Appeals of Michigan reversed summary judgment for the pharmacist and allowed recovery by the Troppis for lost wages, medical and hospital expenses, pain and anxiety of pregnancy and childbirth as well as for the economic burden of child rearing.[106] The ruling allowed application of the benefits rule in mitigation of the damages, but did not dismiss the suit by that rule's application.[107] It held that dismissing a case by blanket application of the view that the benefit of a healthy child always outweighs the burden is "unsound."[108]

The case law in the field of wrongful life/wrongful birth may conveniently be grouped into three categories. First are the cases which deny recovery for damages for malpractice when the result is the birth of a healthy child.[109] These cases show the dependence of the courts on the benefits rule, holding that the benefits of a child, as a matter of law, outweigh the burdens. The second grouping involves those cases which allow the parents a cause of action, but either do not discuss damages[110] or limit the recovery to costs related to a defect or expenditures related to the pregnancy.[111] In neither situation has there been recovery for expenditures related to the raising of the child. The final cluster of cases recognizes the parent's cause of action and awards damages according to normal tort principles.[112] There are no restrictions of public policy hampering these recoveries.

As can be seen readily, tort liability does exist for wrongfully causing one to be born,[113] yet this recognition of liability has not yet developed into an equal recognition that a child "wrongfully born" may sue in its own right for that wrongful life. At this juncture it is both appropriate and important to consider the standing of infants to maintain suit for pre-natal torts. An investigation of this issue will illustrate the only barrier left to a suit grounded in a theory of wrongful life. That obstruction is one of not allowing the child to sue its parents for the very act which gave it life, and will be shown to be no longer valid. If a claim has merit, it should not be disallowed because of the identities of blood relationships of the parties--even though they are parents and child.

Allowing Suits In Tort For Pre-natal Injuries

In the area of pre-natal injuries, the law has not been clear regarding the time a cause of action accrues--as well as to whom such an action inures. It has traversed the arc of a swinging pendulum. In 1884, Justice Holmes then of the Massachusetts Supreme Judicial Court, ruled in Dietrich v. Northampton[114] that there could be no recovery of damages for injuries suffered by fetus while in its mother's womb. With few exceptions,[115] the law retained this holding until 1946. Then, a United States District Court in Bonbrest v. Kotz[116] held that injuries to an unborn child which is viable are compensable in an action brought by the child after it is born and thus sounded the eventual death knell for Dietrich.

One area of considerable concern still remained after Bonbrest. A child could not bring an action unless the injury occurred when it was in a viable state.[117] This distinction was highly artificial and its weaknesses were soon discovered both by commentators[118] and courts.[119] By allowing suit by the infant if born alive for all prenatal injuries, the courts soon buried this distinction. Now, in order to prevail, a plaintiff must direct his attention to very difficult problems of proof in cases of this nature. There must be negligence shown on the part of the defendant, injury suffered by the child and causation between the two established. For example, a pregnant woman may fall down some icy steps and several months later her child could be born defective or deformed. There will be no recovery unless it can be shown that the injuries were caused by the negligence of the person in control of the step's condition. Difficulties in proof also arise as the point of injury occurs earlier in pregnancy.[120] The Statute of Limitations has been held to run from birth rather than from the date of the injury.[121] This is only logical, since the injury will not be known in all probability until birth.

Once the question is answered as to whether the infant can maintain the action, the important consideration is whether recovery is allowed. For actions involving negligence of physicians, recovery has been allowed in a number of cases. In a suit brought on behalf of a three-year-old girl who suffered serious brain damage due to a congenital anomaly or damage at birth, or both, it was held that the procedure used during birth (forcing the child out by putting pressure on the abdomen) was not good practice.[122] An obstetrician was held liable for injuries sustained by a child during a premature forced delivery in Korman v. Hagen.[123] The forced delivery caused a break in the femur and lacerated nerves in the right arm which caused paralysis of that limb. There was testimony that the baby's life was not in jeopardy when it was forcibly delivered. In yet another instance, recovery was allowed in a suit brought on behalf of a child born with cerebral palsy.[124] The condition was alleged to have been caused by the physician's negligence in failure to discover the mother's anemic condition during pregnancy.

As previously indicated, the most difficult problem once the cause of action is allowed is establishing causation. Womach v. Buchhorn[125] was an action for prenatal brain damage suffered during the fourth month of pregnancy. The injury was suffered in an automobile accident. The court allowed that a common-law negligence action could be brought on behalf of the child. The decision cited Smith v. Brennan[126] and adopted its rationale:

> "And regardless of analogies to other areas of the law, justice requires that the principle be recognized that a child has a legal right to begin life with a sound mind and body. If the wrongful conduct of another interferes with that right, and it can be established by competent proof that there is a causal connection between the wrongful interference and the harm suffered by the child when born, damages for such harm should be recoverable by the child."[127]

Causation has been the central issue in several other cases as well, involving such topics as drugs,[128] disease[129] and death.[130]

Recapitulation

With a child having the right to maintain an action for prenatal injuries, all that remains for a suit in wrongful life to lead to a recovery is the recognition by the courts of the complete interest of the child on being born with a sound mind and body. There are those who would argue against such a recovery,[131] using as their support policy arguments which would focus upon two concerns other than the difficulty in ascertaining damages and public attitudes against abortion. Specifically, they would emphasize the effect the allowance of such suits on parents and the family unit, and the effect on the medical profession. Such arguments do not take into consideration the very positive factors of allowing such a suit. Parents and the medical profession, knowing of the risk of incurring liability for wrongful life, will become much more conscious of the child's right to a healthy mind and body and be less hampered with notions of the sanctity of human life. This will reap benefits for society in that such concern by parents and medical practitioners will greatly decrease the number of persons dependent on society for their support and existence. The debate between existence with defects and nonexistence will become a matter of the past as recognition of the right to exist without defects becomes accepted.

This type of approach places a burden of some dimension upon the medical profession. For example, a doctor--because of the very real danger of a lawsuit--may elect to take one of two options. He may either abort all borderline cases in an effort to immunize himself from suit or, resign from the practice of pediatrics. For obvious reasons, neither decision is a satisfactory one for the general public. The burden is, however, mitigated as a consequence of the doctrine of informed consent. Since a parent is entitled to full disclosure before consenting to surgical or medical treatment, it should not be regarded as a drastic change to require that pregnant women be advised as fully as possible of the condition of their child. In this way, an intelligent decision can be made at the proper time whether to continue or terminate the pregnancy.[132]

As with any action in negligence, the parameters of the cause of action must be defined with clarity and precision--either by the courts or by the legislatures. Difficulty in performing such a task should not be used as a reason for failing to meet the burden of responsibility. Although the courts have shown themselves, in the past, to be slow to react to new challenges within the law, the fact of the matter is they have--on the average--eventually responded in a far more positive manner than the legislatures. Meritorious causes of action are, in the course of time, established and recognized. Within the guidelines of causation--again, in the course of time-- total recognition can be true for an action termed, "Wrongful Life."

Epilogue: A Long Awaited Resolution?

On December 27, 1978, a final determination of Park v. Chessin was made by the New York Court of Appeals in a decision rendered by Associate Justice Matthew Jasen.[133] The Court combined Chessin with Becker v. Schwartz in reaching this decision and--consequently--the facts of Schwartz are important to a full understanding of the related issues in both cases.

Dolores Becker, a thirty-seven year old woman, conceived a child in September, 1974. During the pregnancy, she asserted that she was never advised of the increased risk of Down's syndrome in offspring born to women over thirty-five years of age and, furthermore, of the availability of amniocentesis in order to aid in the determination of whether the fetus might be afflicted with the syndrome. After the birth of their mongoloid child on May 10, 1975, Mr. and Mrs. Becker maintained this action not only seeking damages on behalf of their daughter, Barbara, for her "wrongful life," but in their own right for the various sums of money they would be forced to expend for the long term institutional care of their retarded child. The complaint also sought damages for emotional and physical injuries suffered by Dolores Becker as a result of her child's birth and for additional damages to her husband, Arnold, for the injury he suffered by the loss of his wife's services and the medical expenses stemming from her medical treatment following pregnancy. While the Special Term dismissed the complaint as failing to state a cause of action, this order was modified by the Appellate Division of the Supreme Court of the State which found for the plaintiff to the extent they sought recovery of damages "for psychiatric injuries or emotional distress of plaintiff Dolores E. Becker and to the extent that plaintiff Arnold Becker's claim for loss of services and medical expenses is based upon some psychiatric injuries."[134] The appeal was taken to the New York Court of Appeals based upon this finding.

Judge Jasen found that since one does not have what may be termed as a fundamental right to be born as a whole or functional human being, no legally recognizable determinable injury was suffered by the infants in both cases and--accordingly--damages were not ascertainable as to those infants, Barbara Becker and Lara Park. Regarding the claims of the parents based upon the extraordinary care and treatment of their children necessitated because of their "wrongful life" caused as such by the physicians' alleged negligence in failing to inform the plaintiff parents accurately of the risks involved in pregnancy which in turn resulted in the parents' decision to conceive or not terminate the pregnancies as the case may be, the court found the claims for money damages suffered as a consequence of birth valid. Thus in Becker, the court found the complaint made by the plaintiffs to be invalid and dismissed it except to the extent that it sought recovery of the sums expended for the long term institutional care of their daughter, Barbara. In Park, the complaint was dismissed again-- except to the extent that it sought a recovery for the sums expended for the care and treatment of Lara until her death which would be allowed.[135]

Judge Jasen takes considerable pains in his majority decision to distinguish those actions based on theories of "wrongful conception" and "wrongful diagnosis" where some courts have allowed recovery, from cases based on theories of "wrongful birth" where no judicial validation has been forthcoming.[136] Here, under the latter theory, the Judge observes a healthy child--although stigmatized as such by its illegitimacy--seeks recovery in its own right for an alleged injury it has sustained owing to its birth under what might be recognized as untoward circumstances. To the contrary, however, in cases of "wrongful conception," the parents of a healthy and normal yet unplanned child, because of what was thought by them to be a successful birth control precedures, seek damages not on behalf of the child but by themselves for expenses attributable not only to the birth, but for expenses in rearing the child. Similarly, with actions for "wrongful diagnosis," damages are sought for injury incurred as a consequence of the birth of a child whose birth it is maintained, is the direct cause of an inaccurate diagnosis of an existing pregnancy thereby resulting in the deprivation of the mother's choice to terminate the pregnancy within the permissible time period. Noting the mixed judicial reaction to these theories, Judge Jasen states the plaintiffs in the instant cases premise their actions upon "the birth of a full intended but abnormal child for whom extraordinary care and treatment is required."[137] The basic argument made by the Parks and the Beckers is that had they been properly advised by their physicians of the risks of abnormality, they would never have allowed their children to be born.[138]

"Irrespective of the label coined, plaintiff's complaints sound essentially in negligency or medical malpractice. As in any cause of action founded upon negligence, a successful plaintiff must demonstrate the existence of a duty, the breach of which may be considered the proximate cause of the damages suffered by the injured party."[139] Judge Jasen then continues his analysis of the case by observing the two central flaws for recovery by the infants under a theory of "wrongful life:" failure to show that they--the infants--suffered a legal injury and an inability to calculate damages.

There is no precedent either at common law or in statute which allows the judiciary to recognize the very birth of a defective child as an injury to the child. To hold a contrary position would bring staggering consequences. "Would claims be honored, assuming the breach of an identifiable duty, for less than a perfect birth? And by what standard or by whom would perfection be defined?"[140] Since the core of the remedial relief granted in negligence is an effort-- through financial compensation--to place the injured party in that position he would have occupied but for the defendant's negligence, it is readily apparent that the damages recoverable under a wrongful life theory by an infant would be limited directly to that which is necessary to restore the infant to the position it occupied had it not been for the failure of the defendant physicians to render advice to the parents of the infant in a non-negligent manner. Thus, the second central weakness of the case is that, "a cause of action brought on behalf of an infant seeking recovery for wrongful life demands a calculation of damages dependent upon a comparison between the Hobson's choice of life in an impaired state and non-existence."[141] ". . . . creation of a hypothetical formula for the measurement of an infant's damages is best reserved for legislative, rather than judicial attention."[142]

Regarding the claims of plaintiff parents maintained in their own right for negligence--and based upon the failure of the defendant physicians to advise of the full consequences of a completed pregnancy, Judge Jasen states wisely that ascertainable damages (e.g., the expenses associated with the care and treatment of their infants) may be properly claimed and awarded, since their calculation is relatively simple.[143] The invalidity of wrongful life claims does not negatively affect the claims of the plaintiff parents for pecuniary loss. A determination of the particular items of expense to be considered in computing damages "properly await consideration and resolution presumably on trial, after liability has been proved, if it can be."[144]

Drawing upon the court's previous determination of Howard v. Lecher[145] and asserting consistency with the subsequent holding by the same court in Johnson v. State of New York,[146] Judge Jasen states that the plaintiffs in the instant case may not recover damages for psychic or emotional harm allegedly caused as a consequence of the birth of their genetically defective infants because of their speculative nature.[147] Restating the Johnson case rule that once a duty flowing directly from a defendant to a plaintiff is breached, recovery by the plaintiff may be allowed for "the proven harmful consequences proximately caused by the breach," the Judge--while sympathetic to the anguish caused the parents by the birth of a defective child--opines that "certainly dependent upon the extent of the affliction, parents may yet experience a love that even an abnormality cannot fully dampen. To assess damages for emotional harm endured by the parents of such a child would, in all fairness, require consideration of this factor in mitigation of the parents' emotional injuries."[148] The cognizability of an action for emotional harm "is a question best left for legislative address."[149]

Waiting for legislative action to resolve major uncertainties in this area--as Judge Jasen suggests--means that little positive direction will be forthcoming. Perhaps judicial passivity and legislative inactivity may be attributed to a yet-to-be-determined, concrete societal perspective on matters of this nature. The fluidity of social mores, then, means continued fluidity for the law itself.

Complications

On February 17, 1979, The New York Times reported a perturbing news story. Dolores and Arnold Becker had--during the course of their lawsuit--placed their retarded daughter, Barbara, up for adoption and, subsequently, the Nassau County Department of Social Services had advised an adoption of the child was finalized before the New York Court of Appeals reached its decision.[150] No disclosure of any nature had been made by the Beckers during the legal proceedings relative to their daughter's adoption. This matter was disclosed just as a new trial was to be scheduled in the New York Supreme Court in Mineola on the issue of the provable liability of the physicians for their alleged negligent treatment of Mrs. Becker and the factors to be used in computing damages for the wrong itself once proved. If negligence is in fact proved, the physicians would be liable to the natural parents for the lifetime special costs of caring for the child.

Once this news story was released, the defense counsel for the physicians immediately termed the Beckers' action reprehensible and a travesty upon the court and called for a dismissal of the case since, as a consequence of the adoption, the Beckers no longer had any legal obligations to the child. The lawyers for the Beckers met these contentions by maintaining that the case should go forward--especially in light of the physical injuries sustained by Mrs. Becker from the pregnancy and the money previously expended by the plaintiffs for Barbara's care. It was, furthermore, noted by the lawyers that any money judgment arising from a favorable disposition of the case for the plaintiffs would be used solely for Barbara.[151]

Notes--Chapter 4

1. Sylvia v. Gobeille, 220 A. 2d 222, 224 (R. I. 1966). The case held that a child born alive has an action in tort for prenatal injuries that are the result of the negligence of another. Yet, reliable proof of causation is still necessary to sustain the action.

2. See generally, Keeton, "Creative Continuity in the Law of Torts," 75 HARV. L. REV. 463 (1962). There Professor Keeton observes that: "Unhampered by a rigid theory of precedent like that adhered to in England, American courts have a great responsibility for participation in the creative adaptation of law to current needs. . . . Where a need for reform is clear but no reforming statute has been enacted, courts must choose among the unsatisfactory precedent and other rules open to judicial adoption, even though the range of choice may be not as wide as that open to a legislature." Id. at 484.

3. See Comment, "A Cause of Action for 'Wrongful Life,'" 55 MINN. L. REV. 58 (1970). See also Ploscowe, "An Action for 'Wrongful Life,'" 38 N.Y.U.L. REV. 1078 (1963).

Interesting amplifications of this basic problem are to be found in an awareness of the controversy over whether an action will lie for acts causing injury to or death of a child in the womb. While most courts are now recognizing this as a valid cause of action (See Annots., 15 A.L.R. 2d 992 (1967), 40 A.L.R. 2d 1222 (1971), the Supreme Court of Illinois has extended the blanket of liability to a child not even conceived when the wrongful act was done. In Renslow v. Mennonite Hospital, 67 Ill. 2d 348, 10 Ill. Dec. 484, 367 N. E. 2d 1250 (1977), a hospital negligently transfused a thirteen year old girl who was Rh-negative with Rh-positive blood--thereby sensitizing her blood. When a number of years later, she conceived a child, its life was threatened by her condition. Premature birth had to be induced and the child suffered various hemolytic-related injuries. After an elaborate discussion of duty and causation, the court concluded that the harm was reasonably foreseeable and therefore a breach of duty to the child. Justice Ryan, in a strong dissent, observed that the majority abandoned the concept of foreseeability and accepted the notion that where causation is shown, all results are foreseeable.

4. The earliest case using the theory of wrongful life was Zepeda v. Zepeda, 41 Ill. App. 2d 240, 190 N. E. 2d 849 (1963), cert. den. 379 U. S. 945 (1964).

See notes 24-29 infra, and accompanying text.

See generally, Annot., Right to Die, Wrongful Life, AM.JUR. 2d (New Topic Service) 1 (1979).

5. 88 Misc. 2d 222, 387 N.Y.S. 2d 204 (Sup. Ct. 1976), aff'd 60 App. Div. 2d 80, 400 N.Y.S. 2d 110 (1977).

The New York Court of Claims in Case of Riveria v. New York, 46 U.S.L.W. 1169 (Mar. 9, 1978), allowed recovery for what it determined to be the wrongful life of a health child born after the mother had undergone unsuccessful operation for sterilization. The court held the parents had a simple medical malpractice action for damages, medical expenses, pain and suffering incident to the suffering and for anticipated costs of raising the child. The court observed that the anticipated costs of raising the child were not too speculative inasmuch as such compensations are made by estate planners and insurance companies--for example--and that not every birth was an ultimate good.

"The notion that individuals should be compensated for the negligence of a physician in facilitating the birth of an unwanted child is no more offensive to such beliefs than is the concept of birth control." Id.

6. E.g. Terrell v. Garcia, 496 S.W.2d 124 (Tex. Civ. App. 1973).

7. E.g. Troppi v. Scarf, 187 (N.W. 2d 571, 31 Mich. App. 240 (1970).

8. See Annot., Tort Liability for Wrongfully Causing One to be Born, 22 A.L.R.3d 1441, 1443 (1968).

Presently, some seventeen states have either abrogated, limited or rejected the parent-child immunity doctrine. Under this doctrine, a shield of protection is extended to parents thereby protecting them from suits brought by their minor children for acts of ordinary parental negligence resulting in injury to such children. The immunity also extends to children by providing them with as strong a shield for their acts of negligence which cause injury to their parents. The seventeen states having taken judicial action are: Emmert v. United States, 300 F.Supp. 45 (D.D.C. 1969); Xaphes v. Mossey, 224 F. Supp. 578 (D. Vt. 1963); Hebel v. Hebel, 435 P.2d 8 (Alaska 1967); Streenz v. Streenz, 106 Ariz. 86, 471 P.2d 282 (1970); Gibson v. Gibson, 3 Cal. 3d 914, 479 P.2d 648, 92 Cal. Rptr. 288 (1971); Petersen v. City & County of Honolulu, 51 Hawaii 484, 462 P.2d 1007 (1970); Schenk v. Schenk, 100 Ill.App. 2d 199, 241 N.E.2d 12 (1968); Rigdon v. Rigdon, 465 S.W.2d 921 (Ky. 1971); Plumley v. Klein, 388 Mich. 1, 199 N.W.2d 169 (1972); Silesky v. Kilman, 281 Minn. 431, 161 N.W.2d 631 (1968); Rupert v. Stienne, 90 Neb. 397, 528 P.2d 1013 (1974); Briere v. Briere, 107 N.H. 432, 224 A.2d 588 (1966); France v. A.P.A. Transport Corp., 56 N.J. 500, 267 A.2d 490 (1970); Gelbman v. Gelbman, 23 N.Y.2d 434, 245 N.E.2d 192, 297 N.Y.S.2d 529 (1969); Nuelle v. Wells, 154 N.W.2d 364 (N.D. 1967); Falco v. Pados, 444 Pa. 372, 282 A.2d 351 (1971); Smith v. Kauffman, 212 Va. 181, 183 S.E.2d 190 (1971); Goller v. White, 20 Wis. 2d 402, 122 N.W.2d 193 (1963).

The effect which liability insurance has had upon family immunities in tort law (and especially automobile collision cases) is difficult to evaluate. A number of courts hold that liability insurance does not create liability--but only recompenses it when it otherwise exists. The fact that a defendant may have insurance coverage does not automatically change the rule which traditionally denies a remedy to either a spouse, or to parent or child. Although it becomes more difficult to maintain the usual argument against allowing intra-familial recovery for tortious acts where insurance exists, the danger of collusion between the injured party and the insured is magnified considerably within every familial unit. W. PROSSER, HANDBOOK OF THE LAW OF TORTS 868, 874 (4th ed. 1971). See also, McCurdy, "Torts Between Parent and Child," 5 VILL L. REV. 521 (1960; McCurdy, "Torts Between Persons in Domestic Relations," 43 HARV. L. REV. 1030 (1930).

See generally, Annot., Wrongful Death, Interspousal Tort Liability, 92 A.L.R.2d 901 (1979).

It is interesting to speculate, given what appears to be a steady erosion of intra-familial immunity, whether this movement will be a precursor to actions by children to sue their parents when they are born defectively. If a child has a right to be born physically and mentally sound--as some courts maintain--do the parents incur a duty to ensure that the fetus they allow to come to livebirth be without significant defect, genetic or otherwise? See Seaberry, "Children Asserting Rights--Suits Against Parents, Guardians Increase," Wash. Post, Feb. 26, 1978, at 1, col. 1; "Parents Beware--Your Child May Want to Sue You," TIME Mag., May 22, 1978, at 68.

An eminent physician, with an additional law degree, who serves as Director of the Medical Genetics Center at The University of Texas Health Science Center at Houston, has opined that "once a pregnant woman has abandoned her right to abort and has decided to carry her fetus to term, she incurs a 'conditional prospective liability' for negligence acts toward her fetus if it should be born alive. These acts could be considered to be negligent fetal abuse resulting in an injured child. A decision to carry a genetically defective fetus to term would be an example. Abuse of alcohol or drugs during pregnancy could lead to fetal alcohol syndrome or drug addiction in the infant, resulting in an assertion that he had been harmed by his mother's acts. Withholding of necessary prenatal care, improper nutrition, exposure to mutagens and teratogens, or even exposure to the mother's defective intrauterine environment caused by her genotype, as in maternal PKU, could all result in an injured infant who might claim that his right to be born physically and mentally sound has been invaded." Shaw, "The Potential Plaintiff: Preconception and Prenatal Torts," in GENETICS AND LAW II (A. Milunsky, G. Annas eds. 1979); Note, "Intrafamilial Tort Immunity in New Jersey: Dismantling The Barrier to Personal Litigation," 10 RUTGERS-CAM.L. REV. 661 (1979).

9. 49 N.J. 22, 227 A.2d 689 (1967).

In the Civil Division of the Court of Common Pleas of Allegheny County, Pittsburgh, Pennsylvania, the case of Speck et al. v. Finegold & Schwarts, No. GD 76-7752, decided July 21, 1976, by Judge Silvestri presents an intriguing application of the failure of a court to recognize--based upon a "public policy of necessity"--the right of a child wrongfully born to maintain damages for being born. Relying heavily upon Gleitman v. Cosgrove, Zepeda v. Zepeda and Williams v. State of New York, infra, and other similarly related cases, the Court stated, "Believing, as we do, that policy and social necessity mandates the continuation of life, we hold that the birth of a child, whatever its physical or mental condition or its ethnic, social, racial and economic status, is not a wrong resulting in injury or damage to the parent or parents, the child or siblings of the child. It also follows that the period of gestation and the physical act of birth of the child is not a wrong resulting in injury or damage to the parent or parents." (Opinion at 20, 21). The Court continued to observing that, "The fact that some societies do not proscribe contraceptive devices, sterilization and abortion, leaving it to the determination of the individual members in their utilization as to whether to conceive life or terminate a pregnancy, such freedom of choice does not create rights superior to the necessity of the continuation of life."

The facts of the case are rather bizarre. Surgery in the nature of a vasectomy for purposes of sterilization was performed on Frank Speck by a Dr. Finegold on April 28, 1974. This action was undertaken because Mr. Speck, who is afflicted with a genetic disease known as neurofibromatosis, had passed that disease to his two daughters born by his wife, Dorothy. After his vasectomy, Speck was assured of its successful nature and advised he could engage in sexual intercourse without the use of any contraceptive device. His wife became pregnant on April 29, 1975, and gave birth to another daughter who was born prematurely afflicted with neurofibromatosis. Dr. Schwartz, the co-defendant, was retained to perform a therapeutic abortion to terminate Mrs. Speck's unanticipated pregnancy on December 27, 1974, and warranted the success of the surgery. Following the procedure, Dorothy advised Dr. Schwartz of her belief that the pregnancy was continuing--but she was assured of the success of the abortion. Thereafter, she gave birth to her third child.

The five count complaint was in trespass and assumpsit (contract), tortious negligence and malpractice. It alleged injury and damage suffered by each of the plaintiffs--Mr. and Mrs. Speck, their two previous children and the third "unanticipated" child (a daughter, Francine)--was the direct and proximate cause of negligence, misrepresentation and breach of contract by Drs. Finegold and Schwartz. The court sustained the motions of the defendants for more specific pleadings. The case was subsequently appealed to the Superior Court of Allegheny County, April Term, 1977, Docket No. 7. Telephone Conversations, Judge Silvestri and Thomas Hollander, Esquire, July 27, 1978.

As of May 15, 1979, a decision had not issued--although argument was heard in April, 1977. Apparently President Judge Cercone of the Superior Court was circulating a draft of the Speck decision among his colleagues on the Court. Telephone Conversation, Thomas Hollander, Esquire, Pittsburgh, Pennsylvania, May 15, 1979.

A letter dated August 2, 1979, from the chambers of Judge Cercone enclosed a copy of the Court's decision in Speck decided July 25, 1979. In denying the claims of the daughter Francine to be made whole, regardless of a claim for wrongful life or otherwise, the court concluded there was a failure to state a legally cognizable cause of action because of two reasons: a failure of judicial precedent to hold that a child has a fundamental right to be born as a whole, functional human being; and an inability of the court to accept the placement of the child in a position it would have occupied if the defendants had not been negligent when to do so would make her non-existent. While the recovery in negligence "is intended to place the injured party in the position he would have occupied but for the negligence of the defendant," the court observed that, "a cause of action brought on behalf of an infant seeking recovery for a 'wrongful life' on grounds she should not have been born demands a calculation of damages dependent on a comparison between Hobson's choice of life in an impaired state and non-existence. This the law is incapable of doing." (Opinion at 19, 20).

The court allowed recovery for the parent's claim of negligence, holding that but for the breach of the defendant's duty to both treat and correctly advise the plaintiff-parents of their true medical situation, they would not have been required to assume the large expenditures connected with the wrongful birth of their daughter, Francine. But as to the claim by the parents for emotional disturbance and mental stress, the court referred to the decision in the consolidated cases of Park v. Chessin and Becker v. Schwartz rendered by the New York Court of Appeals and, as did the New York Court, held that these claims were to be denied. 408 A.2d 496 (1979). The Supreme Court of the Commonwealth of Pennsylvania will issue its decision in the Speck case in early 1981.

See generally, Robertson, "Civil Liability Arising from 'Wrongful Birth' Following an Unsuccessful Sterilization Operation," 4 AM.J. L. & MED. 131 (1978); R. VEATCH, DEATH, DYING AND THE BIOLOGICAL REVOLUTION (1976).

One of the largest out of court malpractice settlements in the District of Columbia--in the sum of $1.3 million--was reached in the United States District Court in Judge June L. Green's chambers. Plaintiffs, Mr. and Mrs. Ira Biggar sued Dr. B. J. Mundell and Georgetown University Hospital for malpractice in supervision the birth of their son. Because of a serious infection of her womb, it was maintained that Mrs. Biggar should have delivered at once instead of waiting for fourteen hours after the infection was discovered. Severe brain damage resulted as a consequence of the delayed birth to the child and it will need a lifetime of care. The defendants asserted the child's physical and mental retardation was due largely to its premature birth--not to medical malpractice. Whitaker, "Malpractice Settlement is $1.3 Million," Wash. Post, May 11, 1979, at B1, col. 1.

See generally, Annot., Right to Die, Wrongful Life, AM.JUR. 2d (New Topic Service) 1 (1979).

10. Gleitman v. Cosgrove, 49 N.J. 22, 26, 227 A.2d 689, 691 (1967).

11. Id. at 28, 227 A.2d at 692.

12. Id. at 31, 227 A.2d at 693. This reason is clear evidence of the court's conceptualization of abortion as being against public policy. Both courts did not want to adopt a line of reasoning which would tend to encourage abortions. See, Comment, "A Cause of Action for 'Wrongful Life,'" 55 MINN.L.REV. 58, 59 (1970).

13. 58 Misc. 2d 432, 296 N.Y.S.2d 41 (1968).

14. The child's parents each brought actions for negligence for failure to inform them of a divided panel's (of 4) opinion as to whether or not there should have been an abortion and for their doctor's statement that an abortion was not needed and should not be sought elsewhere. The jury aware of damages for the parents was upheld, although the child's cause of action was dismissed. Stewart v. Long Island College Hospital, 58 Misc. 2d 432, 296 N.Y.S.2d 41 (1968), modified 35 App.Div. 2d 531, 313 N.Y.S.2d 502 (2d Dept 1970), aff'd as modified, 30 N.Y.2d 695, 238 N.E.2d 616, 332 N.Y.S.2d 640 (1972).

The Stewart decision was found of precedental control by the same Appellate Division of the New York Supreme Court in Howard v. Lecher, 42 N.Y.2d 109, 366 N.E.2d 64, 397 N.Y.S.2d 363 (1977), aff'g 53 App. Div. 2d 420, 386 N.Y.S.2d 460 (1976). In Howard, the Court granted the motion of the defendant obstetrician to dismiss a claim for damages for emotional disturbance suffered by the parents and for various expenses relating to their child's medical, hospital, nursing and final funeral expenses incurred as a consequence of their child, Melissa, being wrongfully born with the debilitating and subsequently fatal genetic disease--Tay Sachs. The complaint asserted that had the defendant taken proper genealogical histories of the parents and conducted those tests available for determining the existence of the parents' carrier status (i.e., amniocentesis), this information would have aided the parents in making a decision whether to allow an abortion.

In a 3-2 decision, the majority opinion listed the reasons for denial of the parents' claim for emotional harm resulting from the "wrongful birth" of their daughter as being: the harm which the parents alleged they suffered was not a harm the consequence of any direct injury to them; to recognize such a claim--heretofore not recognized in New York--would lead to a flagrant extension of malpractice liability and the measure of any such damages would be highly speculative and, thus, impossible to prove. 53 App.Div. 2d at 424.

In a persuasive dissent, Justice Margett submitted that if the plaintiffs could establish the proofs required to recover in "the usual negligence of malpractice case" (i.e., duty, breach, proximate cause and damages), a recovery should not be denied simply because the measure of damages is not clearly ascertainable or a policy to limit a physicians liability is believed desirable. Id. at 428 passim.

On the issue of calculating damages, see Comment, "Liability for Failure of Birth Control Measures," 76 COLUM.L.REV. 1187, 1195 (1976). See also, Steitz v. Gifford, 280 N.Y. 15, 20, 19 N.E.2d 661, 664 (1939) where the rule in New York was stated that mathematical certainty was not required in computing an award for damages. See also, Kalina v. General Hospital of the City of Syracuse, 13 N.Y.2d 1023, 245 N.Y.S.2d 599 (1963), aff'd 18 App.Div. 2d 757, 235 N.Y.S.2d

808 (Fourth Dept. 1963), where it was held that parental mental suffering caused by a child's illness is not recoverable.

15. 282 U.S. 555 (1931), the Court stated: "Where the tort itself is of such a nature as to preclude the ascertainment of damages with certainty, it would be a perversion of fundamental principles of justice to deny all relief to the injured person, and thereby relieve the wrongdoer from making any amends for his acts." Id. at 563.

16. Griswold v. Connecticut, 381 U.S. 479 (1965); Roe v. Wade, 410 U.S. 113 (1973).

17. Griswold, supra.

18. See, Roe v. Wade, supra.

The vitality of the Stewart v. Long Island College Hospital case, supra note 14 was seriously undermined in Ziemba v. Steinberg, 357 N.Y.S.2d 265, 45 App.Div. 2d 230 (1974). There, a medical malpractice action was brought alleging that due to the physician's lack of care, the plaintiff's wife was not informed of pregnancy so that she could terminate it within a reasonable time. The wife had sought medication to prevent pregnancy but did become pregnant. In allowing recovery the court distinguished Stewart by pointing out that when that case was decided, abortions were not legal. This is no longer true in light of the new legal environment engendered as a consequence of Roe v. Wade.

19. 69 Wis. 2d 766, 233 N.W.2d 372 (1975).

20. The charges alleged that the hospital and physician were both negligent in four ways: 1) failing to diagnose the rubella in the mother, 2) failing to inquire as to the possibility of either rubella or pregnancy, 3) failing to perform clinical tests in light of the mother's manifestation of the symptoms of rubella, and 4) failing to inform the mother of the effects rubella could have on her unborn child and further failing to advise her of the availability of a therapeutic abortion.

21. 233 N.W.2d at 376.

22. 64 Wis. 2d 514, 219 N.W.2d 242 (1974). The complaint was denied on public policy grounds: "We have no hesitancy in concluding that to hold that the allegations of this complaint constitute a cause of action for recoverable damages would open the way for fraudulent claims and would enter a field that has no sensible or just stopping point." 219 N.W.2d at 245.

23. Completing the analysis for recovery in tort would involve not only showing the duty (standard of care) and its breach, but also injury resulting from the breach under a theory of causation. The person affected must be a "person within the risk" and the injury occasioned by the act must be a "result within the risk" of the negligence being asserted. See KEETON, LEGAL CAUSE IN THE LAW OF TORTS (1963); cf. Palsgraf v. Long Island R.R., 248 N.Y. 339, 162 N.E. 99 (1928).

24. 41 Ill.App. 2d 240, 190 N.E.2d 849 (1963), cert. den. 379 U.S. 945 (1964).

25. 18 N.Y.2d 481, 276 N.Y.S.2d 885, 223 N.E.2d 343 (1966).

26. One difficulty which would arise in allowing this tort action would be the apparent anomaly of a plaintiff complaining of his parents' intercourse, "the very act which gave him life." Note, 77 HARV. L. REV. 1349, 1351 (1964).

27. 41 Ill. App. 2d 240 at 247, 190 N.E.2d 849 at 852.

28. See supra, note 26 where it is recognized that this is a "common judicial worry," subject to the counter argument that courts should place more confidence in their "ability as reasoning institutions to find valid distinctions" where relief would be proper. See also, Note, 49 IOWA L. REV. 1005, 1013 (1964).

The third argument by the Zepeda court obviously anticipated claims brought by persons who are born poor or of a racial minority status.

29. Zepeda v. Zepeda, 41 Ill.App. 2d 240, 190 N.E.2d 849, 853-54 (1963).

30. 18 N.Y.2d at 484, 223 N.E.2d at 344, 276 N.Y.S.2d at 887. The next year a Florida court held that to recognize such a claim would be to ignore "our understanding of both the laws of men and the law of nature." Pinkney v. Pinkney, 189 So. 2d 53, 54 (Fla.App. 1967). It is interesting to note that in Cessna v. Montgomery, 28 Ill.App. 3d 887, 329 N.E.2d 861 (1975), the Illinois court invalidated a state statute that barred paternity suits after two years. The statute was invalidated because it was an invidious discrimination against those born out of wedlock, since their legitimate counterparts could bring suit for support at any time. Zepeda and Williams however, did not recognize such a right in being legitimate. The Supreme Court in Gomez v. Perez, 409 U.S. 535 (1973) held that illegitimate children are to be accorded all rights and privileges that legitimate children have. Recognition of the ability to sue for a tort committed during the prenatal period will bring greater weight on the side of those arguing for allowing suits for wrongful life.

31. 18 N.Y.2d at 484, 223 N.E.2d at 344, 276 N.Y.S.2d at 887. Judge Keating, in a concurring opinion stated that what bothered him about the case was the logic (?) of permitting recovery for the very act which caused existence. See note 25, supra; see also Note, "Wrongful Life--A New Tort?--Williams v. State," 17 HAST.L.J. 400 (1965).

32. The report of the decision at 387 N.Y.S.2d 204 omits most of the court's discussion of all but the sixth cause of action. The remaining causes are included in the unpublished portions of the decision.

33. Park v. Chessin, unpublished portion of decision, at 22.

34. Id. at 26.

35. The court was required in the face of the motion to accept the facts alleged in the complaint as true and to allow the causes of action to stand if it found that the plaintiff was entitled to recover on the basis of any facts alleged. Park v. Chessin, 387 N.Y.S.2d 204, 206 (1976).

36. 24 N.Y.2d 478, 301 N.Y.S.2d 65, 248 N.E.2d 901 (1969).

37. 248 N.Y. 339, 162 N.E. 99 (1928).

38. Park v. Chessin, unpublished portion of decision at 25.

39. Park v. Chessin, 387 N.Y.S.2d 205

40. Id. at 207.

41. Id.

42. 400 N.Y.S.2d 110, 113 (App.Div. 2d 1977), modifying 88 Misc. 2d

221, aff'g 387 N.Y.S.2d 204 (1976).

43. 387 N.Y.S.2d 204, 206.

44. 387 N.Y.S.2d 294, 207 quoting Zepeda v. Zepeda, supra. But cf. Roe v. Wade, 410, U.S. 113, 160 (1973) (discussion of fetal viability).

45. But cf. Roe v. Wade, 410 U.S. 113, 159 (1973) (no pre-natal application of term "person" for purposes of constitutional rights).

46. Park v. Chessin, 387 N.Y.S.2d 204, 210 quoting Endresz v. Friedberg, 24 N.Y.2d 478, 483, 301 N.Y.S.2d 65, 68, 248 N.E.2d 901, 903 (1969).

47. 387 N.Y.S.2d 204, 211.

48. Id.

49. Two aspects of the damage question have proved troublesome in the cases. First, courts have traditionally held that the birth of a child is not a compensable injury (e.g., Shaheen v. Knight, 11 Pa.D. & C. 2d 41 (Lycoming County Ct. 1957); but see Troppi v. Scarf 31 Mich. App. 240, 187 N.W.2d 511 (1971). Second, courts have been reluctant to allow a cause of action to stand on the basis that damages, if warranted, are too indefinite and difficult to calculate (see Gleitman v. Cosgrove, 49 N.J. 22, 227 A.2d 689 (1967).

50. 387 N.Y.S.2d 204, 209.

51. Id., at 211.

52. Park v. Chessin, 400 N.Y.S.2d 110, 114.

53. Id.

54. As to the sixth cause of action, the Appellate Division voted 4 to 1 to deny defendants' motion to dismiss. Justice Cohalan would--however--have granted defendants' motion as to the first (mother's pain and suffering) and seventh (father's claim for costs incurred in caring for Lara) causes of action. 400 N.Y.S.2d 110, 115.

55. 400 N.Y.S.2d 110, 114.

56. N.Y. Penal Code § 125.05, Subd. 3 (Supp. 1972-73).

This section of the Code was revised in 1975. The substance of the revised section is that any offense defined in the article, whether bearing a homicide label or an "abortion" label, is technically classified as "homicide" if it involves either the death of a born human being or the death or destruction of an unborn child of more than twenty-four weeks' pregnancy. "Justifiable abortional acts" are those a pregnant female commits upon herself with the advice of a duly licensed physician, within twenty-four weeks from the commencement of her pregnancy, such an act which is considered necessary to preserve her life. Id., § 125.05.

57. 400 N.Y.S.2d 110, 114.

58. These reasons included the conceptual difficulty of comparing the existence of life with its non-existence experienced by the court in Karlsons v. Guerinot, 394 N.Y.S.2d 933 (1977), the difficulty in calculating damages found in Gleitman v. Cosgrove, 49 N.J. 22, 227 A.2d 689 (1967), and the fact that prior to the abolition of criminal prohibitions, the mother in a wrongful life suit could not legally abort.

The last reason was the basis for the holding in Stewart v. Long Island College Hospital, supra note 14.

59. Supra note 57.

60. Id. at 115.

61. Justice Titone cited the cases of Zepeda v. Zepeda, 190 N.E.2d 849, cert. den. 379 U.S. 945, Gleitman v. Cosgrove, supra, and Aronoff v. Snider, 292 So. 2d 418 (1974) as foreign precedent for granting defendants' motion to dismiss the sixth cause of action. The Justice's reliance on Zepeda is misplaced, however, in that the case is easily distinguished from Park. In Zepeda, which is frequently cited as a proto-type wrongful life case, a healthy child sued its natural father for causing it to be born a bastard. The only commonality shared by Zepeda and Park is the term "wrongful life." Aronoff is likewise distinguishable from the case at bar. There, the birth complained of was unplanned but did not involve injuries to the infant. Moreover, Aronoff was couched in terms of wrongful pregnancy and wrongful birth. The action was brought by the parents and the unexpected infant's siblings, not by the child itself as in Park. The facts in Gleitman bear the greatest similarity to those in Park. There, a defective infant was born following its mother's contraction of rubella during pregnancy. The child and his parents sued the obstetrician, claiming that on specific inquiry, he had advised Mrs. Gleitman that her illness would have no adverse impact on the fetus which she was carrying. The case floundered on the court's reluctance to compare the value of life with defects and that of no life at all. As it was a New Jersey decision, the Appellate Division was not bound in any way by Gleitman.

62. 400 N.Y.S.2d 110, 115.

63. At 400 N.Y.S.2d 116, the dissent traces the legislative creation of a cause of action for wrongful death from its English antecedents in the Death Act (Stat. of Aug. 26, 1846, 9 & 10 Vict., Ch. 93) to its adoption by the New York Legislature in 1847 (L. 1847, Ch. 450). He advocates following this method of creating a cause of action for wrongful birth. In so doing, he adopts a view which has blocked causes of action for wrongful life in the past.

64. The three cases considered, when read closely, observed the court, as constituting no bar to the recognition of a wrongful life cause of action in New York were: Howard v. Lecher, 42 N.Y.2d 109, 366 N.E.2d 64, 397 N.Y.S.2d 363 (1977), aff'g 53 App.Div. 2d 420, 386 N.Y.S.2d 460 (1976); Williams v. State of New York, 260 N.Y.S.2d 953 (Ct. Cl. 1965), rev'd, 269 N.Y.S.2d 786 (1966), aff'd, 276 N.Y.S.2d 885 (1966); and Stewart v. Long Island College Hospital, supra note 14.

65. See, Kass & Shaw, "The Risk of Birth Defects: Jacobs v. Theimer and Parents' Right to Know," 2 AM.J.L. & MED. 213, 226 n. 55 (1976-77).

66. See Waltz & Thigpen, "Genetic Screening and Counselling: The Legal and Ethical Issues," 68 NW.U.L. REV. 696 (1973).

67. See, "Is Fetal Monitoring Safe?" Wash. Post, April 16, 1978, at B3, col. 3.

68. Roe v. Wade, 410 U.S. 113 (1973).

69. Sherlock v. Stillwater Clinic, ___Minn.___, 260 N.W.2d 169 (1977).

70. Id. at 176.

71. Id. at 175.

72. Id. at 172, 174.

73. Id. at 172.

74. Id. at 175.

75. Id.

See generally, Seaberry, "Children Asserting Rights--Suits against Parents, Guardians Increase," Wash. Post, Feb. 26, 1978, at 1, col. 1; "Parents Beware--Your Child May Want to Sue You," TIME Mag., May 22, 1978, at 58.

The Appellate Court of Illinois, First District, Second Division ruled in Wilczynski v. Goodman, 391 N.E.2d 479 (1979) that ". . . a public policy which deems precious even potential life while yet in the womb, at such cost and expense that condition may entail, does not countenance as compensable damage to its parent or parents those additional costs and expenses necessary to sustain and nuture that life once it comes to fruition upon and after successful birth." Id. at 487. Here, the defendant osteopathic physician, licensed under the laws of Illinois, undertook to perform a therapeutic abortion for the purpose of terminating the plaintiff's pregnancy. The pregnancy was in fact not terminated and an "apparently healthy" child was born. Plaintiff sought to recover the medical and hospital expenses associated with the birth--together with the costs of raising and educating her child. The court determined there was a cognizable action for negligent performance of an abortion, but limited damages to pregnancy and birth related costs and expenses. No recovery was allowed for costs incurred in raising or educating the child.

A state appeals court in California recently upheld a child's right to sue for damages because it was conceived and born with a severe genetic defect (Tay Sachs). Shauna Temar Curlender's father was allowed to sue in her behalf for negligence--alleging that her parents were improperly tested for Tay Sachs and erroneously told they were not carriers of the genetic defect. A California Superior Court judge had held earlier that two year old Shauna had no right to file the suit. A Court of Appeals panel presided over by Justice Bernard S. Jefferson sent the case back to the lower court with direction that Shauna could maintain the action in her own right.

Justice Jefferson stated that under California case law, all persons are required to use ordinary care in their conduct with others so as to prevent injury. Thus, the Court found no problem in finding such a duty of ordinary care owed by medical laboratories engaged in genetic testing to parents as well as to their yet unborn children. Noting that the very reality of a "wrongful life" claim is "that such a plaintiff exists and suffers, due to the negligence of others," and that this cause of action is "based upon negligently caused failure by someone under a duty to do so to inform the prospective parents of facts needed by them to make a conscious choice not to become parents," the Court concluded that for every wrong committed there should be a remedy. Again, acknowledging that in California, infants were presumed to experience pain and suffering when injury was established, the Court totally rejected the idea that a cause of action based on a "wrongful life" involves an evaluation of a claimed right not to be born. "In essence, we construe the 'wrongful-life' cause of action by the defective child as the right of such child to recover damages for the pain and suffering to be endured during the limited life span available to such a child and any special pecuniary loss resulting from the impaired condition." Curlender v. Bio-Science Laboratories, 100 Cal.App. 3d 811, 831 (1980).

See generally, Capron, "Tort Liability in Genetic Counseling," 79 COLUM.L.REV. 618 (1979).

76. Custodia v. Bauer, 251 Cal.App. 2d 303, 59 Cal.Rptr. 463 (1967).

77. Id.

78. Restatement of Torts § 920 (1939).

79. See Christensen v. Thornby, 192 Minn. 123, 255 N.W. 620 (1934); Shaheen v. Knight, 11 Pa.D. & C. 2d 41 (Lycoming County Ct. 1957); Ball v. Midge, 64 Wash. 2d 247, 391 P.2d 201 (1974). However, the birth of a child may be regarded as something less than the blessed event referred to in these cases when the parents are unable, financially, to afford another child or had not, in the first instance, planned on that child. Cf. Note, 19 U.PITT.L.REV. 802, 805 (1958).

80. See West v. Underwood, 132 N.J.L. 325, 40 A.2d 610 (1945) where recovery was allowed for all pain and suffering, mental and physical, together with loss of services and any other loss or damage proximately resulting from such negligence, when the physician negligently performed a sterilization operation. The negligence required a subsequent operation to effect the sterilization and that operation resulted in a complication which required yet another operation.

81. Custodia v. Bauer, 59 Cal.Rptr. 463, 476 (1967).

82. Id. This is in direct contrast to one defense that has been used for such an action. It is repugnant to social ethics with respect to family establishment to cause emotional injury to the child involved. For the child to find out that it was unwanted may be a traumatic experience. See Note, 9 UTAH L.REV. 808 (1965).

83. Supra, note 81.

84. Jackson v. Anderson, 230 So. 2d 503 (Fla.App. 1970). Recovery was also allowed in Bishop v. Byrne, 265 F. Supp. 460 (S.D.W. Va. 1967), where seventeen months after a sterilization operation, the wife became pregnant and gave birth to a healthy child. The court allowed that the plaintiff could recover for all pain and suffering, mental and physical, together with loss of services and any other loss of damages proximately resulting from the damages.

85. 281 A.2d 616 (Del. Super. 1971).

86. See note 17, supra.

87. Using the Benefits Rule, Restatement of Torts, § 920 (1939).

88. 281 A.2d at 619. The court said: "By attempting to have Mrs. Coleman sterilized, the Colemans sought to maintain a certain standard of living, a certain amount of love and affection, and a certain amount of parental protection for their then four children. The pregnancy and impending birth of a fifth child will upset the portion allotted to each child. Should such change in the family structure be measurable economically, it should be compensable." Id. at 619.

89. Coleman v. Garrison, 349 A.2d 8 (Del. 1975).

90. See note 78, supra and accompanying text.

91. 349 A.2d at 13-14. The court commented that since the damages were not cognizable at law and hence, not recoverable, citing Stewart v. Long Island College Hospital. It appears as if the Colemans became greedy after their initial success in 1971. The earlier case held

there could be recovery for pain and suffering, medical expenses and loss of consortium; the later case, however, shows the reluctance on the part of courts to extend recovery to the normal expenses of raising a child to maturity.

    Also, the court noted, with respect to the suits on behalf of the siblings, that recovery had not been allowed for such a claim prior to this case, citing Arnoff v. Snider, 292 So. 2d 418 (Fla.App. 1974). This is not exactly complete in that Custodia v. Bauer, 251 Cal.App. 2d 303, 59 Cal.Rptr. 463 (1967) recognized that upon proof of a breach of a duty, recovery could be had. The court there said: "Where the mother survives without casualty there is still loss. She must spread her society, comfort, care, protection and support over a larger group. If this change in the family status can be measured economically, it should be compensable as the former losses."

92. 496 S.W.2d 124 (Tex.Civ.App. 1973) cert. den. 415 U.S. 927 (1973).

93. Id. at 127.

94. The language used by the court shows a certain amount of emotional bias. It said: "Despite such holdings, a strong case can be made that, at least in an urban society, the rearing of a child would not be a profitable undertaking if considered from the economics alone. Nevertheless, as recognized in Hays and Troppi, the satisfaction, joy and companionship which normal parents have in rearing a child make such economic loss worthwhile. These intangible benefits, while impossible to value in dollars and cents are undoubtedly the things that make life worthwhile. Who can place a price tag on a child's smile or the parental pride in a child's achievement? Even if we consider only the economic point of view, a child is some security for the parents' old age. Rather than attempt to value these intangible benefits, our courts have simply determined that public sentiment recognizes that these benefits to the parents outweigh their economic loss in rearing and educating a healthy, normal child. We see no compelling reason to change such rule at this time." Id. at 128.

95. Id. at 131.

96. There have been values (monetary) placed on the companionship, etc. of children. Cf. Wardlow v. City of Keokuk, 190 N.W.2d 439 (Iowa 1971); Lockart v. Besel, 71 Wash. 2d 112, 261 P.2d 605 (1967); Fussner v. Andert, 261 Minn. 347, 113 N.W.2d 355 (1962). See also, Story Parchment Co. v. Paterson Parchment Paper Co., 282 U.S. 555 (1931).

97. 136 N.J. Super. 69, 344 A.2d 336 (1975).
98. See notes 77-83, supra and accompanying text.

99. 55 Cal.App. 3d 698, 127 Cal.Rptr. 652 (1976). For an excellent discussion of treatment of damages by the courts, see 127 Cal.Rptr. at 657.

100. 357 N.Y.S.2d 265, 45 App.Div. 2d 230 (1976).

101. See notes 13-14, supra and accompanying text.

102. 410 U.S. 113 (1973). See also, supra, notes 60 and 86, Chapter 3, which discuss the new legal environment created by Roe and three new Supreme Court cases; Oelsner, "High Court Bars Husband's Power to Veto Abortion," N.Y. Times, July 2, 1976, at A1, col. 8.

103. 519 S.W. 2d 846 (Texas 1975).

    Jacobs holds directly opposite to the rulings in Gleitman and Stewart cited supra at notes 10 and 14, respectively. In view of the previously acknowledged "new legal climate" allowing abortions rather freely as a personal choice of a pregnant woman as evidenced by Roe v. Wade and the new cases of Planned Parenthood of Missouri v. Danforth, Singleton v. Wulff and Bellotti v. Baird (see supra, notes 60 and 87), Jacobs is probably better law for the simple reason that it is not shackled with anachronistic public policy considerations.

In Jacobs--although eugenic abortions were illegal under Texas law at all material times in this case--recovery was allowed for expenses reasonably calculated to cover the treatment and care of the child's impairment. Mr. Justice Reavley, in his majority opinion in Jacobs uses the concepts of wrongful birth and wrongful life synonymously. It would be strained, in any event, to consider Jacobs as grounded in a theory of a wrongful life, however.

Justice Reavley observes that, "Previous Texas cases have indicated this distinction between the cause of action which seeks damages for wrongful birth or life and the cause of action seeking recovery of those expenditures required because the child is deformed--even though the tort is causally related to birth itself and not to deformation alone." Id. at 849 (emphasis added). As noted, this case is properly considered as but a suit for recovery of expenses reasonably necessary for care and treatment of a child's physical impairment.

For an in dept discussion of Jacobs v. Theimer see Kass & Shaw, "The Risk of Birth Defects: Jacobs v. Theimer and Parents' Right to Know," 2 AM.J.L. & Med. 213 (1976-77) where the authors assert this case appears to be the first reported one which recognizes a cause of action for damages by the parents of a defective child based upon a claim that the physician in charge had negligently failed to inform the mother of the risk that her child would be born defective and, thus, foreclosed the opportunity of the parents to obtain an abortion. The authors also assert that since the court cited a number of cases considered "wrongful life or wrongful birth cases," this should be taken as indicating the court's intention "to categorize Jacobs as a case of wrongful life or wrongful birth." Id. at 241. By such assertions, the authors follow the general trend of perpetuating the confusion in the area by treating the two legal doctrines synonomously.

104. See note 102, supra.

105. 31 Mich.App. 240, 187 N.W.2d 511 (1971).

106. In doing so, the court distinguished Christensen v. Thornby, 192 Minn. 123, 255 N.W. 620 (1934) where recovery was denied. There, the suit was for the husband's anxiety over his wife's health, which was not damaged by the pregnancy. The husband did not allege that the child or the economic burdens were unwanted.

107. The court criticized Shaheen v. Knight, supra note 79, and held that the benefits always outweigh the burdens of raising a child.

The court cited Milde v. Leigh, 75 N.D. 418, 28 N.W.2d 530 (1947) for authority for allowing damages. There a negligent sterilization leading to pregnancy was grounds for recovery for loss of services, society and companionship (consortium), as well as for medical expenses. The court in Troppi recognized that the damages are not unascertainable, but that each would have to be determined on a case-by-case basis. 187 N.W.2d 511 at 519.

108. 187 N.W.2d 511 at 518.

109. Accord, Rieck v. Medical Protective Society, 64 Wis. 2d 514, 219 N.W.2d 242 (1974), where the court said: "Every child's smile, every bond of love and affection, every reason for parental pride in a

child's achievements, every contribution by the child to the welfare and well-being of the family and parents is to remain with the father and mother. For the most part, these are intangible benefits, but they are nonetheless real." 219 N.W.2d at 244. See also Shaheen v. Knight, supra note 79, Christensen v. Thornby, 192 Minn. 123, 255 N.W. 620 (1934); Ball v. Modge, 64 Wash. 2d 247, 391 P.2d 201 (1964).

110. See Stewart v. Long Island College Hospital, supra note 14; Jackson v. Anderson, 230 So. 2d 503 (Fla.App. 1970).

111. See Jacobs v. Theimer, 519 S.W.2d 846 (Texas 1975); Terrell v. Garcia, 496 S.W.2d 124 (Tex.Civ.App. 1973), cert. den. 415 U.S. 927 (1973); Coleman v. Garrison, 327 A.2d 757 (Del. Super. 1974).

112. See Ziemba v. Sternberg, 45 App.Div. 2d 230, 357 N.Y.S.2d 265 (1974); Bishop v. Byrne, 265 F. Supp 460 (S.D.W.Va. 1967); Troppi v. Scarf, 31 Mich.App. 240, 187 N.W.2d 511 (1971); Custodia v. Bauer, 251 Cal.App. 2d 303, 59 Cal.Rptr. 463 (1967).

113. See, Annot. Tort Liability for Wrongfully Causing One To Be Born, 22 A.L.R.2d 1441 (1968).

Non recognition of standing for infants in "wrongful life" actions is tied, as has been shown, to a basic reluctance of the judiciary to recognize life itself as a negative (or an injurious act) and an inability to attempt to structure a formula for measuring life in an impaired state v. no life at all. Recovery for pre-natal injuries, although allowed in some jurisdictions, finds difficulty of uniform recognition and proof with the basic issue of causation. Injury during the pre-natal period which in turn causes a defect upon birth is--of course--tied to proof of negligence. Basically, the plaintiff in such an action does not claim that he should not have been born at all (as in wrongful life), but rather, but for the negligence of the defendant he would have been born without being in an injured condition.

It is difficult to draw a strong line of distinction between the two judicial postures re allowance for recovery in some pre-natal injury cases and the reluctance to countenance recovery under a wrongful life theory. Yet, if there is a more ready acceptance by some courts to a basic malpractice argument in a pre-natal injury case, it would appear a better legal strategy to follow than maintain a more precarious posture with a wrongful life theory of recovery.

114. 138 Mass. 14 (1884). This case was the guide for all such decisions for the next 60 years. The Restatement of Torts § 869 adopted this view as well. However, Tentative Draft No. 16 of the Restatement of the Law of Torts, 2d (April 24, 1970) provides that one who tortiously causes harm to an unborn child is subject to liability to the child for such harm if it is born alive.

115. See Annot., Liability for Pre-Natal Injuries, 40 A.L.R.2d 1222, 1226 (1971).

In all jurisdictions, except Alabama, there has been a complete rejection of the Dietrich holding in favor of that of Bonbrest. Id. at 1227, n. 14.

116. 65 F. Supp. 138 (D.D.C. 1946).

The Court specifically addressed itself to the problems of bad faith claims and inseparable difficulties of proof, ruling that those reasons should not be sufficient in order to defeat a cause of action. This is directly opposite to the position taken by courts faced with novel wrongful life suits--as has been shown previously.

The Court quotes Mr. Justice Holmes:

> "'The life of the law has been not logic, it has been experience,'. . . . and here we find a willingness to face the facts of life rather than a myopic and specious resort to precedent to avoid attachment of responsibility where it ought to attach, and to permit idiocy, imbecility, paralysis, loss of function and like residuals of another negligence to be locked in the limbo of uncompensable wrong, because of a legal fiction, long outmoded." 65 F. Supp. at 142.

117. Viable state meaning that the infant is capable of being born alive and living independently of the mother. According to Mitchell v. Couch, 285 S.W.2d 901 (Ky. 1955), this generally comes between the 6th and 7th month of pregnancy.

118. See Note, "The Impact of Medical Knowledge on The Law Relating to Prenatal Injuries," 110 U.PA.L.REV. 554 (1962). The author argues against the viability distinction, giving medical authority that "congenital structural defects occasioned by environmental factors can be sustained only within the earliest stages of the previable period." Id. at 563. It appears, therefore, that some of the most deserving claims stand to be denied by the practice of requiring viability at the time of injury.

119. E.g., Smith v. Brennan, 31 N.J. 353, 157 A.2d 497 (1960); Sylvia v. Gobielle, 101 R.I. 76, 220 A.2d 222 (1966); Cardwell v. Welch, 213 S.E.2d 382 (N.C. 1975); Padillow v. Elrod, 424 P.2d 16 (Okla. 1967); Toth v. Goree, 65 Mich.App. 296, 237 N.W.2d 297 (1975). See also, Gordon, "The Unborn Plaintiff," 63 MICH.L.REV. 579 (1965).

120. See Puhl v. Milwaukee Auto Ins. Co., 8 Wis. 2d 343, 99 N.W. 2d 163 (1959) for an excellent discussion of the difficulties in proving causation in a prenatal injuries case.

121. See Simmons v. Weisenthal, 29 Pa. D. & C. 2d 54 (1962). Cf. Berry v. Branner, 245, Or. 307, 421 P.2d 996 (1966), where it was held that a cause of action for malpractice was not barred by the Statute of Limitations when a surgical needle was discovered in the plaintiff nine years after the operation.

122. Libby v. Conway, 192 Cal.App. 2d 865, 13 Cal.Rptr. 830 (1961).

123. 165 Minn. 320, 206 N.W. 560 (1925).

124. Seattle-First Nat. Bank v. Rankin, 59 Wash. 2d 288, 267 P.2d 835 (1962).

125. 384 Mich. 718, 187 N.W.2d 218 (1971).

126. 31 N.J. 353, 157 A.2d 497 (1960).

127. Id. at 364, 365, 157 A.2d at 503. The decision contains an excellent compilation of the cases allowing recovery for prenatal injuries in the differing jurisdictions to that date.

128. See Jorgensen v. Meade-Johnson Laboratories, Inc., 483 F.2d 237 (10th Cir. 1973), where an action was allowed for children being born Mongoloid as a result of their mother taking birth control pills. The court stated that in order to recover, causation must be shown.

In a case involving thalidomide damaged babies, the Third Circuit refused to allow a suit, applying Quebec's 12 month Statute of Limitations. The court there said that the state did not have a substantial interest in the case because it was one of 41 states where the drug was tested. Henry v. Richardson-Merrell, Inc., 508 F.2d 28 (3rd Cir. 1975). It is interesting to note that not one reported decision on injuries as a result of this drug has been found by this author. This is peculiar in light of the injury which was caused by these children being born without limbs. For a discussion of the drug and its effects on children (during the pregnancy of their mother), see Cluiner, "Trauma to the Unborn Child," 5 TRAUMA 80 (1963); SUFFER THE CHILDREN: THE STORY OF THALIDOMIDE (Sunday Times, London, ed. 1979).

See generally, McCormick, "To Save or Let Die: The Dilemma of Modern Medicine" in CONTEMPORARY ISSUES IN BIOETHICS 331 (T. Beauchamp, L. Walters eds. 1978); S. NICHOLSON, ABORTION AND THE ROMAN CATHOLIC CHURCH (1978).

129. See Dillon v. S.S. Kresge Co., 35 Mich.App. 603, 192 N.W.2d 661 (1972). Here, a company was sued--with the principal allegation against it and its owners being that due to unsanitary conditions in which the company kept one of its stores, a pregnant employee subsequently contracted rubella and passed it on to her child who suffered serious, permanent injuries. A trial was ordered by the Michigan Court of Appeals on the issue of causation.

130. See the cases collected in 5 TRAUMA 40 (1965).

131. See e.g., Tedeschi, "On Tort Liability for Wrongful Life," 1 ISRAEL L. REV. 513 (1966).

But see, Friedman, "Legal Implications of Amniocentesis," 123 U.PA. L.REV. 92 (1974).

132. The question of personal morals will obviously become involved in such considerations. Placing the burden of proof upon the person claiming the moral convictions would not allow termination of the pregnancy will, no doubt, ensure its use as an acceptable or justifiable reason for inaction.

133. Becker et al. v. Schwartz, Park et al. v. Chessin, et al., 413 N.Y.S.2d 895 (1978).

134. 60 A.D.2d 587, 400 N.Y.S.2d 119 and 60 A.D.2d 80, 400 N.Y.S.2d 110 (1977).

135. See generally, Cohen, "Park v. Chessin: The Continuing Judicial Development of The Theory of Wrongful Life," 4 AM.J.L. & MED. 211 (1978).

136. Supra note 133 at 898, 899.

137. Id. at 899.

138. Judge Fuchsberg, in a concurring opinion, states the parents alleged the direct consequence of the medical malpractice here was to frustrate their legal right to have avoided parental responsibilities by arranging to have the pregnancy terminated in one case and, in the other, not even allowing the conception. Id. at 903.

It has been proposed that parents of a genetically defective child should have a cognizable legal action when they show that the attending physician knew, or acting within the standard of care of similarly situated practitioners, should have known of the risk of the particular disorder in question; that such information would have been relevant to a reasonable person and that had the complaining parents known of this possibility, they would not have had their child. Once these requirements have been satisfied, it is maintained that the parents should be awarded substantial damages for their emotional anguish and economic injury. Note, "Father and Mother Know Best: Defining the Liability of Physicians for Inadequate Genetic Counseling," 87 YALE L. J. 1488 (1978).

139. Supra, note 133 at 899.

140. Id. at 900.

141. Id.

142. Id. at 901.

143. Id.

144. Id. at 901.

145. 42 N.Y.2d 109, 397 N.Y.S.2d 363, 366 N.E.2d 64 (1977).

146. 37 N.Y.2d 378, 372 N.Y.S.2d 638, 344 N.E.2d 590 (1975).

147. 46 N.Y.S.2d 895 at 901 (1978).

148. Id. at 902.

149. Id.

Judge Wachtler, in his dissent, agreed with the majority that the wrongful life action brought on behalf of the infants was properly dismissed. He would, additionally, dismiss the collateral suit brought by the parents for the expense of rearing an unwanted child. Chiding the majority for creating a new tort, with no discernible limits of liability, a kind of medical paternity tort if you will, he objected to holding a doctor responsible for the birth of a genetically defective child--thus obligating the doctor to pay most, if not all of the costs of lifetime care and support. Id. 904 at 907.

Judge Wachtler sees the heart of the problem in these two instant cases as being the failure by the majority to fully comprehend the fact that the physician is not the one causing the genetic defect. "The disorder is genetic and not the result of an injury negligently inflicted by the doctor. In addition, it is incurable and was incurable from the moment of conception. Thus, the doctor's alleged negligent failure to detect it during prenatal examination cannot be considered a cause of the condition by analogy to those cases in which the doctor has failed to make a timely diagnosis of a curable disease. The child's handicap is an inexorable result of conception and birth." Id. at 904.

150. Oelsner, "Baby in Malpractice Suit was Put Up for Adoption," N.Y. Times, Feb. 17, 1979 at L 23, col. 1.

151. Id.

In Berman v. Allan, decided by the New Jersey Supreme Court on June 26, 1979, it was held a cause of action for "wrongful birth"--but not "wrongful life"--would be recognized. This is, indeed, a forward posture for this Court to take--especially so in light of the fact that some twelve years ago in Gleitman v. Cosgrove, 49 N.J. 22 (1967) the same Court refused to recognize as valid causes of action for either a claim of "wrongful life" asserted on behalf of a physically deformed infant or a claim for "wrongful birth" asserted by the infant's parents.

In Berman, an allegation was made by the parents of an infant born with Down's syndrome that attending physicians were guilty of negligence in failing to inform the infant's thirty-eight year old mother--during her pregnancy--of the avilability of tests (i.e., amniocentesis) to determine the risks of giving birth to a congenitally defective child. Had such information been made available, the mother alleges she would have learned of the high incidence of mongolism and aborted the fetus. The parents sought damages in their own right for not only the emotional anguish which they experienced because of their child's birth defect, as well as for the medical and other expenses which they expect to incur in order to properly raise, educate and supervise the child. The child, through her father as guardian ad litem, sought compensation for the physical and emotional pain and suffering that she would endure, as a mongoloid, throughout her life.

While acknowledging the previous reasons for not allowing a suit for the parents' injuries in cases of this nature--namely, the difficulty of measuring damages and the policy against abortion--should no longer be regarded as road blocks to validation of claims for wrongful birth, the court nonetheless notes the "troublesome" nature of structuring a proper measure of damages. Stating the parents have nevertheless presented an actionable claim for relief, the court allows that if their allegations are in fact proved at trial, they will be entitled to compensation for the mental and the emotional anguish they have suffered and will continue to suffer. Regarding the claim of the genetically defective child, the court concluded:

> "Our premise for holding that the child fails to state an actionable claim for relief is that she has not suffered any damage cognizable at law by being brought into existence. . . . Although we recognize that the child's abilities will be more limited than those of normal children and that she will experience a great deal of physical and emotional pain and anguish, we cannot say that she would have been better off if she had never been brought into the world." 48 U.S.L.W. 2026, 2027 (July 10, 1979), 80 N.J. 421, 404 A.2d 8.

For a contrary view regarding a defective child's right to maintain suit, see Curlender v. Bio-Science Laboratories, 100 Cal.App. 3d 811 (1980), supra note 75.

See generally, Trotzig, "The Defective Child and the Actions for Wrongful Life and Wrongful Birth," 14 FAM.L.Q. 15 (1980).

## CHAPTER 5

## THE NEW BIOLOGY AND A PROGRAM FOR POSITIVE EUGENICS

Artificial Insemination

Artificial insemination, referred to as AID or heterologous insemination, is the process of inseminating a woman with the sperm of a donor. Although AID was developed to provide a child to a married couple that could not reproduce due to a physical impediment of the husband, the method today has a new vitality and purpose as a technique for implementing a program of positive eugenics.[1] Sperm banks have been established to maintain semen from "distinguished" persons even beyond their lifetime.[2] Positive eugenicists advocate superior sperm banks to develop the population to a position of genetic strength and to assure the survival of the human race in the event of an insufficient number of acceptable male members to allow normal reproduction.[3] The ultimate goal of positive eugenics is to assure eutelegenesis, mass insemination with superior sperm.[4] This suggestion for use of AID practices to implement a program of positive eugenics should encounter little resistance because these practices infringe upon individual rights only minimally, neither restricting nor prohibiting marriage or reproduction.[5]

In Vitro Fertilization and Embryo Implants

In 1974, Dr. Douglas Bevis of Leeds University announced that out of approximately thirty attempts to conceive human embryos in vitro, or in test tubes, and then implant them in utero, or into the wombs of women, he had achieved three successful implants that resulted in the birth of three babies.[6] The three mothers had been infertile because of diseased, blocked, or missing Fallopian tubes. Dr. Bevis had removed ova from each woman, fertilized the ova in the test tubes with sperm taken from the women's respective husbands, and then implanted the fertilized eggs into the women's wombs.[7] Dr. Bevis' announcement has been subject to considerable doubt because he was unwilling to document his research fully.[8]

Some scholars credit Dr. Landrum Shettles, formerly of Columbia University, as the first scientist to achieve test tube fertilization of human eggs in 1953.[9] Dr. Shettles also claims to have achieved an embryo implant in 1963, but it was not allowed to mature.[10] In 1973 he attempted another embroy implant following in vitro fertilization, but his superior terminated the experiment before implantation, asserting that further work should be performed with lower primates "to establish whether the number of malformations is acceptable;[11] his superior also raised the ethical question of who would obtain permission for the fetus to be part of the experiment.[12]

As a consequence of the unauthorized termination of Dr. Shettle's 1973 experiment, the prospective parents filed a $1.5 million lawsuit against Dr. Shettle's superior, claiming the procedures were terminated without their consent. The prospective mother stated:

"I can't see why some people believe a baby conceived in this fashion isn't as sacred as a baby conceived in the normal fashion. There's even more care, more desire, more intent involved here--because so much time, energy, skill and emotion had to be invested in its conception."[13]

Whatever the success of these experiments in the past, in vitro fertilization and embryo transplants in humans probably will be fully documented scientific achievements in a few years.[14] For a variety of reasons, women might rely on these techniques to have a child. If a woman is infertile due to a blocked or missing Fallopian tube, an ovum could be taken from one of her ovaries, fertilized in a test tube with her husband's sperm, and implanted in her uterus. If a woman cannot produce normal egg cells, a donor's egg, already fertilized by the husband's sperm through artificial insemination or fertilized in vitro with the husband's sperm, could be implanted into her uterus.[15] A woman who cannot carry a baby to term because of a physical disability could enter into a contract with a surrogate or host mother to do so,[16] and an egg fertilized either in vitro or in vivo could be implanted into the host mother. A career woman, such as a professional athlete, who has no physical disability also may seek the services of a surrogate mother if she does not wish to miss valuable time from her professional interests to carry a baby for the full term.[17].

Married couples also may benefit from successful implantation of in vitro fertilization techniques. A number of embroys could be developed from a couple's best available ova and sperm cells, and the couple then could select one embryo for implantation.[18] Successful in vitro fertilization also may lead to the development of in vitro gestation, enabling a fetus to develop to term completely outside the womb.[19] These two examples would foster a married couple's wish to have a child that is as much genetically their own as is possible physically as a substitute for adopting a child. Married couples also could rely on in vitro fertilization techniques to have a child that is not genetically their own. An unmarried person desiring a child also might wish to utilize these methods. Since an unmarried individual would need a donor's egg or sperm in the procedure, such a program could introduce positive eugenic concepts to create children with a stronger genetic heritage. As in the case of AID programs, the incorporation of positive eugenic concepts would infringe individual rights minimally because they neither restrict nor prohibit marriage or reproduction as eugenic programs generally do.

Asexual Reproduction: Cloning and Parthenogenesis

The word cloning, which derives from a Greek root meaning cutting, is generally defined as asexual propagation[20] and is a common practice to develop new varieties of plants.[21] In 1966 a team of Oxford University biologists, headed by Dr. John Gurdon, announced that they had grown seven frogs from the intestinal cells of tadpoles.[22] What had been routine in the garden now existed for one group of animals: a new organism produced from a single parent.[23]

Several steps would be required to clone a human. First, the nucleus of a donor's egg cell would be destroyed. A nucleus from any convenient cell of the person to be cloned would be inserted into the enucleated egg by microsurgical techniques not yet fully developed. The new cell, placed in a nutrient medium, would begin to divide and embryo implantation would follow in approximately four to six days.[24] The cloned individual would be the identical twin of the person who contributed the body cell.[25] Significantly, the establishment of banks of tissue cultures would permit the production of genetic copies of deceased persons through cloning.

Parthenogenesis, commonly referred to as virgin birth, is another form of asexual reproduction.[26] The French-American biologist, Jacques Loeb, achieved parthenogenesis in sea urchins in 1899.[27] More recently, scientists have reported laboratory parthenogenic experiments for frogs and mice.[28] If this process is perfected for humans, a woman one day may produce the necessary egg cell for conception, jolt the egg by pulling an electric switch or administering a necessary drug, thereby enabling it to split, and then have it implanted in her woumb for gestation and ultimate birth, all without physical contact with man sexually or with his sperm artificially.[29]

Not enough is known, either technically or ethically, about human cloning or parthenogenesis to allow dogmatizing concern whether it should or should not be undertaken.[30] Present standards of medical ethics require that a researcher be reasonably confident about the outcome of his research, that he undertake research for reasonably humanitarian purposes, and that he obtain the informed consent of the research subjects.[31] These factors do not force the conclusion that cloning is or is not proper. If the rate of pollution of the human gene pool continues to increase through uncontrolled sexual reproduction, however, efforts to produce healthier people may be required to compensate for the spread of various genetic diseases.[32] In that event, one could make a strong ethical argument to justify cloning of healthy individuals on the ground that it could achieve the greatest good for the greatest number.[33]

Legislation that embodies positive eugenics concepts for permitting only individuals with superior genetic endowments to clone raises a serious constitutional issue. Such a statute would require safeguards against the large scale cloning of particular types of individuals. To do otherwise would decrease the genetic variation that is so vitally necessary to natural selection[34] and would even threaten man with his own eventual extinction.[35] By discriminating between those with superior genetic traits and all others, however, legislation of this nature would be subject to equal protection challenges. Under standard equal protection analysis, if a court determined that the statute affected a fundamental right, the state would need to show that the legislation served a compelling state interest.[36] The right to procreate is a fundamental right,[37] but the denial of cloning methods to individuals who are capable of reproducing in the normal manner may not be a sufficient infringement of this fundamental right to trigger the compelling interest requirement.[38] If it were not such an infringement, the state would be required only to show a rational relation between the legislation and a legitimate state interest.[39] A court might determine that the state's interest in the propagation of superior traits is constitutionally impermissible because it violates the Constitution's nobility clause or the Thirteenth Amendment's prohibition of involuntary servitude.[40] If a court determined that the state has a legitimate interest in the propagation of superior traits, it probably would find that the legislation is rationally related to that purpose.

Persons who carry recessive traits might succeed in claiming that permitting only genetically superior people to clone infringes upon their right to procreate--with that claim thus triggering strict judicial scrutiny of the cloning law and requiring the state to show a compelling interest for its action.[41] Under this type of judicial scrutiny, at least two attacks on a statute, itself, could be made in addition to challenging the state's purpose for action as constitutionally impermissible. It is doubtful whether scientific evidence can provide a rational basis for classification of individuals having superior genetic traits.[42] Moreover, the state may be able to achieve its objective through a less intrusive program: its interest in the propagation of superior traits through a positive eugenics program is probably less compelling than its interest in the diminution of inferior traits through a negative eugenics program.[43]

Notes--Chapter 5

1.  See Smith, "Through A Test-Tube Darkly: Artificial Insemination and the Law," 67 MICH.L. REV. 127 at 148 (1968).

Theoretically, one donor could produce enough semen to create 20,000 babies a year. The actual record for production by one donor was, however, only seventeen until a recent study showed a donor had in fact produced fifty offspring. See Ch. 6, n. 37.

Conception through AID is not expeditious. The average time for a "take" is approximately six months: or two or three sessions a month with a charge of anywhere from $75.00 - $100.00 for each insemination.

Many physicians encourage intercourse within a short time after artificial insemination primarily to create an element of doubt about the prospective child's paternity and increase the husband's sense of participation in the child's creation.

It is generally agreed that it is best for any AID baby not to know of its origins. The donor should not be told if his donation of semen resulted in a successful impregnation and birth. Attalah, "Report from a Test Tube Baby," N.Y. Times Mag., April 18, 1976, 16 at 17, 51.

See generally, A. ETZIONI, GENETIC FIX, Ch. 2 (1973).

2.  See Smith, supra at 145, 146.

The student newspaper at Columbia University has advertised for sperm donors, who would be paid for their semen "for artificial insemination for couples who cannot have children due to male infertility." Columbia Daily Spectator, Oct. 28, 1974, at 5, col. 4.

In addition to AID heterologous, there are two other possible types of artificial insemination--depending upon the initial source of spermatozoa: A.I.H. homologous and A.I.D.H. with homologous insemination, where a woman's husband's semen is introduced artificially into her reproductive tract. This is necessitated due to the fact the husband suffers from sexual impotency during the act of intercourse. Semen from the husband is obtained through massage of his prostate gland if not through normal masturbatory techniques. With A.I.D.H., a husband's semen is mixed with that of a donor. Although often explained to the husband as a way of strengthening his sperm count, the real reason A.I.D.H. is used is to give him a form of psychological reassurance that his "manliness" is not--in a total sense--negative. Id., at 128.

See Appendix C for a model statute for artificial insemination.

A proposal surfaced in the British Parliament to allow "eugenical privileges to the aristocracy of blood and mind." Under this proposal, polygamy would have been reinstated for the nobility and the Knights of the Realm would have been asked to make available their semen to a "Bank of Superior Genotypes" for the use of women with mates of less talent. Landauer, "Aristogenics," 149 SCIENCE 816 (1965).

It was also suggested that the benefits of AID--including the legitimization of children born therefrom--be made "available to those members of the British peerage who enjoy hereditary titles." Kilgrandon, "The Comparative Law of Genetic Counseling," in ETHICAL ISSUES IN HUMAN GENETICS, 255 (B. Hilton et al eds. 1973).

See generally, Frankel, "Human-Semen Banking: Implications for Medicine and Society, 39 CONN. MED. 313 (May, 1975).

3. Smith, supra note 1 at 146-147.

4. Id. at 147. See generally, S. PICKENS, EUGENICS AND THE PROGRESSIVES (1968).

5. See Vukowich, "The Dawning of the Brave New World--Legal, Ethical and Social Issues of Eugenics", 1971 U. ILL. L.F.189 at 230-31.

6. See Rorvik,"The Embryo Sweepstakes,"N. Y. Times Mag., Sept. 15, 1974 at 17.

See generally, George, "Life in the Lab", National Observer, July 7, 1973, at 1, col. 1. At the time Dr. Bevis made the announcement, the babies ranged in age from 12 to 18 months. See also, Revillard, "Legal Aspects of Artificial Insemination and Embryo Transfer in French Domestic Law", 23 INT'L COMP. L.Q. 383 (1974).

Dr. Patrick Steptoe, a British Gynecologist, announced the first validly documented laboratory conception of a test-tube baby and of its forthcoming birth. He and Dr. Robert Edwards, a Cambridge University physiologist, have been working in this area of experimentation for a number of years. Cohn, "Test-Tube Baby Reported Near", Wash. Post, July 22, 1978, at 1, col. 1; "Test-Tube Baby conceived in a Laboratory", TIME Mag., July 24, 1978, at 47; Weintraub, "First Test-Tube Baby Born in British Hospital", Wash. Post, July 26, 1978, at 1, col. 5.

See also, Sullivan, "New Ear in Reproduction Seen in British Laboratory's Embryo", N. Y. Times, July 15, 1978, at 1, col. 1.

The Ethics Advisory Board of the Department of Health, Education and Welfare went on record as approving both basic research and the actual creation of so-called test tube babies as ethically acceptable not only for the purpose of aiding couples unable to bear children but also as laboratory experiments which may end in the destruction of an embryo. The Board did not tell HEW whether it should or should not fund such research. Cohn, "Ethics Board Gives Backing to Test Tube Research", Wash. Post, Mar. 17, 1979, at 1, col.5.

Virginia became the first state to approve the establishment of a test tube baby fertilization center--located at Norfolk General Hospital. See TIME Mag., Jan. 21, 1980 at 58; Wash. Post, Jan. 9, 1980, at C1, col. 4.

7. See Rorvik, supra, at 17.

To obtain eggs from a female, researchers insert a laparscope near or through a woman's umbilicus into her peritoneal cavity to obtain a direct view of the ovaries. A suction device is employed to pull selected ova from follicles on the ovarian surface. Various hormones are used to regularize the menstrual cycle to ensure that the egg is at the proper stage of maturation when withdrawn, and then the egg is exposed in a test tube. Fertilization normally is achieved within 12 hours, with cleavage into two cells evident within 38 hours. By the end of the fifth day, an embryo reaches the blastocyst stage having 64 or more cells. In order to avoid rejection of the embryo by the uterus, both the uterus and the embryo must be in a biochemical synchronization that often is different to achieve. The embryo may be introduced into the womb surgically through the walls of the uterus or non-surgically by use of a catheter that is passed through the vagina and cervical canal into the uppermost part of the uterus. See Id. at 54-56.

8. Id.

9. Id. at 50, 54.

But see, Smith supra note 1 at 127 (Dr. John Rock of Harvard fertilized human ova in the laboratory in 1944).

10. See Rorvik, supra note 6 at 56-59.

Reportedly, the first fully documented and successful embryo transfer in a primate occurred with an infant baboon at the Southwest Foundation for Research and Education in San Antonio, Texas. The baboon's genetic mother provided the egg which had been fertilized by a male baboon. The five day embryo was then transferred surgically to a "foster" mother who carried it to gestation. 6 THE HASTINGS CENTER REPORT 2 (Aug. 1976).

11. See Rorvik, supra note 6 at 59.

12. Id.

See generally, McCormick, "Proxy Consent in the Experimental Situation", 18 PERSPECTIVES IN BIOLOGY & MED. 2 (1974).

13. Id.

The lawsuit was first filed in 1974 and was against Columbia University, Columbia Presbyterian Hospital and the director of obsterrics and gynecology, Dr. Ramon Vande Wiele who stopped the procedure. Conrad, "Trial Opening Into Loss of Test-Tube Embryo", Wash. Post, July 17, 1978, at A 11, col. 1. A jury subsequently awarded $50,000.00 to Doris Del Zio and a token $3.00 to her husband. The award was made in order to compensate Mrs. Del Zio for mental anguish she suffered because of the destruction of the developing test tube fetus. The panel determined she should receive $12,500.00 from Columbia Presbyterian Hospital, $12,500.00 from Columbia University and $25,000.00 from Dr. Vande Wiele. Hastings, "U. S. Jury Gives Pair $50,003.00 in Suit over Test Tube Baby", Honolulu Advertiser, Aug. 19, 1978, at B 4, col. 1.

See also, Powledge, "A Report from The Del Zio Trial", 8 THE HASTINGS CENTER REPORT 15 (Oct., 1978).

In vitro fertilization provides new opportunities for obtaining knowledge about the etiology of chromosomal abnormalities--which are the cause of a significant number of birth defects--and for advancing the general level of diagnostic knowledge of the development of genetic disease before later implantation--thus alleviating the need for reliance upon amniocentesis. P. REILLY, GENETICS, LAW AND SOCIAL POLICY 195 (1977); Glass, "Human Heredity and Ethical Problems", 15 PERSPECTIVES IN BIO. & MED. 237, 249 (Winter 1972); Edwards, "Problems of Artificial Fertilization", 233 NATURE 23, 25 (1971).

Today, the list of heterozygous "inborn errors of metabolism" which can be detected biochemically or by karytoype (chromosomal constitution) analysis is rapidly increasing each year. Glass, id. at 239. It may thus become possible in the near future to undertake corrective measures designed to alter the chemistry of the fertilized egg or--for that matter--structuring safeguards in the pre-natal environment in which the fetus develops; or by developing procedures designed to correct ailing or deficient genes directly with chemical additives injected into the cell or the fetus. With the control of the sex-determination process, could also come a significant reduction in the incidence of serious sex-linked conditions. For example, if a man having hemophilia could be treated, and a separation of his X-bearing sperms carrying the harmful gene accomplished so that only his Y sperms would be used in procreation, he would thus sire only sons--free from the threat of the disease. With no daughters, his X-linked "wayward" genes would be taken out of circulation. A. SCHEINFELD, YOUR HEREDITY AND ENVIRONMENT 704 (1965).

14. Cattle embryo transfer techniques already have resulted in the birth of calves. See Brody, "Embryo Transfer Aids Cattle Breeding", N. Y. Times, Sept. 15, 1974, at L 40, col. 4. Researchers are now attempting to refine and perfect these techniques to achieve more efficient breeding methods. If the techniques are perfected, one cow could "mother", genetically speaking, hundreds of calves. See id. The cost of the transfer operation at present prohibits mass application of these techniques. See id.

Cattle embryo transfer techniques begin with stimulating a genetically superior cow with certain hormones so that it will superovulate or produce a greater quantity of eggs in an estrous cycle than it normally would. See id. After the cow is inseminated with prize bull semen, its uterus is "washed" to collect the fertilized eggs. A healthy appearing fertilized egg, identified by microscopic equipment, is removed from the washing liquid and transferred to the uterus of an ordinary cow. This cow, in turn, acts as a surrogate mother, nourishing the embryo until birth. The transfer of the fertilized eggs frees the original, genetically superior cow from nine months of pregnancy and permits it to be superovulated rather soon again. The technique thus greatly multiplies the opportunities for the birth of high quality offspring. See id.

An interesting, novel corollary to the rented womb or surrogate mother concept in cattle breeding, may be found in the rented nest theory for peregrine falcons. By taking eggs laid by rare nonprolific peregrines and placing them in the nests of prairie falcons (known for the proclivity to procreate), the female peregrine falcon will immediately lay three or four more eggs to replace those taken from her nest. Careful performance of this egg theft and the creation of foster parents makes it mathematically possible in one act to double the number of baby peregrine falcons that would oridinarily have been produced. N. Y. Times, Oct. 20, 1974, at L 41, col. 2.

Experiments have shown that the mammary glands of men can develop and produce milk -- with proper stimulation. There are even reports from China that men have functioned through the ages as wet nurses. It is theoretically correct to assume that with proper scientific preparation a pregnancy could be gestated "in a man's abdomen and thrive to term, in a transplanted uterus or other suitable spot" and thereupon delivered by Caesarian section. There is further evidence that wherever a burrowing placenta finds an adequate blood supply there will be an adequate area for it to grow the embryo. Gorney, "The New Biology and the Future of Man", 15 U.C.L.A. L. REV. 273, 284, 285 (1968).

15. See Gaylin, "We Have the Awful Knowledge to Make Exact Copies of Human Beings", N. Y. Times Mag., Mar. 5, 1972, 11 at 48; Rorvik, supra note 6 at 50.

See generally, R. McKINNELL, CLONING: NUCLEAR TRANSPLANTATION IN AMPHIBIA (1978).

Ova transplanting might be undertaken for eugenic reasons similar to those prompting the use of AID. If it is the wife instead of the husband whose germ cells are infertile or carry the threat of transmitting some serious X-linked genetic condition, she can be implanted with eggs from a healthy donor. The results and the parentage problems would then be analogous to those in cases of artificial insemination--with one important difference: instead of the child of a couple not being the husband's genetically, the child in the ova transplant cases would not be the wife's. REILLY, supra note 13 at 217; SCHEINFELD, supra note 13 at 701.

Just as sperm may be frozen and stored for later use, a woman's eggs or ovaries might also be preserved in a similar manner. Reproductive cells could, for example, be taken around the age of twenty years and preserved for use at a later time. Such preservation might serve to prevent the accumulation of detrimental mutations with advancing age. From a eugenics standpoint, parents should have their children while they are young. Older parents may, to be sure, produce genetically healthy children; but the odds for such success are not favorable. By banking reproductive cells under those conditions where mutation is reduced to the lowest possible level, and using implantation of an embryo produced by artificial fertilization, older parents may have children as free of defect--on the average--as when they were young. Glass, supra note 13 at 250. Indeed, the most available and ready source of woman users of in vitro fertilization would come from those having occluded oviducts, those sterilized as a consequence of tubal ligation and those known to carriers of X-linked genetic disorders. REILLY, supra note 13 at 196.

Other conditions might prevail and prompt the use of either artificial insemination or donor egg implants in order to attempt to circumvent particular genetic problems. An Rh situation might exist--with the husband being Rh positive, and the wife Rh negative with such condition ending previous conceptions with miscarriages. Here, donor insemination might be a positive alternative. SCHEINFELD, supra note 13 at 663.

In vitro fertilization holds particular promise for women over the age of thirty-five, known carrier of X-linked disorders, and even couples at risk for having children with autosomal recessive disorders. Thus, for a woman known to be a carrier of hemophilia, an egg could be obtained from a healthy noncarrier, fertilized in a test tube with her husband's sperm, and--a few days later--it could be implanted in the uterus of the carrier female. Predictably, a normal pregnancy would follow and the birth of a child with a very slight risk of hemophilia (since there is always some risk of a new mutation) would be achieved. REILLY, supra note 13 at 191.

Scientific advances may be soon perfected which allow for yet another means of avoid genetic disease. A female carrier of X-linked disease can avoid bearing children with that disorder only if she has daughters. While the daughters may be carriers, as was their mother, they--themselves--will not be affected. This can be achieved modernly only by sexing the fetus and aborting all males, thus sacrificing the healthy fetuses with the diseased. The carrier mother can, however, be virtually guaranteed a female child when it becomes possible to separate X-bearing from Y-bearing sperm. REILLY, id., 192.

Within the last several years, an important scientific discovery was advanced: namely, that human Y-bearing sperm could be differentiated from X-bearing cells by use of a fluorescent stain. This discovery has promoted serious study of the methods of separation of sperm by sex. The ultimate goal of success here would be the development of a method of X-sperm selection which would, in turn, eliminate the need to abort a fetus with a fifty percent risk of being born with a genetic disease. REILLY, id.

In general, it may be said--then--that when a dominant condition exists, depending upon which parent is affected, donor sperm or donor egg should be considered as a genetic alternative to normal procreation. If there is a recessive condition, however, AID would be better utilized than in vitro fertilization. Finally, when there is a sex-linked recessive condition, a donor egg should be used or separation of the X and Y sperm undertaken--thus assuring that only daughters would be conceived.

16. See Gaylin, supra note 15 at 48; cf. Rorvik, supra note 6 at 50 (eggs from one cow can be implanted in womb of another).

17. Gaylin, supra note 15 at 48.

18. Id.

19. Id.

20. D. RORVIK, BRAVE NEW BABY 109 (1971).

21. See G. TAYLOR, THE BIOLOGICAL TIME BOMB 23-25 (1968); Kuhn, "The Prospect of Carbon Copy Humans", 60 CHRISTIANITY TODAY at 11 (Apr.9, 1971).

   Agricultural researchers in the state of Washington, which ranks second in asparagus production, have successfully cloned asparagus. Perfection of the process means that because of a stronger nature of cloned vegetable, it may in turn be harvested by machine instead of by hand--thus effecting a considerable savings in the costs of production. "The Cloning of a Vegetable", Wash. Post, April 8, 1978, at 14, col. 5.

22. See G. LEACH, THE BIOCRATS 94 (1970).

23. Id. Professor James Watson, who has long pioneered in genetic research and who discovered DNA, has observed:

> "A clone is the aggregate of the asexually produced pregnancy of a single cell; for example, all the descendants of a single bacteria present as a colony upon a petri dish. The genetic identicalness of all members of a clone arises from the fact that the normal process of cell division, termed mitosis, produces two daughter cells with idential chromosomal complements. The nuclei of the cells found in the frog's intestine are thus identical to those which could be found in its liver or brain. In contrast, the cell division process, termed meiosis, which generates the sex cells, reduces the chromosome number in half. Only one of each pair of homologous chromosomes enters a sperm or egg. Moreover, a completely random event determines whether the given chromosome is of male or of female origin. Consequently, no two eggs (or sperm) arising in a given individual are ever genetically equivalent. No two sexually produced frogs, having the same two parents, thus will be identical unless they arise by the rare splitting of an already divided fertilized egg into two daughter cells, each of which goes on to develop into a complete embryo. This is the process by which identical human twins are produced. In contrast, all the members of a clone produced by mitosis will be identical, except for the occasional mutant cell resulting from rarely occurring somatic gene mutations . . . it is now routinely possible to produce mature plants starting from highly specialized somatic cells of diploid chromosome number . . . it is highly likely that the embryological development of most higher animals, including man, involves the creation of countless numbers of totipotent somatic nuclei each capable of serving as the

complete genetic material for new organism. This means that, theoretically, all forms of higher animal life may in effect be capable of clonal reproduction."

J. Watson,"Potential Consequences of Experimentation with Human Eggs," Jan. 28, 1971 (Papers 1, 3, 4, Harv. Univ. Biological Labs).

The following is an amusing, yet somewhat frightening scenario for genetic manipulation:

"Although both Alvin and Doctor Pfizer were wholly ignorant of the matter, the boy was in fact the brainchild (along with the identical brothers) of the aforementioned Professor Miriam Poynter, O.B.E., D.Sc., F.R.S., P.L.S., etc. It was she who, with her own two lips, had pipetted the long frozen sperm and introduced it to the chosen egg. She, and she alone, had observed with bated breath as it lashed its frenetic way into the zona pellucide and came at last to rest, all passion spent, with its weary little head flat on the surface of the vitellus. Forty-eight hours later, when the zygote was already busily dividing, here was the finger which, at the critical moment, had touched the switch and administered a one minute electric shock to the blastocyst which caused it to have second thoughts and induced the inner cell mass to split into identical halves. Twelve hours later a second shock had induced a second division. Within a matter of days, Alvin, Bruce, Colin and Desmond as yet sexless and anonymous, were at last free to grow their separate ways, having been transferred, one by one, into four placentas which had once belonged to four Berkshire sows. There in the cosy, uterine darkness laved by synthetic placental fluids, the genes which carried the peculiar inheritance of Alvin's progenitors were able at last to transmit their mysterious messages unimpeded to their embryonic cells." R. COWPER, CLONE 17 (1972).

For consideration of the problems of cloned individuals procreating among themselves, see Batt, "They Shoot Horses, Don't They?: An Essay on the Scotoma of One-Eyed Kings", 15 U.C.L.A. L. REV. 510 (1968).

It was asserted recently by David Rorvik, in his book, IN HIS IMAGE: THE CLONING OF A MAN (1978), that human cloning has already been accomplished. Lescaze, "A Baby Book of a Different Nature", Wash. Post, Nov. 8, 1978, at A 3, col. 4.

But see, Schmeck, "Author of Book on Cloning Offers No Direct Proof Test Succeeded", N. Y. Times, Mar. 23, 1978, at B 10, col. 1; Hellegers, "Book on Cloning is a Hoax", Wash. Post, June 7, 1978, at B 3, col. 1; "Costly Hoax, Scientists Sue Over Clone Book", TIME Mag., July 24, 1978 at 47; Cohn, "Research Shows Human Cloning Doubtful", Wash. Post, June 1, 1978, at A6, col. 1.

Dr. Landrum Shettles reported a successful first step in the cloning of a human at Gifford Memorial Hospital, Randolph, Vermont. Although past research undertaken by Dr. Shettles has been open to serious question, he nonetheless maintains that he withdrew eggs from the ovaries of women with a syringe, placed them in lab dishes and using a microscope and ultra-thin glass needles, teased out the nucleus of each egg. He then proceeded to remove spermatogonia from the testes of male volunteers, detached their nuclei which contain a full complement of 46 chromosomes and inserted them into the incubating ova. Three successes were recorded before Shettles stopped further experimentation. These successes consisted of recording the division of the eggs and their development into multicelled blastocysts--or that state where a normally fertilized ovum leaves the Fallopian tube and becomes implanted in the womb. NEWSWEEK, Feb. 12, 1979 at 99.

24. See D. RORVIK, supra note 20 at 114; Lederberg, "Experimental

Genetics and Human Evolution", 100 AM NATURALIST 519, 562 (1966); Watson, "Moving Toward the Clonal Man", ATLANTIC MONTHLY 50, 51 (May 1971).

See also, R. ETTINGER, MAN INTO SUPERMAN (1972).

See generally, Louisell, "Biology, Law and Reason: Man as Self-Creator", 16 AM. J. JURISPRUDENCE 1, 3 (1971); Note, "Asexual Reproduction and Genetic Engineering: A Constitutional Assessment of the Technology of Cloning", 47 SO. CAL. L. REV. 476 (1974). Cloning could be a useful technique in many situations. If a husband and wife carry a debilitating recessive genetic disease and are unwilling to risk, through normal procreative processes, the birth of a child who in turn might carry their disease, cloning is a viable alternative for them. If the husband is opposed to adoption and the wife has a strong, natural desire to bear her own child, a clone of either the husband or wife could be developed. EXPERIMENTATION WITH HUMAN BEINGS 977 (J. Katz ed. 1972). A couple who loses their only child and is incapable of having another could rely on cloning techniques to produce a duplicate of their lost child: one of the child's cells could be cloned and implanted in its mother's or a substitute womb for the nine month term of gestation. Gaylin, supra note 15, at 48. See also G. LEACH, supra note 22, at 96.

One author has gone so far as to suggest that a government might be forced into a program of state controlled cloning in order to compete with a despotic enemy using cloning to produce more rugged soldiers, more brilliant scientists, and more skilled workmen. G. TAYLOR, supra note 21, at 26. But see G. LEACH, supra note 22, at 94-95 criticizing Taylor's suggestion as unrealistic). See generally Fletcher, "Ethical Aspects of Genetic Controls", 285 N. ENG. J. MED. 776 (1971).

Indeed, it has been suggested that until public assessment and control have been established in cloning, government supported research of clonal techniques be reduced or stopped. Green, "Genetic Technology: Law and Policy for the Brave New World", 48 IND. L. J. 554, 576-580 (1973).

See generally, 117 Cong. Rec. 7670 (1971); GENETIC ENGINEERING, REPORT TO HOUSE SUB COMMITTEE ON SCIENCE AND RESEARCH AND DEVELOPMENT, COMMITTEE ON SCIENCE AND ASTRONAUTICS, 92d Cong. 2d Sess. (1972).

Perhaps the only real value of cloning is to be found in animal husbandry where -- using the process -- would allow an equality to be reached with plant biology where such techniques have been used for a considerable time. With cloning, a chance would be provided to make, for example, 10,000 identical copies of champion race horses or breed the best of any animal. Endangered species would also be protected by cloning. All matters considered equal, there is no legitimate human end served by cloning--unless ego trips are considered legitimate by some. Stated simply, experiments in cloning are dangerous because man runs the risk of coming against an evolutionary dead end as a consequence of such experimentation. Prof. Laurence H. Tribe, Rose F. Kennedy Lecture, "Clones, Cyborgs, and Chimeras", April 3, 1978, Georgetown University, Washington, D.C.

See also, Lederberg, "Law and Cloning--The State as Regulator of Gene Function", in GENETICS AND THE LAW 377 passim (A. Milunsky, G. Annas eds. 1976).

25. See D. RORVIK, supra note 20 at 109. Because women could contribute body cells as easily as men, every offspring could be a cloned female. The same woman could donate the cell, contribute the body-cell nucleus, and carry the baby to term, achieving a monopolization

of the entire reproductive process. See id. at 114.

26. See G. TAYLOR, supra note 21 at 29. When human eggs are formed they receive from the parent cells only a half complement of chromosomes, which are the structures carrying the genetic message; the same occurs when sperm are manufactured. Consequently, when egg and sperm fuse, the full complement of chromosomes then is present. Offspring thus receive genetic instructions from both parents. When eggs develop, in contrast, they have only half the normal number of chromosomes. In complex organisms, this condition often is sufficient to abort development of the embryo. Egg cells may be stimulated into doubling themselves, without sperm fertilization, however, by jolting them with chemicals or even by pricking them with a pin. The offspring formed is not genetically identical to its mother because the unfertilized sex cell is haploid. A cloned cell, which results from a body cell and not a germ cell, suffers no such handicap. The problem of how to nurture a cloned cell, or the artificially launched egg, may soon be resolved completely as a consequence of new techniques in artificial ovulation. See id. at 30-32.

See also, A. SCHEINFELD, YOUR HEREDITY AND ENVIRONMENT 702-704 (1965).

27. G. TAYLOR, THE BIOLOGICAL TIME BOMB 30 (1968).

28. See D. RORVIK, supra note 20, at 95 (mice); Supra note 27 at 29 (frogs).

29. See D. RORVIK, supra note 20, at 94.

30. See Lederberg, "Genetic Engineering or the Amelioration of Genetic Defect", 34 PHAROS 9, 12 (1971).

31. See id. at 12.

32. Fletcher, "Ethical Aspects of Genetic Controls", 285 N. ENG. J. MED 776, 779 (1971).

Gene therapy, or the cure of hereditary diseases by transplanting negative, defective genes with stronger ones, appears to be "far in the future". McElheny, "Gene Transplant is Still Remote", N. Y. Times, May 16, 1975, at 74, col. 5 (quoting Dr. John Morrow of the Carneige Institution).

33. Fletcher, id.

34. SO. CAL. Note, supra note 24 at 560.

35. Id. at 561.

36. Id. at 550, 556.

If one were required to submit a list of goals to be sought in breeding or re-shaping a genetic line, the following might well be listed: freedom from gross physical or mental defects; sound health; high intelligence; general adaptability; integrity of character and nobility of spirit. An alternative list might include: genuine warmth of "fellow felling" and a cooperative disposition; a depth and breadth of intellectual capacity; moral courage and integrity; and appreciation of nature and art and aptness of expression and of communication. On the physical side: to better the genetic foundations of health, vigor and longevity; to reduce the need for sleep; to bring the induction of sedation and stimulation under more voluntary control and to develop increasing physical tolerances and aptitudes in general. Glass, "Human Heredity and Ethical Problems", 15 PERSPECTIVES IN BIO. & MED. 237, 246-247 (Winter, 1972). The first group of goals are those of Professor Glass and the second group belong to the late Nobel laureate, Professor Herman Muller of Indiana University.

Accepting genetic engineering as a proposal for the prevention of suffering, one distinguished authority has suggested its use and feasibility be compared with other available methods in preventive genetic medicine: genetically-conditioned marriage licenses; genetically motivated voluntary sterilization; using three contraceptives at once; abstinence or sequestering oneself in an "old-fashioned Catholic monastery or nunnery". Ramsey, "Genetic Engineering" in HUMAN GENETICS: READINGS ON THE IMPLICATIONS OF GENETIC ENGINEERING at 235 (T. Mertens ed. 1975).

The personal happiness of man is strongly limited by his biological heritage. M. STRICKBERGER, GENETICS 822 (1968).

37. Skinner v. Oklahoma, 316 U. S. 535, 541 (1942).

38. Supra, note 24 at 550-52.

39. Id. at 556; see Shapiro v. Thompson, 394 U. S. 618, 638 & n. 20 (1969).

40. See, supra note 37 at 581-82; U.S. CONST., art. I, § 9, cl. 8; Amend. XIII.

41. See supra note 37 at 556.

42. See id. at 579.

Interestingly, proposed legislation entitled, "The Adoption Identification Act of 1980", (S.2561) was introduced in the second session of the 96th Congress and is designed to provide a voluntary system whereby the natural parents, siblings or other natural relative of an adoptee can locate each other through a centralized computer system. On April 2, 1980, Senate Resolution 401 was introduced which expressed disapproval of this proposed legislation.

See generally, Smith, "Artificial Insemination: Disclosure Issues", 11 COLUM. HUMAN RIGHTS L. REV. 87 (1979).

43. See Vukowich, supra note 5, at 222. If the challenged legislation incorporated negative, rather than positive, eugenic concepts so that it only restricted carriers of recessive debilitating defects from cloning, the constitutional problems would be minimized. The legitimacy of the state interest could not be challenged on the ground that it creates an elite group and therefore violates the nobility clause. See Note, supra note 24, at 581-82. A court could readily find that the statute is rationally related to a legitimate state interest diminishing the propagation of inferior traits. Scientific evidence more readily can provide a rational basis for the classification of those carrying debilitating defects than for those possessing superior genetic traits. See Vukowich, supra at 198-201. Whether the state's interest in a negative eugenics program is sufficiently compelling to sustain the statute under the strict scrutiny test, however, is uncertain. See id. at 208.

What usually triggers the strict scrutiny test is legislation which threatens a fundamental civil right or contains a suspect classification regarded as such because of the nature of the group classified and thereby disadvantaged. See e.g., Shapiro v. Thompson, 394 U.S.

618 (1969); McDonald v. Board of Election Comm'rs, 394 U.S. 802, 807 (1969); Harper v. Virginia Board of Elections, 383 U.S. 663 (1966). Outside the parameters of racial discrimination, the Court has rarely used the strict scrunity test. In non racial classifications, the Court appears to use a balancing process wherein, "...the character of the classification in question, the relative importance to individuals in the class discriminated against of the governmental benefits that they do not receive, and the state interests in support of the classification," are weighed. Dandridge v. Williams, 387 U.S. 471, 520-521 (1970) (Marshall, J., dissenting).

When classifications based on individual qualities which are wholly or largely beyond the control of the individual himself are found to exist, they are regarded as suspect. See e.g., Korematsu v. United States, 323 U.S. 214, 216 (1944) (classification disfavoring Japanese); Levy v. Louisiana, 391 U.S. 68 (1968) (classification disfavoring illegitimate children); Takahashi v. Fish & Game Comm'n, 334 U.S. 410 (1948) (classification disfavoring persons ineligible for citizenship). See generally, Note, "Developments in the Law -- Equal Protection", 82 HARV. L. REV. 1065 (1969).

The purpose of government is to make life tolerable for those who are governed. B. RUSSELL, THE SCIENTIFIC OUTLOOK 278 (1931). Thus, in the "area of economics and welfare, a State does not violate the Equal Protection Clause merely because the classifications made by its laws are imperfect," observed Mr. Justice Stewart for the majority in Dandridge, supra at 485. Indeed, "the problems of government are practical ones and may justify if they do not require, rough accommodations, illogical though they may be, and unscientific." Metropolis Theater Co. v. City of Chicago, 228 U.S. 61, 69-70, (1913), per Justice McKenna.

# CHAPTER 6

## THE LEGAL RESPONSE

The social and legal complexities of biomedical research raise issues that must be resolved. Courts cannot refuse to confront these issues or the law may become immobilized or estranged from science.[1] To deal adequately with these issues, courts must educate themselves in the research being conducted and its potential applications; their failure to develop this knowledge fully will result in a chaotic patchwork of decisions, each based upon a consideration of only a limited aspect of a particular biomedical technique. Legislative consideration of these issues should begin soon to give courts guidance.[2] Legislation dealing with specific techniques that remain largely unexplored, such as in vitro fertilization and cloning, is premature because the available options are unknown.[3] As scientific knowledge of these techniques expands, however, the states can be expected to draft appropriate legislation to protect the health, safety, and welfare of their citizens.[4] The role of the federal government at this stage will be limited to using its taxing and spending powers to condition the grant of federal monies on incorporation of minimum technological standards into state statutes.[5]

A review of the judicial and legislative treatment of family law issues arising from use of artificial insemination (AID) will serve to illustrate the type of challenges that will confront the more exotic methods of reproduction and will suggest the appropriate legal response. In vitro fertilization and embryo implantation raise few issues of family law that AID has not raised already.[6] Until recently, the law treated a child born by AID as the child only of its mother, not that of the mother's husband or putative father.[7] In Gursky v. Gursky,[8] for example, a New York trial court held that even though a husband consents to his wife's use of AID, the child is illegitimate.[9] In People v. Sorenson[10] the California Supreme Court rejected the approach of Gursky and held that a husband who consents to his wife's use of AID cannot disclaim his lawful fatherhood of the child for the purpose of child support.[11] The court construed the penal nonsupport statute to incorporate liability for a consenting father of the AID child, finding a genetic relationship unnecessary to establish the required father-child relationship.[12]

Since the Sorenson decision, several states have passed legislation legitimizing the offspring of AID when the husband consents to the procedure.[13] These developments indicate that courts no longer equate AID with adultery, and they may signal the public's willingness to sanction more startling genetic developments.[14]

Those who oppose legitimization of the new reproductive methods might make two interesting sociological objections: that separation of sexual love from procreation leads to depersonalization of the marriage bond, and that the biological family unit is the one most likely to provide the proper motivation for child rearing.[15] The first argument does not distinguish the newer methods of reproduction from AID since they all separate sexual love from procreation. This threat has not stopped legislatures and courts from sanctioning AID procedures. The second argument, which raises the importance of maintaining genetic continuity in the lineage of the family,[16] does separate AID from the newer methods of reproduction since these methods do present a possibility of genetic discontinuity.

The child resulting from AID can be the genetic child of the mother but not that of the father. The newer methods of reproduction, in vitro fertilization, embryo implantation, cloning, and parthenogenesis, all could be used to create children who also are the genetic children of one of the parents but not the other. Perfection of in vitro fertilization and embryo implants also will allow the birth of a child who is neither the genetic child of the woman who bore him nor that of her husband; the ovum and sperm of donors can create an embryo that is later implanted in another woman.[17] Cloning of persons outside of the marriage partners and developing a donor's ovum parthenogenetically would result in similar genetic discontinuity.

To the extent that newer reproductive techniques do not differ significantly from AID, family law should treat the two similarly; these techniques eventually will allow a more significant departure from the traditional family concept, however, and the law should be ready to respond to new legal problems.[18] Statutory legitimization of children born as a consequence of artificial insemination, invitro fertilization, ovum or embryo implantation, cloning, or parthenogenesis would resolve the basic family law problems that the New Biology will create.[19] Such legislation would bring a radical new dimension into the concept of hte traditional family. One reason for society to approve such a change in the concept of marriage will be to allow a couple who, for physical or professional reasons, cannot have children without artificial or scientific manipulation to bless their marriage with children. To protect the interests of the child, legislation that permits marriage partners to use these techniques following a full and knowing consent provide further that the resulting child is the legitimate child of the consenting marriage partners.[20] "Little else would be needed to establish firmly the mutual responsibilities of the husband, wife, and child, to each other, both with respect to duties of support and the right to inherit."[21]

The reasons for allowing a married couple to give birth to a child by artificial insemination, in vitro fertilization, or ovum transplants do not apply when an unmarried woman seeks to have a child; the unmarried woman cannot assure her child a typical family environment.[22] If this argument is convincing, legislation prohibiting physicians from performing these procedures for the unmarried effectively could limit their use to married couples.[23] The liberalization of certain state adoption laws that now permit single individuals to adopt child[24] and the growing recognition of the liberation of women, however, raise doubts about the rationale of such a prohibition. A society that accepts the spirit of female liberation that motivates some women to conceive with no wish for formalized family relationships or ties should allow a single woman who does not conceive because of physical inabilities or professional interests to conceive with bioengineering techniques. The "Do Your Own Thing" philosophy of the 1970's would embrace such preferences of a single woman.[25]

Once the public accepts AID and other techniques of the New Biology, the incorporation of positive eugenic principles into these procedures should not be objectionable.[26] Although mandatory mating for eugenic purposes is patently offensive, using genetically superior semen or ova in the application of these techniques should be viewed favorably. "The use of frozen semen or frozen ova from long-dead men and women whose genetic heritage should not be lost would involve an interesting variation . . . ."[27] Before implementing positive eugenic methods, however, society must grapple with the moral dilemmas resulting from their implementation and then, hopefully, develop a carefully formulated decision making process to determine which programs will benefit society as a whole. The need for a social system for controlling biomedical research and its application comes into conflict with the traditional notion that scientists alone should make those decisions and raises a fundamental ethical issue: who decides?

Confidentiality

Not only has wider use of genetic testing encouraged the increased potential for collection and storage of such data,[28] but the expansion of computer programming of medical information by insurance companies and the heavy reliance by the federal government and its many agencies upon data banks have all brought with their development very real potentials for misuse and invasions of privacy.[29]

Two examples will serve to highlight rather typical dilemmas a physician faces. Suppose a physician, in performing a routine medical examination of a corporate employee discovers that his thirty year old patient while presently in "outwardly" good health, is carrying an autosomal or single-gene dominate genetic disease--Huntington's chorea --distinguished, as it is recalled, by its late onset. Among the symptoms which will eventually become manifest between the ages of thirty and forty are quickness of temper, diminished memory and judgment, dementia and--finally--complete loss of physical control leading to death within ten to fifteen years. The attending physician knows his examination records will be stored in a centralized computer data bank of medical records which are, in turn, shared by the insurance industry and--further--that some eighty-five per cent of all insurance medical records are currently included in the bank. Not only will the physician's disclosure of his patient's position as a carrier of Huntington's chorea jeopardize his ability to get any insurance coverage, but he would also be excluded from the company's group health plan which would be possible under a "prior conditions exclusion clause" of the contract. Thus, the patient's family would be totally unprotected. It is also conceivable that the very position, itself, would be denied the patient and with the critical information being placed in a data bank, it might well mean that positions of managerial responsibility would never again be within his grasp. Yet, failure to include the family history and the patient's physical condition by the doctor would not only have the possible effect of increasing the risk in the corporate insurance pool, but also be a disservice to other employees insured by the company. Does a physician owe a primary moral obligation to his individual patient or--if employed by industry-- does that obligation become a shared one with industry?[30]

Now, suppose that a genetic counselor knows that a new born baby has an extra Y chromosome. There is, as has been observed, a fair degree of evidence pointing to the fact that an extra Y chromosome predisposes an individual to aggressive behavior. Should a disclosure be made by the counselor of this fact--it thereby becoming part of the child's medical record? What impact might that knowledge have on a school principal if the child with an extra Y becomes rather obstreperous in school? Should the parents be advised of the genetic condition; or would that knowledge have an adverse effect on the way in which they raise the child?[31]

The very physician-patient relationship, expressing as it does a respect for the patient's right of privacy, dictates a high standard for observing and preserving the principle of medical confidentiality. Although a most important obligation, some would note that it is not an absolute obligation or, for that matter, one's sole obligation. Thus, when serious competing interests and values are brought into consideration, the principle may be violated. The most basic of such overriding interests would be the need to protect either the life of the patient, itself, or the lives of other persons.[32]

While clearly the state has, under its broad police powers, the right to compile and store genetic information which may well serve as a potential benefit to the public health, its administrative agencies and other non-state affiliated businesses need to be guided by a definitive legislative policy which enumerates the guidelines for data collection and dissemination and which structures a system of sanctions--civil and criminal--for violations thereof.[33]

The issue of confidentiality also comes into distinct focus and conflict when it is placed under the rubric of what has become known as the right to know or, freedom of information. Indeed, the extent of the right to know one's roots--genetic or otherwise--has become a current topic of study and investigation. The very parameters of this right to know one's genetic roots are yet to be determined. Nonetheless, the problems encountered in AID of continuing to assure donor confidentiality frame the truly quintessential dilemma. Even though such a confidentiality has always been maintained, and for the most part respected, recent studies reveal a critical need for action of some type by some authority to impose standards for donor selectivity. New York City remains the only jurisdiction to have taken steps to impose actual standards for donor selectivity.[35] What rules or guidelines that may exist elsewhere appear to be those informally developed by the individual physician administering AID. From a recent survey it appears that such self-regulation is sorely wanting.[36]

An investigation of seven hundred and eleven physicians likely to perform AID yielded a response from four hundred seventy-one, three hundred seventy-nine of whom reported that they had in fact performed this procedure. Their work accounted for approximately three thousand give hundred and seventy-six births during 1977. The reasons given for administering AID were: to correct infertility; to prevent transmission of genetic disease (here, twenty-six per cent of the doctors sought to accomplish this end) and to assist single women in their desire to have a child (with ten per cent of the doctor's being directed toward the achievement of this purpose). The astonishing statistics from the survey were found in the disclosure that: a mere twenty-nine per cent of the doctors tested the donors of the semen--and then primarily only for communicable diseases; only thirty-seven per cent kept records on the subsequent issue born of the artificial union; thirty per cent kept records of the donors they used; seven per cent of the physician's used the same donor for a given recipient and thirty-two per cent used multiple donors within a single cycle. One donor had been used to produce fifty children--thereby raising markedly the danger of accidental incest among offspring who unknowingly have the same father.[37]

The unprofessional and seemingly cavalier attitude exhibited by certain members of the medical profession in their indescriminate administration of AID, will mean greater control of the whole process and bring greater demands for the revelation of donor identity. Regarding the selection of donors, it has been suggested that perhaps a board of physicians privately or in association with a hospital board deal with the management of AID procedures at the local level and thereby seek to assure the highest standards of genetic screening and medical record keeping be utilized. The panels could even be mixed as to their membership--admitting, as such, non-professionals. Of course, panel composition of this nature--while assuring collective judgment--would also open a greater possibility for invasion of the donor's anonymity.[38] As an alternative, judicial control over the process could be undertaken by designating perhaps a division within a Family Court to handle all matters concerning donor suitability.[39] The unfettered exercise of discretion by the physician is no longer a suitable alternative to decision making.

The Nevada Supreme Court has docketed what may well be a landmark case regarding donor anonymity, the liability of a donor for the birth of a defective child that was sired through AID, and the obligations of a physician to screen all such donors before allowing them to participate in the procedure.[40] Very basically, a substantial part of this case turns on the physician's admission that in selecting the donor in question, who was acknowledged to a medical student, he spent but a few minutes talking to him about his background. The plaintiff-appellants maintain that the doctor was guilty of malpractice in failing to properly screen the donor and ascertain his genetic profile and his compatability with the co-plaintiff's wife. Furthermore, it is asserted that since the supervising physician failed admittedly to advise the co-plaintiff wife of the "chance" of "something going wrong" in her use of AID, she was not properly advised and was, therefore, unable to give an informed consent to the procedure. Query: was the statistical probability of genetic error occurring so insignificant that the physician should not be regarded as being negligent? The issue born of AID died before the age of two of "a failure to thrive." The plaintiffs argue there was a mismatch between the donor and the co-plaintiff's wife and, additionally, that no admissible evidence points to anything other than an inherited defect from the donor; thus the donor is liable for providing defective or incompatible semen and the doctor is liable for malpractice.

The lower trial court ruled in favor of the defending physician and refused to allow the identity of the donor to be disclosed or to impose liability upon him. On appeal, the Supreme Court of Nevada may well wish to consider remanding the case for a determination of the genetic compatability of the donor with the co-plaintiff wife. This could accomplished "in camera" by the judge, with the assistance of a qualified geneticist--thereby providing a greater degree of assured donor confidentiality. Should such incompatability be established, then the doctor's liability would appear to have been established under a theory of negligent malpractice. It would be antithetical to all concepts of equity and public policy consideration to impose liability upon a donor in cases of this nature. The physician is the dominant or controlling force here and should be held liable for his negligent mis-use of that force.

The opinion of the New York Court of Appeals in the closing days of 1978 in the consolidated case of Becker v. Schwartz and Park v. Chessin,[41] may serve as a pertinent influence or vector of force for the Nevada court in its decision making. There, it will be remembered, the New York Court held that where a physician fails to apprise a married couple with a given genetic deficiency (profile) of the risks of bearing a handicapped child, and--furthermore--neglects to advise of the availability of tests for detecting the disorder, if a defective child is subsequently born, the physician, under basic malpractice law, is liable to the parents for the special costs of raising the handicapped child. Could not an expansive, yet rational, reading of this holding lead the Nevada High Court to conclude: that the doctor is guilty of malpractice in failing to work up a full and careful genetic profile of the donor and his compatability with the co-plaintiff wife and, furthermore, failed to disclose all relevant information to the wife in order to enable her to give an informed consent to hte AID procedure?

New directions in the field of adoption law have significant import for the resolution of conflicts associated with donor anonymity in AID cases. Recently, a District of Columbia Superior Court Judge, June Green, ruled that a twenty-two year old mother of two children living in Tacoma, Maryland, who herself was adopted as a child, should be granted permission to see her sealed birth records and thus learn the identities of her natural parents. The plaintiff in this case asserted her basic right to know her total historical identity and to also discover whether possible hereditary diseases or other health problems were a part of her genetic inheritance.[42] By analogy, a similar argument could obviously be made by the progeny of AID.

As the courts begin to recognize "a best interests of the child test," it would surely appear that in cases where genetic heritage is brought into question, the confidential files (if such are maintained) of a participating physician to an AID intervention, should be examined by a judge "in camera" and--where necessary--with the assistance of a geneticist. Public disclosure of the donor's identity should not be revealed [43] nor, for that matter, should liability be imposed upon him for "error" that might follow as a consequence of his participation. As has been asserted, it is the doctor who must be held liable for error.

Interestingly, the Uniform Parentage Act provides that all records involving AID interventions are to be kept "confidential and in a sealed file." Inspection of them is only sanctioned when a court order acknowledges the existence of "good cause."[44] It is certainly not without the realm of possibility to anticipate a court deciding the genetic identity of a plaintiff AID offspring is of such importance that "good cause" has, indeed, been shown and that the confidentiality of the original AID papers must be compromised.

It is both regrettable and distasteful for judicial supervision and intervention to be contemplated or urged here. Lacking professional self-regulation and a recognition of a professional (i.e., ethical) duty to carefully evaluate the genetic profile of prospective AID donors and to maintain through records of the donors and of the families to which they have been of assistance, little else can be recommended.

Notes--Chapter 6

1. Louisell, "Biology, Law and Reason: Man as Self-Creator," 16 AM. J. JURISPRUDENCE 1, at 3 (1971). See R. ETTINGER, MAN INTO SUPERMAN (1972).

Oliver Wendell Holmes, Jr., lists the first requirement of a sound body of law as being its correspondence "with the actual feelings and demands of the community, whether right or wrong. If people would gratify the passion of revenge outside of the law, if the law did not help them, the law has no choice but to satisfy the craving itself, and thus avoid the greater evil of private retribution." O. W. HOLMES, JR., THE COMMON LAW 41, 42 (1881).

Sir James Fitzjames Stephen has suggested that "Legislation ought in all cases to be graduated to the existing level of morals in the time and country in which it is employed. You cannot punish anything which public opinion, as expressed in the common practice of society, does not strenuously and unequivocally condemn...." J. STEPHEN, LIBERTY, EQUALITY, FRATERNITY 159 (1873).

2. See Tunney & Levine, "Genetic Engineering," SATURDAY REV. Aug. 5, 1972, at 24-28. Senator Tunney lists several considerations necessary to effective legislative action: whether a technique or technology assists individual needs or should be designed for a greater societal good; whether there is a real difference between social genetic engineering and individual gene therapy; how the words normal, abnormal, health, disease, and improvement should be defined; whether research for genetic improvement should be of a continuous nature or should be confined to a single generation; and to what degree genetic engineering would affect the diversity among men. Id. at 24-26.

See also Gaylin, "Symposium -- The Law and the Biological Revolution," 20 COLUM. J.L. & SOC. PROBS. 47, 48 (1973) (in attempt to anticipate possible legal problems, law students at Columbia University have developed a Surrogate Mothers Act, Uniform Human Gamete Storage Act, Uniform Controlled Fertilization Act, and General Utilization of Tissue Statute).

3. Grad, "New Beginnings in Life -- A Lawyer's Response" in THE NEW GENETICS AND THE FUTURE OF MAN 75-76 (M. Hamilton ed. 1972); Grad, "Legislative Responses to the New Biology: Limits and Possibilities," 15 U.C.L.A. L. REV. 480, 485 (1968).

4. See Grad, Legislative Responses, supra at 485.

5. See id., at 485-86.

6. Id. at 501.

See Smith, "Through A Test Tube Darkly: Artificial Insemination and The Law," 67 MICH. L. REV. 127 at 134-42 (discussion of A.I.D. and illegitimacy and inheritance).

See also, Atallah,"Report from a Test Tube Baby,"N. Y. Times Mag., April 18, 1976, 16 at 52.

Recently a married man, upon discovering his wife's infertility, advertised as follows in The San Francisco Chronicle: "Childless husband with infertile wife wants test tube baby." The husband in question believed it to be immoral for him to have sexual relation outside of his marriage in order to provide him with the offspring he desired. He therefore, consequently, gave samples of his semen to a doctor and it was--in turn--introduced artificially into the reproductive tract of an unmarried (childless) woman donor chosen from among 181 applicants. The husband and donor never met. Total expenses--including legal and medical--were approximately $10,000.00. The donor received a flat $7,000.00 for her services. The "regular" wife was initially dubious over the whole procedure. After the issue worn of the "union" and the surrogate mother turned over the baby to the husband and his legal wife, the wife accepted the child. This is an interesting variant of artificial insemination. See, "Ad for Mother by Test Tube Brings a Baby," N. Y. Times, Nov 19, 1976, at A13, col. 1.

See Harris, "Stand-In Mother: Maryland Woman to Bear Child for Couple," Wash. Post, Feb 11, 1980, at 1, col. 2; Comment,"Contracts to Bear a Child," 66 CAL. L. REV. 611 (1978).

A report from Stuttgart, Germany, shows the sexual complexities that arise from efforts to conveive in a rather strained relationship. Here, a breach of contract action was maintained by a Demetrius Soupolos against a Frank Maus. It appears that the plaintiff, although declared sterile, nonetheless wanted his wife to conceive. Thereupon, he hired a neighbor who physically resembled him to be his "substitute in the marriage bed." The substitute, Mr. Maus, was married and the father of two children. He was offered $2,500.00 to perform his job three evenings a week for six months. At the end of the six month period a startling examination revealed Mr. Maus to be sterile and thereby conclusively proved his own lack of paternity as to the children of his own marriage. Mr. Maus' defense was that he did not guarantee conception--only an honest, industrious effort. Wash. Post, PARADE Mag., June 25, 1978 at 17.

7. Smith, supra note 6 at 134-45.

8. 39 Misc. 2d 1083, 242 N.Y.S. 2d 406 (Sup. Ct. 1963).

9. Id. at 1088-89, 242 N.Y.S. 2d at 411-12 (although consent does not vitiate wife's adulterous act, husband liable for support on implied contract or equitable estoppel grounds).

10. 68 Cal. 2d 280, 437 P. 2d 495, 66 Cal. Rptr. 7 (1968).

11. Id. at 283-84, 437 P. 2d at 498, 66 Cal. Rptr. at 10. The court expressly left the issue of legitimacy unanswered, however, suggesting it should be dealt with by the legislature. Id. at 284, 437 P. 2d at 501, 66 Cal. Rptr. at 13.

See also, Smith, "Artificial Insemination--No Longer A Quagmire," 3 FAM. L. Q. 1, 2 (1969); Smith, "For Unto Us A Child Is Born--Legally," 56 A.B.A.J. 143, 144 (1970); Smith, "A Close Encounter of the First Kind: Artificial Insemination and an Enlightened Judiciary," 17 J. FAM. L. 41 (1979); Grobstein, "External Human Fertilization," 240 SCIENTIFIC AMERICAN 57 (June, 1979).

12. 68 Cal. 2d at 284, 437 P. 2d at 498-99, 66 Cal. Rptr. at 10-11.

13. See e.g., ALASKA STAT. § 20.20.010 (1975); ARK. STAT. ANN §61-141 (c) (1971); CAL. CIV. CODE § § 7004, 7005 (West 1975); COLO. REV. STAT. § 19-6-106 (1977); CONN. GEN. STAT. § 45-69(f) (1975); FLA. STAT. ANN. § 742.11 (West 1973); KAN. STAT. § 23-219 (1977); LA. CIV. CODE ANN., art. 188 (West 1975); MD. EST. & TRUSTS CODE ANN §1-206(b) (1975); MONT. REV. CODE ANN. § 61-306 (1975); N.Y. DOM. REL. LAW § 75 (McKinney 1974); N.C. GEN. STAT. § 49A-1 (1974); OKLA. STAT. tit. 10, § 552 (1974); OR. REV. STAT. §109.243 (1977); TEX. FAM. CODE ANN., tit. 2, § 12.03 (Vernon 1977); VA. CODE § 64.1-7.1 (1977); WASH. REV. CODE ANN., § 26.26.050(1) (West 1976); WYO. STAT. §14-2-,03(a) (1978).

14. See Grad, "Legislative Responses," supra note 3 at 490.

15. Note, "Asexual Reproduction and Genetic Engineering: A Constitutional Assessment of the Technology of Cloning," 47 SO. CAL. L. REV. 476 at 502 (1974).

16. Id.

17. Grad, Legislative Responses, supra note 3 at 502.

18. Id. at 502-503.

   See Appendix C for a Model A.I.D. Statute.

19. See id. at 506-07.

   Statutes that legitimize issue born of consensual A.I.D. procedures also force a reconsideration of established tenets of the laws of inheritance that normally define the issue for inheritance purposes at meaning "issue of the loins." Id. at 507. Such a reconsideration would be useful and could be accomplished without major difficulty.

20. Id.

21. Id. at 509.

22. Id. at 506.

23. Artificial insemination of unmarried women may be more difficult to regulate than the other modes of reproduction because medical expertise is not required to perform the procedure. See Id. at 506.

   See also, Smith, "A Close Encounter," supra note 11.

24. See N.Y. DOM. REL. LAW § 110 (1964); 2 AM. JUR. 2d, Adoption § 10 (1963).

25. A Harris poll conducted in 1969 that surveyed attitudes toward the New Biology throughout the country found that 19 percent approved the use of A.I.D. and 56 percent disapproved; 35 percent approved the use of A.I.D., however, assuming it was the only means by which a married couple could have a child. See Smith, "For Unto Us A Child Is Born," supra note 11 at 143.

26. Grad, Legislative Responses, supra note 3 at 508.

27. Id.

28. See Riskin & Reilly, "Remedies for Improper Disclosure of Genetic Data", 8 RUTGERS-CAMDEN L. J. 480 (1977). Only four states--Kansas, Massachusetts, Virginia and Iowa provide for the confidentiality of test results of genetic screening. Id. at 498-503.

29. See Miller, "Computers, Data Banks and Individual Privacy: An Overview", 4 COLUM. HUMAN RIGHTS L. REV. 1 (1972); Fried, "Privacy", 77 YALE L. J. 475 (1968).

   The Federal Privacy Act of 1974 provides for the person about whom an improper disclosure is made to recover a minimum award upon the successful civil prosecution of the action. While the Act prohibits a federal government agency from disclosing personal information it holds about citizens without their consent, it has a central weakness in that it requires a moving plaintiff to establish that the fedenting agency acted in its acts of disclosure with intent. 5 U.S.C. s 552 § (Supp. V, 1975).

30. R. VEATCH, CASE STUDIES IN MEDICAL EHICS 127-129 (1977).

31. Rivers, "Genetic Engineering Portends a Grave New World", in GENETICS AND REPRODUCTIVE ENGINEERING 83, 86 (D. England ed. 1974).

32. Simonsen v. Swenson, 104 Neb. 224, 177 N. W. 831 (1920); Tarasoff v. Regents of the Univ. of Calif., 131 Cal. Rptr. 14, 551 P. 2d 332 (1976). The Tarasoff court observed that when a patient has threatened to harm certain persons--even though expressing the intent in a privileged counseling session--his attending psychotherapist may well be under a duty to warn the persons of the threat.

See T. BEAUCHAMP, J. CHILDRESS, PRINCIPLES OF BIOMEDICAL ETHICS, Ch. 7 (1979); Walters, "Ethical Aspects of Medical Confidentiality" in CONTEMPORARY ISSUES IN BIOETHICS 169, 174 (T. Beauchamp, L. Walters eds. 1978); P. REILLY, GENETICS, LAW AND SOCIAL POLICY 167-275 (1977).

A new and sweeping piece of legislation in Massachusetts not only ensures the confidentiality of all hospital records of every patient, but allows the patient a complete access to his medical records. ACTS, MASS. LEGISLATURE, Ch. 214 (West's Mass. Legislative Service, 1979).

33. REILLY, supra, at 252-255.

34. This question or fascination for the ascertainment of any and all information about one's self has perhaps been heightened and promoted in large part by the passage of The Freedom of Information Act, 5 U.S.C. C. § 552 et seq. (1977); and, at the state level, similar laws are being passed. On a less conventional level, the television dramatization of Alex Haley's book, ROOTS, has quickened and/or popularized the interest in genealogy (and, thus, social genetics).

35. REILLY, supra, note 32 at 201.

Under the New York City Health Code, provision is made for maintaining minimum standards with regard to determining the health of the donor, the compatability of Rh factors between the donor and the donee as well as the maintenance of records. N. Y. CITY HEALTH CODE, Art. 21 (1959), formerly N. Y. SANITARY CODE § 112.

36. Currie-Cohen, Luttrell and Shapiro, "Current Practice of Artificial Insemination by Donor in the United States", 300 N. ENG. J. MED. 585 (No. 11, 1979).

37. Id. at 587.

38. Waddlington, "Artificial Insemination: The Dangers of a Poorly Kept Secret", 64 NW. U. L. REV. 782, 804 (1976).

See also, Smith, "Artificial Insemination: Disclosure Issues", 11 COLUM. HUMAN RIGHTS L. REV. 87 (1979); Annas, "Fathers Anonymous: Beyond The Best Interests of the Sperm Donor", 14 FAM. L.Q. 1 (1980).

39. Id.

40. Fitzgerald v. Rueckl #11433, set for argument September 15, 1981.

The trial court of the second judicial district of Washoe County, Nevada, found for the doctor.

41. See Ch. 4, supra, note 133, passim for a detailed discussion of these two cases.

42. Whitaker, " Birth Data Ruled Open to Adoptee", Wash. Post, Feb. 5, 1979, at C1, col. 5.

43. REILLY, supra, note 32 at 202.

44. The following states have adopted, in essence, these provisions of Section 5 of the Uniform Act: CAL. CIV. CODE § 7005(a) (West 1975); COLO. REV. STAT. § 19-6-106(i) (1977); CONN. GEN. STAT. § 45-69h (1975); KAN. STAT. § 23-130 (1977); MONT. REV. CODE ANN. § 61-306 (1975); N. Y. CITY HEALTH CODE § 21.07 (1959); OKLA. STAT. tit. 10, § 553 (1974); OR. REV. STAT. § 677.365 (3) (1977); WASH. REV. CODE ANN. § 26.26.050 (1976); WYO. STAT. § 14-2-103(a) (1978).

See generally, Note, "The Uniform Parentage Act: What It Will Mean for the Putative Father in California", 28 HASTINGS L. J. 191 (1976).

CHAPTER 7

THE SCIENTIFIC METHOD AND THE NEW BIOLOGY:
AN OVERVIEW

By promoting changes in knowledge that force a re-examination of the ideals, principles, and methods men employ in deciding between alternative moral claims, scientific developments can disrupt and bring into dispute a system of moral values and commitments.[1] The striking incompatibilities between more traditional moral standards and alternative standards supported by significant advances in knowledge can lead to general social anxiety.[2] To ease that anxiety, the reflective man confronts three tasks: to clarify the "bearing of trends in scientific inquiry upon pervasive conceptions of man's place in nature;" to make "explicit the intellectual methods by which responsibly held beliefs are achieved;" and to "interpret inherited beliefs and institutions in the light of current additions to knowledge . . . in order to exhibit the enduring wisdom which may be embodied in them."[3] The freedom of moral decisionmaking carries with it a burden of considerable dimension, since only men can choose the ends to which they will direct scientific research.[4]

Scientific investigations of necessity begin with fact-gathering. Thus, the central consideration becomes whether, as a form of expression, such acts are entitled to constitutional protection. Stated otherwise, is the lynchpin to scientific inquiry tied to a constitutional right-to-know and a right-to-learn? Two commentators have concluded, on the basis of a thorough study of constitutional law, first amendment theory and actual amendment cases heard by the United States Supreme Court, that persuasive evidence exists for asserting a strong argument that scientific inquiry is within the field of protected expression. Drawing heavily upon the masterful exposition of Professor Emeritus Thomas Emerson of Yale,[6] and in turn upon certain cases which speak of man's right to pursue self-fulfillment, to develop character and all human potentialities and thereby attain truth,[7] the authors conclude that the right to read, to listen and to see has constitutional protection.[8] Indeed, the passive right to know--distinguished from the more active right to gather and seek out information--has enjoyed a guaranteed protection by the United States Supreme Court as a necessary coordinate of the right to speak as well as to publish.[9]

Freedom to gain useful knowledge is--additionally--protected within the ambit of substantive due process.[10] To be in free enjoyment of one's faculties is a right protected by not only due process and privacy rights, but by the first amendment as well.[11] In a state court in Michigan, it was held that "there is no privacy more deserving of constitutional protection than that of one's mind."[12]

Not only does the First amendment protect the freedom to acquire useful knowledge, but this freedom is within the very meaning of "liberty" as spelled out in the due process clauses of both the Fifth and Fourteenth amendments to the United States Constitution. Accordingly, notice and an opportunity to be heard must be accorded any scientist or group of scientists whose constitutionality guaranteed freedom of inquiry is either being abridged or denied by governmental action.[13]

Improper legislative classifications could be challenged as unconstitutional by invoking the powers of the equal protection clause of the Constitution. The rationality of the suspect standard for classification could be attacked or judicial review sought by using a strict scrunity standard of analysis of the particular legislative enactment in question. Such a challenge might arise as a consequence of a government ban on recombinant DNA research, for example. Here, a class of scientists might be created who wished to engage in such research and would, thus, arguably, be treated in a dissimilar way from other scientists.[14] This classification would frame the equal protection challenge to the government action.

Judicial review of governmental actions which impinge on an individual scientist's right to undertake basic research and do so on grounds of content instead of valid state concerns within the scope of its police power, should be exercised decisively and on behalf of maintaining the interests of liberty.[15] In structuring the scope of its judicial review of cases of this nature, the courts should consider and balance the extent to which the research in question is regarded as essential to "the generation of new ideas or theories;" the extent to which the scientific inquiry contains as a necessary factor, teaching learning, writing, or other protected activity; the level at which social and political discourse is viewed as being promoted and as essential to self-governance as a consequence of the particular research; and, finally, the fundamental or basic extent of the research (thus providing the state with an opportunity to intervene at some later point if harmful effects appear).

From the government's viewpoint, those actions which are demonstrably predicated on the actual use of the police power or the so-called power of the purse--as opposed to being based on an intent to quell legitimate inquiry--and, as a parallel, are aimed at which might be viewed as negative externalities or consequences of the research, itself, rather than its content, will normally raise less serious constitutional problems for it.[16] It has been postulated that when in fact cases arise where the individual researcher's interest is high in those elements which arguably demand protection--and, correspondingly, the state's interest in intervening is low in those conferring legitimacy--then the courts should not shirk from their responsibility to intercede on behalf of "critical liberty interests."[17]

The possibilities of biomedicine are just beginning to challenge society's moral standards. New scientific developments will allow man to achieve many goals, but someone must make the value judgments of which goals are proper, based on an ethical standard.[18] Who will exercise those judgments is a threshold question. Individual moral dilemmas may not stand in the way of scientists,[19] but a collective moral dilemma confronting society as a whole may halt research or redirect it towards another goal.

The current debate over the limits of experimentation with genetic engineering focuses on this issue of how much power society should exercise over scientific research decisions. Currently, society does not join in that decisionmaking; scientists make research decisions without public involvement. Moreover, the only guide for scientists is an international code of ethics for clinical research[20] accepted by numerous American professional medical societies, including the American Medical Association.[21] Neither that code, nor the first international attempt to develop standards for scientific research. The Nuremberg Code,[22] imposes no radical principle on the medical researcher; each only restates a general standard of conduct, primarily imposing responsibility on the researcher for his subjects' safety.[23]

Several alternatives to the present method of self-regulation exist. Recognizing the inadequacy of self-imposed and sometimes artificial moral burdens on scientists, some scholars have suggested that scientists should share decisions concerning genetic research with the public.[24] The American public's ability to understand and analyze the complex decisions that would be thrust upon them under this proposal is doubtful, however, thus suggesting the need for a different alternative. Perhaps the most satisfactory way to resolve the rather troubled relationship between ethics and life sciences is to promote total cultural revision, or revolution, rather than to focus on individual normative behavior.[25] The development of a contemporary and "fresh" ethic for the life sciences would require an attemtp to build a new culture where all people work for common ends, share binding visions, and agree on a set of shared values.[26] That ideal provides no direction for the immediate ethical problems posed by the New Biology. A more practical means of assuring society's participation in basic research decisions would operate by exerting influence on the sources of funding. The dependence of genetic research on federal funds in particular suggests the leverage society could exert to force scientists to break with tradition and seek society's support for their research.

New Avenues of Regulation: The Federal Government Acts

The federal government already has taken steps to regulate medical research programs supported by federal funds. In 1966, the Surgeon General announced that the United States Public Health Service, a division of the Department of Health, Education and Welfare (HEW), would not grant, renew, or continue to support research programs involving humans unless the institution at which the research is being conducted undertakes a review of the risks and potential medical benefits of the research, the rights and the personal welfare of the research subjects, and the need for their informed consent to participate.[27] In 1973 HEW initiated a public dialogue about biomedical experimentation by proposing regulations for the projection of human subjects in federally funded scientific experiments.[28] Final regulations, which became effective July 1, 1974, apply to all HEW (now the Department of Health and Human Services) grants and contracts supporting research, development, and related activities involving human subjects.[29] Among other provisions, the regulations state that if experimentation may expose research subjects to possible physical, psychological, or social injury, an independent review committee must find that these risks are outweighed by the benefit to the subject and the value of the knowledge to be gained.[30] All subjects must give an informed consent before taking part in the experiment,[31] and any person obtaining an HEW grant or contract monies for research involving human subjects must submit both a written assurance of compliance with HEW policy and a set of implementing guidelines.[32]

Eleven days after the HEW regulations went into effect, Congress established a National Commission for the Protection of Human Subjects of Biomedical and Behavioral Research.[33] Congress directed the Commission, composed of eleven representatives from various fields of professional interest,[34] to identify basic ethical principles for biomedical and behavioral research involving human subjects;[35] the Commission specifically was directed to study risk-benefit criteria in determining the appropriateness of research involving human subjects and the nature of informed consent in biomedical research.[36] The Commission was to make its recommendations to the Secretary of HEW for guidance in developing guidelines for HEW-funded research and to Congress if the need for statutory guidelines for private research arises.[37]

On August 23, 1974, about one month after Congress created the Commission, HEW supplemented its July 1974 regulations with amendments governing biomedical research activities that involve minors, fetuses, abortuses,[38] prisoners, and institutionalized mental defectives.[39] The new regulations also give the Secretary of HEW authority to establish two Ethical Advisory Boards to review biomedical research proposals.[40] Moreover, even after HEW approves a research project for funding, an Institutional Review Board must continue to monitor the project and intervene if necessary.[41] Although these new regulations assure the federal government a role in making decisions concerning scientific experimentation in human research, the Secretary of HEW has adopted the position that it would be contrary to the public interest to impose permanently research restrictions that are based on the successes and limitations of current technology.[42] Without such broad guidelines, the Secretary advised the Ethical Advisory Board to evaluate each proposal involving in vitro fertilization in light of "the state of the art, legal issues, community standards, and the availability of guidelines to govern each research situation."[43] The HEW regulations set a standard of care that provides protection for the rights of human subjects at the earliest level of conception, while also providing for the freedom of scientific inquiry that is vital for research in this area. The establishment of an Ethical Advisory Board assures that ethics will be a factor in scientific decisionmaking.

Another HEW sponsored project, the National Commission for the Protection of Human Subjects and Behavioral Research, recently published its report and recommendations.[44] Based on a finding that human fetal research in the past has assisted in saving many thousands of lives that otherwise would have been lost,[45] the Commission encouraged therapeutic research directed toward the fetus or its mother without the imposition of complex safeguards.[46] In addition, the Commission called for the repeal of the present HEW moratorium on federally funded medical research involving fetuses in utero.[47]

The Commission's specific recommendations establish useful guidelines. Nontherapeutic fetal research undertaken in anticipation of abortion would be allowed if the research were consistent with guidelines for all other nontherapeutic research,[48] but a national ethical review board should review special problem cases.[49] Nontherapeutic ex utero fetal research might be permitted, following abortion, if three conditions are met: the fetus is less than twenty weeks old, the mother gives an informed consent, and the father does not object.[50] HEW also should require an assurance not only of the value of the anticipated research but also of the inability to achieve the research goals by alternative means.[51] These rather stringent restrictions will serve to reduce to a minimum any risks or indignities to the research subjects.[52] Most of the fetal research currently being conducted by American scientists, however, would nto be affected by these standards because researchers presently comply with the guidelines.[53]

The Commission's recommendations satisfactorily balance the need for continued scientific research in the area and the need for an ethical standard of guidance.[54] Although limited to HEW funded research, the recommendations and the accompanying HEW guidelines provide a basis for structuring a uniform standard of scientific research. An international and interdisciplinary inquiry into the ethical considerations surrounding biomedical research could build on these recommendations to develop a broader research standard.

Ethical Tribunals in Nongovernmental Areas

The concept of an ethical tribunal, embodied in the HEW regulations governing biomedical research with human subjects, could easily be incorporated into the procedures governing privately funded research in this area.[55] The Clinical Pathological Conferences, which already are an integral part of major hospital post-mortem procedures, could serve as an ethical tribunal for private research.[56] At these conferences, the clinician who administered to the needs of hte diseased patient describes his diagnosis, its rationale, and therapy he prescribed, and the pathologist presents the post-mortem findings. This process reveals to the clinician any errors in his judgment or the astuteness of his diagnosis.[57] Introduction of an ethical review of the research by adding a third panel member would entail little difficulty.[58]

The most perplexing dilemmas, with imposing ethical considerations, arise in human experimental research, not in therapeutic treatment. The selection and composition of the members of an ethical tribunal that would consider prospective action and the determination of the nature and extent of its responsibility pose serious questions.[59] Should a tribunal be composed completely of doctors, or should an ethicist, philosopher, general member of the community, or atheist be included? The composition of each tribunal could be tied to the nature of the research to assure that the personal composition of the tribunal would change and that each panel could have a different philosophical disposition for each new research problem. Under this proposed structure, a tribunal could approach a cancer research problem differently from a fetal research problem. Whether any tribunal reviewing private research would be legally liable for its judgments is doubtful, but each member would clearly be morally responsible for the consequences of his actions.[60]

Yet another imaginative proposal--somewhat akin to an ethical tribunal--calls for the establishment by Congress of a Council for Societal Informed Consent with a membership of fifty persons (called consentors) drawn broadly from all segments of society. The Consentors, upon selection, would be enrolled for one or two years in a specially designed university program providing intensive training in--among other areas--biomedical ethics, philosophy, law and science. They would serve in a full-time capacity. The Council's function would be to provide, in some cases original policy, in others, advisory--or withhold the informed consent of society to scientific questions, proposals and programs which are clearly of immediate importance to society--as for example the current DNA experimentation. Structured as such, this body would serve as a Naderesque "citizens gorup."[61]

Finding An Equilibrium

Government regulation, international codes of ethical behavior, and rules of internal hospital administration are limited in their attempt to set standards for scientific research in genetic engineering. In the final analysis, the private researcher charts the course of scientific investigation. He will determine the balance between freedom of scientific inquiry and concepts of what is socially good; he will determine whether his research should be totally utilitarian, providing the greatest good to the greatest number even if it may compromise the rights of some individuals, and how his research should accommodate the competing interests of each subgroup in society. The system of self-regulation would be workable, of course, if scientists accepted the fundamental principle that their research must promote the social good by seeking to minimize human suffering for the greatest number; that principle would provide adequate ethical guidance for research decisions.[62] This general standard appears to favor independent scientific study since the research scientist determines both the definition of the social good that the experimentation will promote and the beneficiaries of that research. The standard is based on a broader conception, however, in that it seeks to minimize mental, physical, social, and spiritual suffering throughout the human community. The standard therefore, would allow a scientist to undertake in utero or in vitro fetal research in order to produce children free from debilitating recessive hemophilic or sickle cell genes because successful research ultimately would benefit parents and prospective parents and their offspring by eliminating suffering and also would benefit society by making available for others research resources otherwise allocated to maintain the genetically defective.

Because this ethical standard is future-oriented, one could argue that it does not adequately concern itself with present human suffering. As a branch of applied science, genetics is directed toward the study of the evolutionary aspects of heredity and reproduction, while other branches of medical science exist in part to treat the present infirmities of the genetically inferior. Genetic manipulation holds for the future, not a cure for the present, and this prospective character complicates the ethical considerations.

Notes--Chapter 7

1. E. NAGEL, SOVEREIGN REASON 297 (1954).

Professor Ayer has cautioned that "moral judgments are emotive rather than descriptive; they are persuasive expressions of attitudes and not statements of fact--consequently they cannot be either true or false." Ayer, "On the Analysis of Moral Judgments", in A MODERN INTRODUCTION TO ETHICS 545 (M. Munitz ed. 1958).

See generally, Sinsheimer, "The Presumption of Science", 107 DAEDALUS 23 (1978).

2. E. NAGEL, supra.

See generally, "Symposium, Ethical Aspects of E perimentation with Human Subjects", 98 DAEDALUS 219-594 (1969); Graham, "Concerns about Science and Attempts to Regulate Inquiry", 107 DAEDALUS 115 (1978)

3. E. NAGEL, supra note 1 at 297-298.

The etiology of fear has yet to be determined fully. What is understood, however, is that all of the advances of civilization--by they scientific in origin or otherwise--have been accompanied by an irrational, but almost instinctual fear or anxiety. Some conclude that the magnitude of the anxiety over the particular occurrence or innovation. Others suggest that one is more likely to fear that which he cannot control. Contrariwise, one is more tolerant of developing situations or actual occurrences when they can be controlled.

In order to educate the greater public to the realities of the New Biology and thereby erode the all pervasive anxieties associated with its development, it has been suggested that a panel of "wise and broad-minded citizens be assembled" (perhaps under the aegis of the National Academy of Sciences) to meet the task of collecting evidence from (i.e., the possibility of escape into the environment of an organism containing recombinant DNA molecules). Additionally, the panel would assess both the nature and the magnitude of the damage occurring if the acts were to in fact happen and--furthermore--list the efforts which would be undertaken to counteract, contain or--if possible--correct the catastrophe. Stetten, "What Men Fear", 21 PERSPECTIVES BIOLOGY & MED. 515 passim (1978).

4. See R. HARE, FREEDOM AND REASON 3 (1963).

See also, Bok, "Freedom and Risk", 107 DAEDALUS 115 (1978).

Arguably, a presumption in favor of the validity of technological advancement should be indulged unless the advancement of a particular technology poses a clear and present danger to society or an identifiable minority group within the society.

5. Delgado & Miller, "God, Galileo, and Government: Toward Constitutional Protection for Scientific Inquiry", 53 WASH. L. REV. 349, 353 (1978).

6. T. EMERSON, THE SYSTEM OF FREEDOM OF EXPRESSION (1970).

7. Police Department v. Mosley, 408 U. S. 92, 96 (1972); Whitney v. California, 274 U. S. 372, 375-377 (1927), (Brandeis, J., concurring); Jones v. Opelika, 316 U. S. 584, 618 (1941) (Murphy, J., dissenting), vacated, 319 U. S. 103 (1943).

8. Delgado & Miller, supra note 5 at 383.

9. Id. at 382.

10. Meyer v. Nebraska, 262 U. S. 390, 399 (1923).

11. Delgado & Miller, supra note 5 at 395.

   See Griswold v. Connecticut, 381 U. S. 479 (1965); Stanley v. Georgia, 394 U. S. 557 (1969).

12. Kaimowitz v. Department of Mental Health, Cir. Ct. Wayne County, Mich., July 10, 1973, reprinted in 1 ABA MENTAL DISABILITY L. REP. 147 (1976) as cited by Delgado & Miller, supra at 395.

   See Mackey v. Procunier, 477 F. 2d 877 (9th Cir. 1973) where the court recognized a right to control one's own mental processes based as such on a privacy based right as discussed by Delgado & Miller, supra.

13. Delgado & Miller, supra note 5 at 396.

   See also, Robertson,"The Scientists Right to Research: A Constitutional Analysis,"50 SO. CAL. L. REV. 1203 (1978).

14. Id. at 401.

15. Id. at 403.

16. Id.

17. Id.

18. See Kass, "The New Biology: What Price Relieving Man's Estate", 174 SCIENCE 779, 781 (1971).

19. Rose & Rose, "The Myth of the Neutrality of Science", in THE SOCIAL IMPACT OF MODERN BIOLOGY 215, 219 (W. Fuller ed. 1971).

20. Declaration of Helsinki, 1964, reprinted in Ratnoff & Smith, "Human Laboratory Animals : Martyrs for Medicine", 36 FORDHAM L. REV. 673, 680-81 (1968).

   See Appendix D for the Declaration, infra.

21. See Ratnoff & Smith, supra at 681.

   Interestingly, the American Medical Association has promulgated its Ethical Guidelines for Clinical Investigation (A.M.A. Opinions and Reports of the Judicial Council 9, (1969)) which do not complement the World Medical Association's Declaration of Helsinki in certain key areas of concern. (See Appendix D). For example, unlike the Helsinki Declaration, the AMA guidelines propose only that when mentally competent adults are found to be unsuitable subjects, minors or mentally incompetent subjects are to be used as subjects. The Helsinki Declaration states--while the Guidelines do not--that the subject or his guardian are free to withdraw permission for research at any time during the course of the clinical research. Both documents are quite vague in many respects. The following statement from the Guidelines is illustrative: "Ordinarily consent should be in writing except where the physician deems it necessary to rely upon sent in other than written form because of the physical or emotional state of the patient". (Id. at 11). The criteria for determining the nature of the physical or emotional conditions which allow for the dispensation of the written consent are conspicuously absent.

See Appendix J, infra, The Belmont Report of the National Commission for the Protection of Human Subjects of Biomedical and Behavorial Research for a review of hte various ethical standards which should be used when experimentation with humans is undertaken.

22. See generally, 2 TRIALS OF WAR CRIMINALS BEFORE NUREMBERG MILITARY TRIBUNALS, THE MEDICAL CASE 181-84 (1950), reprinted in Beecher, "Experimentation in Man", 169 J.A.M.A. 461, 472-74 (1959) (text of Nuremberg Code). See Appendix E, infra, for the Code.

23. See Comment, "Non-Therapeutic Medical Research Involving Human Subjects", 24 SYRACUSE L. REV. 1067, 1078 (1973). The researcher must be able to justify the use of human subjects after evaluating the risks and potential benefits; additionally, the participants must give a fully informed, voluntary consent. Moreover, the researcher has a continuing affirmative duty to protect his subjects' safety even to the point of terminating the experiment if he has reason to believe the subjects otherwise would be harmed. Id.

See generally, Brody, "The Problems of Exceptions in Medical Ethics" in DOING EVIL TO ACHIEVE GOOD, Ch. 2 (R. McCormick, P. Ramsey eds. 1978).

24. Rose & Rose, supra note 19 at 219.

The American public is not at this time fully equipped with a level of sophistication to understand or appreciate complex decisions of a nature which would be naturally thrust upon them if biomedical research went "public" in a total sense. As of 1970, only 52.3% of the population of the United States twenty-five years of age and older were high school graduates. STATISTICAL ABSTRACT OF THE UNITED STATES, U.S. Dept. Comm. at 119 (W. Lerner ed. 1974). Of the Negro population 25 years and older, 31.4% were high school graduates. Id. For members of the Caucasion Race twenty-five years or older, 12.3 school years had been completed by March, 1972, with the Negro Race showing a completion record of 10.3 years. THE OFFICIAL ASSOCIATED PRESS ALMANAC 1974 at 736 (L. Urdang ed. 1973).

When political hysteria combines with unsophisticated public emotional reactive powers, a genetic moratorium may result. In Cambridge, Massachusetts, the City Council--in a 5/4 vote--demanded Harvard University and M.I.T. halt for three months their research activities concerned with DNA and, specifically, those experiments which involve inserting segments of DNA from other organisms into E. coli. Some feared such experiments might create new, lethal microbes against which humans have no immunity. The National Institutes of Health then proceeded to issue new and rather demanding rules designed to govern such experimentations. TIME Mag., July 19, 1976, at 67. N.I.H. Proposed Guidelines for Research Involving Recombinant DNA Molecules (1976), 41 Fed. Reg. 27,911 (1976).

See Appendix G, infra, for pertinent part of the Final Guidelines for research conducted with federal grant monies involving Recombinant DNA molecules together with the policy decision of the Director of the National Institutes of Health to release the Final Guidelines dated June 23, 1976. Schmeck, "Guidelines Issued to Curb Genetic Research Hazards", N. Y. Times, June 24, 1976, at 1, col. 2. A summary of the revision of the Guidelines issued by the Secretary of Health Education and Welfare, Joseph A. Califano, Jr., on December 17, 1978, is made a part of Appendix G, infra.

"There is a real danger," observed the President of Columbia University, "that fearful communities will seek to regulate scientific research at the lowest level of our political governance where serious thought and careful analysis are often lacking". McGill, 3 COLUMBIA TODAY 24 (1977). See also, McGill, "Science and the Law", 23 CATH. LAW. 85 (1978).

Following the furor at Cambridge, Massachusetts, the City Manager there structured a nine member (non scientific) experimentation review board to evaluate, pursuant to the N.I.H. safety guidelines, the safety of future DNA experiments undertaken with grant federal monies. The board's views are advisory and affect only the city of Cambridge--yet its decisions and functions will serve as an "international model". See Bennett & Gurin, "Science That Frightens Scientists--The Great Debate Over DNA", ATLANTIC MONTHLY, Feb. 1977, 43 at 61. See also, Jonas, "Freedom of Scientific Inquiry and The Public Interest", 6 THE HASTINGS CENTER REPORT 15 (Aug. 1976); Toulmin, "DNA and the Public Interest", N. Y. Times, Mar. 12, 1977, at C23, col. 3.

See generally, Berger, "Government Regulation of the Pursuit of Knowledge: The Recombinant DNA Controversy", 3 VT. L. REV. 83 (1978); Stich, "The Recombinant DNA Debate", 7 PHILOSOPHY & PUB. AFF. 187 (1978); Dworkin, "Science, Safety and the Expert Town Meeting: Some Comments on ASILOMAR", 51 SO. CAL. L. REV. 1471 (1978); Hollman & Dutton, "A Case for Public Participation in Science Policy Formation and Practice", 51 SO. CAL. L. REV. 1505 (1978).

It has been suggested that Congress create a National Biohazards Commission analogous to the original Atomic Energy Commission, with legal authority to evaluate, license, supervise and inspect all activities that may subject the public to biological hazards and--furthermore--that an international council of the same nature be structured in order to set uniform world-wide policy in this area. Cavalieri, "New Strains of Life--or Death", N. Y. Times Mag., Aug 22, 1976, at 8. See also, "Research at Yale on Combination of Genetic Materials Stirs Fears, But Safety Panel Defends Technique", N. Y. Times, Nov. 7, 1976, at L 23, col. 1. See also, BIOHAZARDS IN BIOLOGICAL RESEARCH (A. Hellman, M. Exman, R. Pollack eds. 1973).

Interestingly, Mack v. Califano, 447 F. Supp. 668 (D.D.C. 1978) determined that federal government research which was to be undertaken at Ft. Detrick, Maryland, and would follow Recombinant DNA research standards, was consistent with preserving the environment.

25. Callahan, "Search for an Ethic--Living with the New Biology", 5 THE HASTINGS CENTER REPORT 4 (July-Aug., 1972).

It has been proposed that a new social role be formed for an individual to be designated as an Informed Outsider. His duties as a non professional member of a peer goup advisory panel on human experimentation would be tied to obtaining a thorough familiarity with all matters of normal concern brought with each case presented to the review group for their consideration. B. BARBER, J. MAKARUSHKA, D. SULLIVAN, RESEARCH ON HUMAN SUBJECTS 196 (1973).

26. Callahan, supra at 6.

27. See Curran, "Governmental Regulation of the Use of Human Subjects in Medical Research: The Approach of Two Federal Agencies" in EXPERIMENTATION WITH HUMAN SUBJECTS 402, 436 (P. Freund ed. 1970). See generally MEDICAL PROGRESS AND THE LAW (C. Havighurst ed. 1969); Jaffee, "Law as a System of Control", 98 DAEDALUS 406, 428 (1969).

See also, Kaplan, "Emotion Versus Objectivity in Funding of Biomedical Research", 196 ANNALS N. Y. ACAD. SCIENCES 274 (1972); Curran, "Ethical Consideration in Human Experimentation", 13 DUQUESNE L. REV. 819 (1975).

Some claim federal regulation of biomedical research was first undertaken in 1953 with a N.I.H. requirement that at its Clinical Center in Bethesda all research involving humans was to be first approved by a review committee responsible for protecting the human subjects. N. HERSHEY, R. MILLER, HUMAN EXPERIMENTATION AND THE LAW 5 (1976). The book, itself, has been prepared specifically for investigators, members of institutional review boards and institutional administrators with responsibility for research with human subjects. It directs its focus on institutional processes and on those procedures by which a research proposal dealing with human subjects comes first under scrutiny, is later assessed and--finally--an institutional decision made as to its merit and feasibility. All in all, it is a most useful book. For another enlightening discussion of the history of biomedical research see P. RAMSEY, THE ETHICS OF FETAL RESEARCH 3-20 (1975).

28. U.S. Dep't.of Health, Educ. and Welfare, Protection of Human Subjects, Proposed Policy, 38 Fed. Reg. 27882 (1973). After public comment, final regulations were adopted on May 30, 1974, to be effective July 1, 1974. 45 C.F.R. §§ 46.1-.22 (1975). See generally 39 Fed. Reg. 18914-17 (1974) (summary of public comments). See also Martin, "Ethical Standards for Fetal Experimentation", 43 FORDHAM L. REV. 547 (1975).

29. See 45 C.F.R. § 46.1 (1975).

30. Id. § 46.2(b) (1).

See Brody, "Study Finds Injury Risk in Medical Research Similar to Normal Living", N. Y. Times, Sept. 19, 1976, at 54, col. 2. See generally, Childress, "Compensating Injured Research Subjects: I. The Moral Argument", 6 THE HASTINGS CENTER REPORT 21 (Dec., 1976); Robertson, "Compensating Injured Research Subjects: II. The Law", 6 THE HASTINGS CENTER REPORT 29 (Dec. 1976).

31. Id. § 46.2(b) (3); see Id. §§ 46.3(c), 46.9, 46.10.

32. Id. § 46.4.

Presently, the National Institute of Health--an arm of HEW--is responding to both recommendations from the former National Commission for the Protection of Human Subjects of Biomedical and Behavioral Research concerning institutional review boards (originally published in 43 Fed. Register 56174 (Nov. 30, 1978) and public comments. Final regulations are expected to be released sometime in early 1981. The Proposed Regulations Amending Basic HEW (and Food and Drug Administration) Policy for Protection of Human Research Subjects which arose from the National Commission's recommendations are to be found at 44 Fed. Register 47688-47729 (Aug. 14, 1979).

At present the primary effect of the proposed rules would be to: 1). Continue to provide protections for human subjects of research conducted or supported by the HEW; 2). Require Institutional Review Boards (IRB's) review and approval of research involving human subjects, even if it is not supported by Department funds, if it is conducted at or supported by an institution receiving HEW funds for research not exempt from the regulations; 3). Require review of human subject research irrespective of risk--unless the research is specifically exempted from coverage; 4). Exempt from coverage certain kinds of social, economic and educational research; 5). Either exempt or require only expedited review of certain kinds of research involving solely the use of survey instruments, solely the observation of public behavior, solely the study of documents, records and specimens, or solely a combination of any of these activities; 6). Require only expedited review for certain categories of proposed research involving no more than minimal risk and for minor changes in research already approved by the particular IRB; 7). Provide specific procedures for full IRB review and for expedited IRB review as well; 8). Designate the basic elements of informed consent which are a necessary pre-requisite to research subject participation and additional elements which, when appropriate, are a necessary pre-requisite to subject participation; 9). Indicate circumstances under which the IRB may approve withholding or altering certain information otherwise required to be presented to research subjects; 10). Require that IRB membership include at least one nonscientist; 11). Establish regulations which to the extent possible are compatible and consistent with proposed Food and Drug Administration standards for IRB's.

See Reinhold, "Guidelines On Studies of Humans Stir Protest", N. Y. Times, April 22, 1980, at C1. col. 6; "Symposium on Institutional Review Boards", 9 THE HASTINGS CENTER REPORT 18--33 (Feb. 1979). Professor John A. Robertson's article, "Ten Ways to Improve IRB,", at 29 passim is most important.

33. National Research Act, Pub. L. No. 93-348, % 201, 88 Stat. 342, 42 U.S.C.A. § 2891-2 (1974).

34. The fields represented were: medicine, law, ethics, theology, biology, physical science, behavorial and social science, philosophy, humanities, health administration, and government and public affairs. Id. § 201(b).

35. Id. § 202(a) (1) (A) (i), 42 U.S.C.A. § 2891-2(a) (1) (A) (i).

36. Id. § 202(a) (1) (B), 42 U.S.C.A. § 2891-2(a) (1) (B).

37. Id. §§ 202(a) (1) (A) (iii), 202(a) (3), 42 U.S.C.A. §§ 2891-2(a) (1) (A) (iii), 2891-2(a) (3).

When the Commission ended its life October, 1978, President Carter signed Public Law 95-622 into law on November 9, 1978, creating the President's Commission for the Study of Ethical Problems in Medicine and Biomedical Behavioral Research (92 Stat. 3437-3442). Given until September 30, 1982, in order to complete its work, the main agenda facing it is to oversee the ethics of research and medical care conducted or financed by all federal agencies and to seek agreement among the states on a uniform definition of death; explore and define the parameters of informed consent to medical procedures and experiments; guarantee protection and confidentiality of individual medical records and extend protection against the advances of computerization and invasions of privacy and chart the scope and effect of genetic screening. See Cohn, "New Presidential Commission to Study Ethics of Gaps in U.S. Health Services", Wash. Post, Jan. 12, 1980, at A4, col. 4; Annas, "All The President's Bioethicists", 9 THE HASTINGS CENTER REPORT 14 (Feb. 1979).

Interestingly, a draft report of this Presidential Commission prepared recently posits a new standard for defining death and will, in turn, be submitted to Congress the first part of 1981. The draft states that: "An individual who has sustained either irreversible cessation of circulatory and respiratory functions or irreversible cessation of all functions of the entire brain, including the brain stem is dead. A determination of death must be made in accordance with accepted medical practice." Hilts, "Revised Definition of Death Suggested", Wash. Post, Sept. 17, 1980, at A8, col. 4.

38. Abortuses are fetuses that are expelled prior to viability, either spontaneously or as a result of medical or surgical intervention.

39. 39 Fed. Reg. 30648 (Aug. 23, 1974) (regulation proposed); see 40 Fed. Reg. 33526 (Aug. 8, 1975) (adoption of final regulations).

See also, EXPERIMENTATION WITH HUMAN BEINGS, Chs. 12, 13 (J. Katz ed. 1972); M. VISSCHER, ETHICAL CONSTRAINTS AND IMPERATIVES IN MEDICAL RESEARCH, Ch. 9 (1975).

The then preliminary proposals of the National Commsision for Protection of Human Subjects of Biomedical and Behavorial Research re medical experimentation with prisoners pointed to what one author concluded would probably result in a moratorium on most medical research presently conducted in prisons. A drug industry spokesman told the Commission that 85% of the first tests of new drugs in humans was done on prisoners who volunteered for such research. Schmeck, "Panel's Proposals May Curtail Medical Experiments in Prisons", N. Y. Times, June 14, 1976, at 58, col. 5. See Appendix F, infra, for extracts of the Final Regulations.

See generally, Burt, "Why We Should Keep Prisoners from the Doctors", 5 THE HASTINGS CENTER REPORT 25 (Feb., 1975); Steele, "U.S. to Keep Testing Drugs on Prisoners", Wash. Post, Aug. 27, 1977, at A6, col. 4.

See also, 3 ENCYCLOPEDIA OF BIOETHICS 1349-1353 (W. Reich ed. 1978); Gobert, "Psychosurgery, Conditioning and the Prisoner's Right to Refuse Rehabilitation", 61 VA. L. REV. 155 (1975); Hatfield, "Prison Research: The View From Inside", 7 HASTINGS CENTER REPORT 11 (Feb. 1977); Branson, "Prison Research: National Commission Says 'No Unless' . . ." 7 HASTINGS CENTER REPORT 7 (Feb. 1977); Supra, note 104, Ch. 3, for a discussion of the Kaimowitz case.

Dr. Alan A. Stone, President of the American Psychiatric Association and a member of the Harvard University Law School, has opined that all prison situations are coercive and--thus--a real difficulty is encountered in attempting to show a voluntary, knowing capacity to give an informed consent exists as to prisoners being considered for research experimentation. Lecture, Psychiatry for Lawyers, July 21, 1979, Harvard University Law School, Cambridge, Massachusetts.

40. 45 C.F.R. § 46.204 (1975).

41. Id. § 46.205 (intervention "as necessary"); see Id. § 46.106.

42. 39 Fed. Reg. 37993 (Oct. 23, 1974).

43. Id.

44. NATIONAL COMMISSION FOR THE PROTECTION OF HUMAN SUBJECTS OF BIOMEDICAL AND BEHAVIORAL RESEARCH, REPORT AND RECOMMENDATIONS--RESEARCH ON THE FETUS, (DHEW Publication No. (05) 76-127, 1975) (hereinafter cited as COMMISSION REPORT AND RECOMMENDATIONS); see 40 Fed. Reg. 33547 (Aug. 8, 1975) (text of Commission recommendations).

See Symposium, "On the Report and Recommendations of the National Commission for the Protection of Human Subjects of Biomedical and Behavioral Research , 22 VILL. L. REV. 297 (1976-77). See generally, Symposium, "Medical Experimentation on Human Subjects", 25 CASE WES. RES. L. REV. 431 (1975).

45. COMMISSION REPORT AND RECOMMENDATIONS App. 15-1 to 15-2 (study of the Columbus-Battelle Laboratories). The conclusions focus on four major medical achievements aided by fetal research: treatment of the Rh hemolytic diseases; the development of a vaccine for rubella; the treatment of newborn humans with respiratory distress syndrome; and development in amniocentesis. See Id. at 15-9 to 19-96. See also Schmeck, "Report on Human Fetal Studies Finds Work Saved Thousands", N. Y. Times, Mar. 15, 1975, at 20, col. 1.

46. See COMMISSION REPORT AND RECOMMENDATIONS 73 (Recommendations 1 and 2); 40 Fed. Reg. 33547 (Aug. 8, 1975).

47. See COMMISSION REPORT AND RECOMMENDATIONS 76 (Recommendation 16); 40 Fed. Reg. 33548 (Aug. 8, 1975). See also 39 Fed. Reg. 30962 (Aug. 27, 1974) (HEW Moratorium). The ban, although limited to HEW funded research, effectively stifled all fetal experimentation in the United States because of scientific researchers' heavy dependence on federal support. See Schmeck, supra note 45 at 20, col. 1.

In its final regulations, adopted after the Commission issued its report, HEW lifted its ban, providing, among other things, that in utero fetal research may be conducted only if required to meet the health needs of the fetus or if the rick to the fetus is minimal and important biomedical knowledge cannot be obtained by other means. 45 C.F.R. § 46.208(a) (1975); see 40 Fed. Reg. 33526 (Aug. 8, 1975) (secretary's statement approving final regulations).

The Ethics Advisory Board of the Department of Health Education and Welfare--a group designed in large part to carry on work and investigation initially undertaken by the Commission for the Protection of Human Subjects upon the Congressionally mandated time for the expiration of the Commission--recently went on record as urging the Department to void its ban on providing grant monies for research on in vitro fertilization. A number of scientists expressed their concern that more research on nonhuman embryos was necessary before going forward with primates. The Board, perhaps with a recognition of objections of this nature, recommended as in vitro research proceeded, the public be informed of any evidence such acts produce a higher rate of abnormal children than natural reproduction. Additional proposed safeguards were that: research be funded only where new information not otherwise obtainable relative to the safety and effectiveness of in vitro would be gained; implanted embryos come only from the sperm and eggs of legally married couples, and research be limited to human embryos in the first 14 days of development after fertilization--or that time when implantation into the womb is completed in a normal pregnancy. TIME Mag., April 2, 1979, at 89.

48. See COMMISSION REPORT AND RECOMMENDATIONS 74 (Recommendation 5); 40 Fed. Reg. 33548 (Aug. 8, 1975). See also, Schmeck, "Members of Panel on Fetal Research Object to Several of its Recommendations", N. Y. Times, May 21, 1975, at 31, col. 4. Non-therapeutic research is research that does not directly aid the individual involved; the purpose of such research is to strengthen the health of persons in that same affinity group or category.

See generally, Ramsey, "The Enforcement of Morals: Non-therapeutic Research on Children", 6 THE HASTINGS CENTER REPORT 21 (Aug. 1976).

49. See COMMISSION REPORT AND RECOMMENDATIONS 74 (Recommendations 5 and 6); 40 Feb. Reg. 33548 (Aug. 8, 1975).

50. See COMMISSION REPORT AND RECOMMENDATIONS 74 (Recommendation 6); 40 Fed. Reg. 33548 (Aug 8, 1975).

The first public debate on the issue of medical experimentation on the human fetus was seen in England and culminated with the so-called Peel Report of May, 1972. This report concluded the area of investigation presented no major ethical difficulty and that no unresolved moral questions needed to be raised over the issue. Under the Peel Report, experiments with the human fetus were to be allowed up to 20 weeks or 140 days. Consequently, a fetus over a 20 week gestational age would be regarded as viable for purposes of research. Any experiment on a viable fetus thus defined was prohibited unless regarded as necessary to promote its life. P. RAMSEY, THE ETHICS OF

FETAL RESEARCH 1, 2 (1974).

51. See COMMISSION REPORT AND RECOMMENDATIONS 74 (Recommendation 6); 40 Fed. Reg. 33548 (Aug. 8, 1975).

52. See COMMISSION REPORT AND RECOMMENDATIONS 68-69.

53. See Schmeck, "Some Research on Human Fetus Backed in Report", N. Y. Times, Apr. 12, 1975, at 55, col. 1.

The impact of the National Environmental Policy Act of 1969 is being felt in the New Biology. The decision of the Director of the National Institutes of Health on guidelines for research on Recombinant DNA molecules issued June 23, 1976, acknowledges the need for--and directs-- the preparation of an environmental impact statement on recombinant DNA research activity consistent with NEPA, See Appendix G, infra. See also, Parenteau & Catz, "Public Assessment of Biological Technologies: Can NEPA Answer the Challenge?", 64 GEO. L. REV. 679 (1976); Comment, "Governmental Control of Research in Positive Eugenics", 7 U. MICH. J.L. REFORM 615 (1974); Wollan, "Controlling the Potential Hazards of Government Sponsored Technology", 36 GEO. WASH. L. REV. 1105 (1968); Epstein, "Medical Genetics: Recent Advances with Legal Implications", 21 HASTINGS L. J. 35 (1969).

The courts have been quick to respond to the need for caution in advancing the new biological technologies. That cautious attitude has found strength in developing the provisions of NEPA for application here. In Scientists' Institute for Public Information, Inc. v. AEC, the court -- through Judge Wright -- held that the preparation of an environmental impact statement prior to the actual implementation of a technology was required by NEPA. 481 F. 2d 1079 (D.C. Cir. 1973). The Judge opined that once technological advances are brought to a stage of commercial feasibility they should be viewed as capital investments and the investment made in their development acts to compel their ultimate application. "To wait until a technology attains the stage of complete commercial feasibility before considering the possible adverse environmental effects attendant upon ultimate application of the technology will undoubtedly frustrate meaningful consideration and balancing of environmental costs against economic and other benefits." Id. at 1089. Continuing, the Judge stated that, once -- as in the terms of NEPA there has been an irretrievable commitment or resources -- in the technology development stage, "the balance of environmental costs and economic and other benefits shifts in favor of ultimate application of the technology." Id. at 1090.

See e.g., National Resources Defense Council, Inc. et al v. Callaway, 524 F 2d 79 (2d Cir. 1975); Jones v. Lynn, 477 F.2d 885, 891 (1st Cir. 1973); Natural Resources Defense Council, Inc., v. Morton, 458 F. 2d 827, 836 (D.C. Cir. 1972); Aberdeen & Rockfish Railroad V. Students Challenging Regulatory Agency Procedures (SCRAP II), 422 U. S. 289 (1975).

See generally, Robertson, "The Scientists Right to Research: A Constitutional Analysis", 51 SO. CAL. L. REV. 1203 (1975).

See also, Mack v. Califano, supra note 24, for the holding that DNA research would have no adverse effect on the environment or jeopardize the public health.

54. See generally, P. RAMSEY, THE ETHICS OF FETAL RESEARCH (1975); Friedman, "The Federal Fetal Experimentation Regulations: An Establishment Analysis", 61 MINN. L. REV. 961 (1977).

55. See Dagi, "The Ethical Tribunal in Medicine", 54 B.U.L. REV. 268, 274-77 (1974).

A survey undertaken at some 293 institutions having peer review group committees for clinical research showed 68% of the committees were not specialized in any way. Twenty-seven percent had membership of from between 1 to 5 members; 48% had from 6 to 10 members; 25% had more than 10 members and 1 committee had 40 members. A further break down of membership on the committes is informative: 9% retained lawyers; 9% members of hospital boards of trustees; 9% were composed of basic scientists; 9% had behavorial scientists; 2% employed pharmacologists or pharmacists; 18% used nurses. Over-all, 93% of the institutions surveyed reported using members of their own staffs to sit on such review committees. As to outsiders, 10% of the responding institutions reported using members on their peer review committees from other institutions who were doing related clinical research; 10% used physicians from other institutions who did no clinical research in the area of specialization; 4% used outside (non-institutional, affiliated) lawyers; 5% used outside behavorial scientists; 4% used clergymen and 1 had a patient sitting on its peer review committee. Altogether, only 22% of the 293 institutions had any kind of outsider as a committee member. An outsider for pusposes of this study was a non member of the institution. B. BARBER, J. LALLY, J. MAKARUSHKA, D. SULLIVAN, RESEARCH ON HUMAN SUBJECTS 253 (1973).

See generally, Gray, "The Functions of Human Subject Review Committees", 134 AM J. PSYCHIATRY 907 (1977); Robertson, "The Law of Institutional Review Boards", 26 U.C.L.A. L. REV. 484 (1979).

56. Dagi, supra at 275.

57. Id.

58. Id.

See Appendix G.

59. Id. at 276.

60. Id.

Interestingly, a modified type of ethical tribunal has been used as a part of the administration of the Federal Social Security Act. The 1972 amendments to this Federal Act structured professional standards review organization. These organizations have, as their purpose, to assist in reducing the number of poor and aged sent to hospitals by doctors. Membership on the boards is composed of doctors, community representatives and specialists in societal and ethical matters. A. ETZIONI, GENETIC FIX 201 (1973). Sometimes doctors will allow an indigent or terminally ill person to occupy a hospital bed indefinitely when individuals of this nature might be better off and able to obtain similar treatment in county nursing homes or in their own homes--thus vacating hospital space for the more seriously ill.

See generally, "Symposium: Professional Standards Review Organizations", 1975 UTAH L. REV. 355; Robertson, supra note 55.

61. "Societal Informed Consent: A Proposal", Unpublished Paper, May 5, 1977, Law, Science and Medicine Program, Yale University Law School, New Haven, Connecticut.

62. Pauling, Foreward, "Symposium--Reflections on the New Biology", 15 U.C.L.A. L. REV. 267, 270 (1968).

Cf. Greenawalt, "A Contextual Approach to Disobedience", 70 COLUM L. REV. 48, 50-51 (definition of concept "social good" in abortion context).

CHAPTER 8

THE BIOETHICAL CONUNDRUM: A CONSIDERATION IN MICROCOSM

The Interaction Of Science And Ethics

Bioethics attempts to develop a philosophy regarding the application of man's biological knowledge in furtherance of the social good.[1] Several philosophers have attempted to structure a general system of bioethics. For Teilhard de Chardin, the "Omega Point," that cultural stage that will occur in the evolutionary process where "the minds of men attain a common language of scientific humanism"[2] was a workable philosophy.[3] Darwin constructed a general ethic dependent on a "Scientific-philosophic" concept of progress.[4] One contemporary scholar has stated that Darwin's concept rests on several premises regarding knowledge. Darwin apparently assumed that the limits of knowledge are infinite, that no single individual can begin to encompass the knowledge that presently exists, that the only effective solution to what may be termed dangerous knowledge is more knowledge, and that knowledge should be disseminated as widely as possible. Indeed, to Darwin, wisdom, the knowledge of how to use knowledge, was by far the most important knowledge of all.[5]

Whether one defines wisdom in Kantian terms, as policy of action that espouses "doing or letting be," or as Darwin did, as the knowledge of how to use knowledge, the principal focus of wisdom is society's competence to act.[6] Moreover, this problem of how society should use knowledge for the social good must be considered in terms of the total volume of information that society can manipulate:[7]

"[M]oral status (our ethical integrity) depends upon two things at least: first, freedom of choice, and, second, knowledge of the facts and of the courses between which we may choose. In the absence of either or both of these things we are, in the forum of conscience, more like puppets than persons. Lacking freedom and knowledge, we are not responsible; we are not moral agents of personal beings. . . .mankind is constantly growing and gaining ground both in knowledge of life and health and in human control over them. This is, indeed, the same as saying that the means to heightened moral stature are available. The appeal of moral idealism is that we take advantage of every opportunity to grow in wisdom and stature, that we assume our responsibility; in short, that we act like human beings."[8]

Human beings ideally will act with rational purpose and design in addressing the ethical problems of biomedical research. Some urge a cessation of all research, observing that we lack total knowledge.[9] Significant dangers do exist in undertaking research and in applying the fruits of that research,[10] and man often chooses the path of ignorance to escape the burdens of responsibility that arise from new knowledge. To end research now, however, will foreclose any opportunity to grow in wisdom and use that wisdom to act with dignity and responsibility. Since man cannot escape responsibility, we should continue research in the New Biology and increase the public debate over the social and legal consequences arising therefrom.[11]

Ethical and scientific factors continuously interact as the scientific process creates new possibilities that influence ethical judgments.[12] The set of values and ordering of commitments to which the scientist ascribes influences not only the research objectives he seeks but also the results he can recognize.[13] Science is descriptive and attempts to resolve the question: What is? Ethics is prescriptive and attempts to resolve the question: What ought to be?[14] Paradoxically, the law is charged with structuring a standard for present behavior and simultaneously remains a step behind science in a reactive capacity.[15] Exclusive reliance should not be placed on legal remedies, however, to resolve the complex ethical problems that biomedical research presents.[16] Indeed, the law should probably not support any one particular scientific ethic, however styled.[17]

Much of the ethical theory surrounding biomedicine attempts to harmonize individual desires with the greater social welfare.[18] Moral dilemmas in biomedicine may be thought of as arising from real or apparent conflicts between perceived obligations to distant generations and to the present generation.[19] In determining whether continued investigations into genetic engineering will jeopardize future life, one should inquire whether an act with uncertain consequences would be harmful to one's own children.[20] Man should not inflict on future generations that which can be disastrous to a present generation.[21]

One scholar has suggested a bioethical creed for individuals. The creed states that "the future survival and development of mankind, both culturally and biologically, is strongly conditioned by man's present activities and plans."[22] The creed encompasses a corresponding commitment to live life and influence the lives of others to promote the evolution of a better world for future generations by avoiding actions that would detrimentally impact the future.[23]

The Metaethical Quagmire

Metaethics examines specifically how normative standards should be structured and what the standards should be for applying genetic rules of research and development of future generations.[24] A uniform core of standards is needed. Individual judgments of scientists, which have proven faulty and inadequate,[25] should be replaced by an ethic that assures collective social responsibility.[26] An a priori ethic, which rests on the faith that certain acts are inherently immoral,[27] does not meet this requirement. A pragmatic ethic, which requires that one make choices that offer a maximum of desirable consequences,[28] does seem to fulfill the goal of collective responsibility. If the results of biomedical research will contribute to human well-being, a practical ethic would sanction the research.[29]

Two types of pragmatic ethics exist within the general category: rule utilitarianism and case utilitarianism. Rule utilitarians stress the need for a weighing of the good that an entire class or category of experiments, such as reproduction in the laboratory, would produce.[30] If they conclude that the research would not provide sufficient benefits, they would disapprove the entire class or category of experiments.[31] Case utilitarians, on the other hand, would weigh the good that each separate case or situation would provide. Under this ethical approach, laboratory reproduction might be proper in certain cases but improper in others.[32] Either type of a practical ethic is consistent with the need to seek a consensus ethic to guide biomedical research that is not aligned with humanism, meta rationalism, or assumptions of faith, but is tied solely to a communion of shared values derived from observable experiences.[33] No condemnation of laboratory reproduction would be made pursuant to a consensus ethic unless either the means or the ends of the research were incompatible with human needs[34] or unless a common consent, achieved through verifiable reasoning, required ending the experiment.[35] One scholar has suggested that, in the final analysis, reason together with imagination can produce a "reasonable guess" and that is about all that can ever be done.[36]

The creation of life and the remaking of man frame the ultimate ethical issues resulting from increased genetic knowledge.[37] Genetic modifications are intermediate expressions of this ultimate capacity, and cloning exemplifies the final consequences. To illustrate the issues that an ethical system must resolve in dealing with biomedical technologies, consider the consequences of surrogate motherhood.38 If donors of sperm have no claim over children born of their sperm through artificial insemination, a donor of an ovum should have no superior rights over the real mother.39 When a physician seeks to implant an ovum into another woman, he should obtain permission from the donor for the transfer or implant. But, what if the donor woman has strong religious or other objections to in vitro fertilization that would have led her to refuse permission if she were told that her ova were to be used for that purpose? Even if the doctor has obtained permission to use a donor's ova for in vitro fertilization, what happens if, after fertilization, an embryo begins to develop abnormally; who should make the decision to discard or to keep a defective embryo: the donor woman, the desiring couple, the geneticist, the obsterician, or all of these individuals together?40 These dilemma may be upon us rather quickly.

The prospect of producing "optimum babies" introduces another issue that bioethics must resolve. Many people might raise objections to the regulation of life beginning in the laboratory rather than in the home. This issue forces consideration of the interests of a new participant-- the scientist. To some, this depersonalization of the procreative process is most undesirable;41 human procreation for them "is a more complete human activity precisely because it engages us bodily and spiritually as well as rationally."42 Acts of laboratory procreation may threaten the sanctity of marriage and the human family.43 A scientific mastery should not, however, drive out a spiritual mystery.

The criticisms of new biomedical research have led some to suggest a professional moratorium on human experimentation with in vitro fertilization and embryo transfer until safety procedures are perfected that will safeguard against "the further dehumanization of man."44 Development of such safeguards could take many forms: studies of the normality of the offspring produced by the technologies of the New Biology among lower order mammals; establishment of intraprofessional organizations to discuss and evaluate work in this area; promotion of interdisciplinary discussion to fully apprise all concerned of the consequences of continued in vitro fertilization and embryo transfer of research in order to minimize any possible negative social consequences; development of an international forum for full exploration of ways in which misdirections in biomedicine may be averted; and, ultimately, the creation of ethical guidelines for the application of the New Biology, drawing on the skills of lawyers, legislators, theologians, philosophers, humanists, social scientists, and laymen.45

Research into the impact of biomedical technologies and consideration of the ethical dilemmas involved does not require a moratorium on human experimentation; the two can continue concurrently. The new HEW regulations regarding human research, which apply to most of the research proposals in the field, provide sufficient guidelines for new research. The successor to the National Commission for the Protection of Human Subjects in Biomedical and Behavioral Research--The President's Commission for the Study of Ethical Problems in Medicine and Biomedical Behavioral Research--hopefully will continue to present valuable insights into this area and provide guidelines for subsequent research. Formalized standards, such as The Nuremberg Code, also encourage scientists to consider ethical factors and supply helpful benchmarks for research. Society should encourage, not stifle, research; for a society unable to accept and encourage either current or future behavioral variations does not promote a hospitable environment for the free development and expression of ideas of any kind.46 Man cannot learn by merely thinking in this area.

Notes--Chapter 8

1. V. POTTER, BIOETHICS: BRIDGE TO THE FUTURE 26 (1971).

See generally, T. DOBZHANSKY, GENETIC DIVERSITY AND HUMAN EQUALITY (1973); R. PAOLETI, SELECTED READINGS: GENETIC ENGINEERING AND BIO-ETHICS (1972).

Ethics is a science of right energizing. F. ADLER, AN ETHICAL PHILOSOPHY OF LIFE, 221 (1919).

See also, 16 NEW CATHOLIC ENCYCLOPEDIA, Supp. 1967-1974 (1974); ENCYCLOPEDIA OF BIOETHICS (W. Reich ed. 1978); CONTEMPORARY ISSUES IN BIOETHICS (T. Beauchamp, L. Walters eds. 1978).

2. V. POTTER, supra at 34.

Chardin may be correctly viewed as having established a form of scientific humanism. P. CHAUCHARD, MAN AND COSMOS 153-156 (1965).

See also, P. de CHARDIN, SCIENCE ET CHRIST (1965); Barbour, "The Significance of Teilhard", in CHANGING MAN: THRUST AND THE PROMISE 130 (K. Haselden, P. Hefner eds. 1968).

See generally, Stent, "The Poverty of Scientism and the Promise of Structuralist Ethics," 6 THE HASTINGS CENTER REPORT 32 (Dec. 1976).

3. V. POTTER, supra note 1 at 31.

4. Id. at 45-46.

5. Id. at 49.

6. Id. at 184-186.

The Supreme Ethical Rule is to act in such a manner so as to elicit the best in others and thus in one's own self; or act in that way which elicits "the sense of unique distinctive selfhood, as interconnected with all other distinctive spiritual beings in the infinite universe." ADLER, supra note 1 at 208.

7. Id. at 186.

8. J. FLETCHER, MORALS AND MEDICINE 35 (1960) (emphasis in original) (footnotes omitted).

See also, J. FLETCHER, SITUATION ETHICS: THE NEW MORALITY (1966); H. ENGELHARDT, D. CALLAHAN, SCIENCE, ETHICS AND MEDICINE (1976); ETHICAL ISSUES IN MODERN MEDICINE (R. Hunt, J. Annas eds. 1977).

See generally, McCormick, "Ambiguity in Moral Choice," in DOING EVIL TO ACHIEVE GOOD, Ch. 1 (R. McCormick, P. Ramsey eds. 1978).

9. See Greene, "Genetic Technology: Law and Policy for the Brave New World," 48 IND. L. J. 559, 576-80 (1973); Kass, "The New Biology: What Price Believing Man's Estate," 174 SCIENCE 779, 786-87 (1971).

10. See Friedman, "Interference with Human Life: Some Jurisprudential Reflections," 70 COLUM. L. REV. 1058, 1076 (1970).

See also, T. SZASZ, THE THEOLOGY OF MEDICINE (1977).

11. Id. at 1077.

See generally, "Symposium--Morals, Medicine and the Law," 31 N.Y.U. L. REV. 1157, 1161 passim (1956).

12. Shinn, "Genetic Decisions: A Case Study in Ethical Methods," 52 SOUNDINGS 229, 308-09 (Fall 1969).

See generally, P. RAMSEY, FABRICATED MAN: THE ETHICS OF GENETIC CONTROL 2--22 (1970); Shinn, "Perilous Progress in Genetics", 41 SOC. RESEARCH 83, 94-103 (1974).

See also, Deutsch, "Scientific and Humanistic Knowledge in the Growth of Civilization", at 121 in EXPERIMENTATION WITH HUMAN BEINGS (J. Katz, ed. 1972); M. VISSCHER, Ch. 2, ETHICAL CONSTRAINTS AND IMPERATIVE IN MEDICAL RESEARCH,(1975).

13. Friedman, supra note 10 at 1077.

14. Fletcher, "Ethical Aspects of Genetic Controls", 285 N. ENG. J. MED. 776 (1971).

See generally, B. DeMARTINO, A SWYHART, BIOETHICAL DECISION MAKING: RELEASING RELIGION FROM THE SPIRITUAL (1975).

15. Burger, "Reflections on Law and Experimental Medicine", 15 U.C.L.A. L. REV. 436, 439 (1968).

David Bazelon, former Chief Judge of the United States Court of Appeals for the District of Columbia Circuit, asserts that the new laws required to accommodate biological advances should not race ahead of community attitudes and understanding. The public must be informed of the problems and of the nature of the new enterprises. The Chief Judge observes that no unique demands upon the law are threatened as a consequence of either present or potential medical technology. "They will require the same sort of adjustments between conflicting values and the same sort of interweaving of moral and scientific data that the law has always taken as its province. Courts cannot alone cope with the tasks. But other lawmaking institutions have each their special skills. If we define the problems posed, confront them before events have imposed their own haphazard solutions, and call upon the right sort of institution to resolve them, the law can insure that medical progress does not carry with it the perils so easily imagined and so rightly feared. In responding to earlier upheavals, society and its legal institutions have not responded quickly enough, or wisely enough, to reap the promise without an accompanying harvest of misery. The lessons learned there, however, and the sophistication in social control acquired in those struggles, may have prepared us for the confrontation with medical technology. If they have not, we all must bear the shame." Bazelon, "Medical Progress and the Legal Process," 32 PHAROS 34, 57, 58 April, 1969).

See also, Canavan," Law and Society's Conscience," 2 HUMAN LIFE REV. 1 (1976); Stent, "The Poverty of Scientism and the Promise of Structuralist Ethics," 6 THE HASTINGS CENTER REPORT 32 (Dec. 1976).

One writer has stated that, "The promise of law is a stable and ordered society; the thrust of science is rampant social upheaval. The two are on a collision course." Miller, "Science vs. Law: Some Legal Problems Raised by Big Science", 17 BUFF. L. REV. 591, 629 (1968). See also, R. VEATCH, VALUE-FREEDOM IN SCIENCE AND TECHNOLOGY (1976).

See generally, Smith, "The Medico-Legal Challenge of Preparing for a Brave Yet Somewhat Frightening New World", 5 J. LEGAL MED. 9 (1977).

16. Grad, "New Beginnings in Life -- A Lawyer's Response," in THE NEW GENETICS AND THE FUTURE OF MAN 77 (M. Hamilton ed. 1972).

17. Id.

The purpose of law should be to maintain the more developed members of society at the level they have reached, and, by educative penalities, bring the backward up to the same level. Law is, thus, designed so that "mankind shall be at liberty to discover new and more significant conditions which in turn are again to become automatic." Adler, supra note 1 at 307.

18. See generally A. CAMPBELL, MORAL DILEMMAS IN MEDICINE 1-14 (1972); Leake, "Changing Concepts in Medical Morals", 37 CONN. MED. 139, 140 (1973).

The formal law of morals is to "Act absolutely in conformity with your conviction of your duty." It is for the individual to find his determined duty. There appears to be a feeling of truth and certainty and this feeling is the sought for absolute criterion of the correctness of one's conviction of duty. J. FICHTE, THE SCIENCE OF ETHICS 173, 177 (1907). See also, S. REISER, A. DYCH, W. CURRAN, ETHICS IN MEDICINE: HISTORICAL PERSPECTIVE AND CONTEMPORARY CONCERNS (1977).

The central problem of ethics--and more especially medical ethics-- has been to find a rule or principle of conduct. The rightness or the wrongness of human action has concerned historians, ethicists and most religions since history was first recorded. Perhaps the central most weakness in today's debate concerning medical ethics is the very lack of a basic yardstick against which the rightness or wrongness of the physician's actions can be measured. "There is no general agreement among physicians and ethicians as to what should be the ethical determinant of what doctors should or should not do in a particular situation." Gunn, "On Medical Ethics," 4 THE HUMAN LIFE REV. 81 (1978).

See READINGS IN BIOETHICS, Parts 4, 6 & 7 (T. Shannon, ed. 1976).

See generally, T. BEAUCHAMP, J. CHILDRESS, PRINCIPLES OF BIOMEDICAL ETHICS (1979); A. BAHM, ETHICS AS A BEHAVORIAL SCIENCE, Ch. 1 (1974).

19. Golding & Callahan, "What Obligations Do We Have to Future Generations?", 164 AM. ECCLESIASTICAL REV. 265, 275 (1971).

20. Id. at 279.

21. Id. at 279-90.

22. V. POTTER, supra note 1, at 196.

The remaining parts of the Creed expres the following beliefs and commitments: a belief in the need for prompt remedial action in a world beset with crises and a commitment to work with others to improve the formulation of such beliefs and unite in a worldwide movement to make possible both the survival and improved development of the human species in harmony with the natural environment; a belief in the uniqueness of each individual and his need to contribute to the betterment of society and a commitment to listen to reasoned viewpoints from either the minority or the majority; a belief in the inevitablity of human suffering resulting from natural disorders in the biological and physical world but not an acceptance of the suffering resulting from man's inhumanity to man and a commitment to face personal problems in a dignified and courageous way and strive for the elimination of needless suffering among mankind as a whole; and, finally, a belief in death's finality adn a re-affirmation of life's values, man's brotherhood, and an obligation to future generations of men and a commitment to live in such a way which will provide benefits for the lives of men presently and those to come and, hence, be remembered in a favorable way by those who survive after death. Id.

23. Id. One author maintains, however, that one cannot argue, solely on the basis of reason, that the survival of mankind should be promoted; if religious convictions are set aside, it cannot be shown that the indefinite continuation of human species is desirable. Heilbroner, "What Has Posterity Ever Done For Me?", N. Y. Times Mag., Jan. 19, 1975, at 14. But see, J. RAWLS, A THEORY OF JUSTICE 284-293 (1971).

24. Callahan, "Normative Ethics and Public Morality in the Life Sciences," 32 THE HUMANIST 6 (Sept.-Oct. 1972); See K. VAUX, BIOMEDICAL ETHICS 51-68 (1974). Metaethical judgments are statements about possible practical judgments, one step removed from actual situations. Such judgments define and appraise the standards, rules, and principles that are sought to justify practical decisions. R. ABELSON, ETHICS AND METAETHICS 4 (1963).

25. Callahan, supra, at 6.

26. See Brody, "Biomedical Innovation, Values and Anthropological Research," 158 J. NERVOUS & MEN. DISORDERS 85, 86-87 (1974).

27. Fletcher, supra note 14, at 777-78; Fletcher, "New Beginnings in Life" in THE NEW GENETICS AND THE FUTURE OF MAN 81 (M. Hamilton ed. 1972).

28. Fletcher, supra note 14 at 778.

29. Fletcher, "New Beginnings in Life", supra note 27, at 86-87.

30. Id. at 82. Utilitarians assert that the right act in any given circumstances is the one act, out of all the possible acts, that, based on available data, will more than likely produce the greatest good. This process reduces ethics to defining good and bad not as means, but as ends. One therefore judges whether any individual action is right by whether the action leads to a happier life for the particular individual. M. MUNITZ, A MODERN INTRODUCTION TO ETHICS 128 (1958). See also A. EDEL, ETHICAL JUDGMENT 132 (1955); S. ZINK, THE CONCEPTS OF ETHICS 93-94 (1962).

31. Fletcher, "New Beginnings in Life", supra note 27, at 86-87.

32. Id. at 82-83.

Fletcher suggests two alternative designations: situation ethics and rule ethics. Situation Ethics state that variables in any case determine what ought to be undertaken. Consequently, what is right sometimes is wrong other times. Rule Ethics maintain some things are inherently wrong -- whatever the mitigating circumstances might be. Fletcher argues that a mature ethics is needed in dealing with the new genetics -- one which is social, not egocentric. He opines that only the morally superficial reject efforts to build better quality fetuses by rejecting low quality ones. He suggests that this whole melodrama is not set with tragic overtones -- "sad, but not agonizing." J. FLETCHER, THE ETHICS OF GENETIC CONTROL 30, 159 (1974).

The dominant current of secular Anglo American ethical philosophy may be regarded as a mixture of utilitarianism and Kantian humanism. Fried, "The Need for Philosophical Anthropology", 48 IND. L. J. 527, 530 (1973).

33. Fletcher, "New Beginnings in Life", supra note 27 at 88-89.

34. Id. at 89.

35. Id. See also Gustafson, "Basic Ethical Issues in the Bio Medical Fields", 53 SOUNDINGS 151, 177 (1970) (maximum possible freedom for

research and self-development should be sought, but values must be enunciated as guidelines for future research).

36. Callahan, "A Philsopher's Response", in THE NEW GENETICS AND THE FUTURE OF MAN 92 (M. Hamilton ed. 1972).

"Today's knowledge is often tomorrow's error." For prophets, then, their drive is found in optimism, hope or great expectations. J. FLETCHER, THE ETHICS OF GENETIC CONTROL 21 (1974).

37. K. VAUX, BIOMEDICAL ETHICS 51-52 (1974).

38. See Kass, "New Beginnings in Life," in THE NEW GENETICS AND THE FUTURE OF MAN 35-38 (M. Hamilton ed. 1972).

See generally, MORALITY AS A BIOLOGICAL PHENOMENON (G. Stent ed. 1978).

39. Id. at 37-38.

40. Id. at 34-35.

41. Id. at 53-54.

42. Id. at 53.

43. Id. at 54.

44. Kass, "Babies by Means of In Vitro Fertilization: Unethical Experiments of the Unborn?", 285 N. ENG. J. MED. 1174, 1178 (1971). An international conference of biologists has structured a set of rules to govern research and experimentation in genetic engineering. Although conceding that there was considerable ignorance concerning the multiple facets of the revolution in molecular biology, scientists from 16 nations recommended that research continue. The conference called for safer biological tools for gene manipulation, however, determining that these tools needed to be tested and thoroughly developed under laboratory conditions before they are used in human experimentation. Two Nobel Prize biologists, Dr. Joshua Lederberg and Dr. James D. Watson dissented from the conference's conclusions on the ground that any "safeguard" is virtually unenforceable because of the difficulty in determining the nature of the exact risk undertaken in specific experiments. McElheny, "World Biologists Tighten Rules on 'Genetic Engineering' Work," N. Y. Times, Feb. 28, 1975, at 1, col. 4.

Dr. Franz J. Ingelfinger, Editor of the New England Journal of Medicine, has observed that existing and proposed new regulations of fetal research may hinder medical progress. "The pendulum swings too far when ethical principles are used to denigrate scientific inquiry and creativity." Altman, "A Medical Editor Discusses Ethics", N. Y. Times, Apr. 1975, at 15, col. 1. He encouraged medical researchers and ethicists to work together more closely with one another in hospitals, laboratories, and classrooms to minimize unethical experimentation on humans and hopefully narrow the large gap in understanding between the purists in medicine and philosophy. Id.

See generally, Ramsey, "Incommensurability and Indetermininancy in Moral Choice" in DOING EVIL TO ACHIEVE GOOD, Ch. 3, (R. McCormick, P. Ramsey eds. 1978).

CHAPTER 9

SCIENCE AND RELIGION: COMPATIBILITIES AND CONFLICTS

Science has been defined as, ". . . intelligence in action with no holds barred."[1] It began as but a simple pursuit of truth but is today fast becoming incompatible with veracity quite simply because a complete veracity leads to a form of complete scientific skepticism.[2] Science was originally recognized, and indeed valued, as a method to know and/or understand the world.[3] Ever since the time of the Arabs, in fact, science has had but two simple functions; to enable us to know and learn about things and to thereby assist us in doing things.[4] Now, as a consequence of the development of scientific method and the triumph of technique, science is viewed as a means of changing the world.[5]

Probabilities are at the center of scientific inquiry. As such, an absolute form of truth is not within its scope of realization. Yet science can yield--in the final analysis--such a high degree of probability to, for all practical purposes, become certainty.[6]

Science is a way of ordering experience. It is ordered knowledge. Its constant testing and referral to the facts of past experience should be viewed as the only valid way which enables man to progressively increase both his knowledge and control of the objective world.[7] This constant reference back to experience in the quest for knowledge is the most significant attribute of the scientific method; for from it comes, "the cosmic side of that intellectual scaffolding of religion we call theology."[8] Yet, there has been a prolonged conflict between religion and science.[9] Perhaps one of the basic reasons for the built-in conflict has been the difference in focus of religious creed and scientific theory. A creed is said to embody both eternal and absolute truth. Scientific theory is always recognized as tentative--with modifications sooner or later found necessary. The scientific method, then, unlike the religious creed, is one which is logically incapable of arriving at an ultimate statement.[10]

Religion, to a considerable extent, consists in a way of feeling sometimes more than in a set of beliefs.[11] The beliefs are secondary or supportive of these feelings. There are some things people believe, then, because they feel as though they are true;[12] and such feelings and beliefs are a source of mystery and incomprehensibility to the scientific mind. Faith is an unknown and rather primitive principle to the scientist.[13]

Religion must--from the standpoint of maintaining its strength, efficiency or power--face change in the same spirit as science does. While its principles may be immutable and eternal, the expression of those principles requires a continual development.[14]

The Roman Catholic religion is a perfect example of a religion which has charted--predominantly under the leadership of the late Pope John XXIII--a new course of contemporary expression particularly in its liturgy. Certain dogma such as the Virgin Birth by Mary, papal infallability, priestly celibacy, the exclusivity of the male priesthood and the sanctity of creation remain inviolate and not subject to modernist revision today. It is on this last point that has been seen --and will be explored further--that has presented problems to the scientific community as it explores eugenic proposals and fetal experimentation.

During the middle and latter half of the 19th century, science made its greatest inroads into religion; for at that time a type of credibility gap was beginning to open between what could be explained within the framework of religion and, contrariwise, what could be explained within the scientific frame of analysis. Some view this gap as continuing to widen simply because the more scientific discoveries about the universe that are made, the less explicable they become. Some thirty years ago it was generally believed that gradually science was attempting quite successfully to explain the entire universe. The more scientific facts presented for understanding the more knowledge of the universe would emerge. Today, however, there is a concern because rationalists and humanists are suggesting that within the foreseeable future science will not be able to say anything fundamental about the true nature of the universe.[15]

The advancement of science is often blamed for loss of religious faith.[16] There is, on the other hand, a belief that the greater understanding of religious truths today has been caused by a work of science more than any other factor.[17]

The overriding fact to be observed is that normally a scientific advance will show that statements of various religious beliefs, if they have contact with or are tied to physical facts, require some sort of modification: either expanded, re-interpreted or entirely restated. If the particular religion in focus as a consequence of the scientific advance is grounded in sound expressions of truth, the required modification will only "exhibit more adequately the exact point which is of importance."[18] A contradiction, in formal logic, is the signal of a defeat. In the evolution of real knowledge, it marks but the first step in progress toward a victory; and this--then--is the principal reason why variety of opinion is tolerated and even encouraged.[19]

What equivocal attitudes that have developed among Christians regarding their religious faith runs much deeper, however, than what has been stated thus far. They are compounded of suspicion, ignorance and misunderstanding: suspicion directed against advancing technology which appears to have a considerable power for good or evidence depending on the technologist who directs it, ignorance from not knowing sufficiently the true nature of science and technology and--finally--misunderstanding of the Christian doctrine of creation which has in turn led to false ideas about materialism.[20]

Viewed modernly, between the statement of theological principle and the scientific method of inquiry of investigation, there is really no real conflict because there is no interrelationship or mutual dependence.[21] Based on revelation and faith, theology presents its concepts and principles totally independent of the scientific theories about nature or of speculations regarding the past.[22] Both science and religion present different phases of human activity and embody distinctive experiences. While religion is fundamentally a spiritual experience, science is based on "sensuous experience."[23] Yet, science and religion are one in the experience of revelation they offer, however, to those who pursue them--the revelation of a supreme fact of mental or progressive spirit and experience.[24]

The scientist as well as the theologian depend, in the final analysis, upon experience and interpretation--and both use their minds. They ask different types of questions and neither expect to receive in return the same types of answer. Science and religion are but reflection of different aspects of man's social experiences. If one can move beyond popular misconceptions regarding the nature and role of both science and religion, he will feel no conflict between their methods of study and practice.[25]

Religion should be devoted to the expression and fulfillment of final values--beyond which no other values can exist.[26] A scientific approach to religion then becomes but a noble effort to study the true story of man--the relation to the source of his being and his duties, privileges and structure of values. Science--pursued within this construct--provides the basic framework for a new dynamic testament: indeed, a new scripture of truth about man and his destiny.[27]

If the administration of science is to be advanced for the betterment of mankind, not only are moral ideals needed, but a spiritual vision as well. The most notable scientific work has flowed consistent with a high conception of social duty and with a spirit of altruism. Science is but a means to an end, with its values being determined by the end.[28]

Societal progress, in the ultimate analysis, must embrace two complementary plans of development--plans which both embrace scientific research as well as increased moral understanding and appreciation.[29]

Theological Considerations

The Catholic View

Roman Catholic dogma teaches that marriage does not bring to the married couple an absolute right to children--only a conditional right. All that may be done is for the couple to avail themselves of the use of legitimate medical processes in order to assure their sexual act be performed in a natural way to attain "its fertile union."[30]

The Church, thus, stresses the fact that coition be recognized solely as an act designed for procreation--between husband and wife only--and that the act, itself, be unimpeded by direct means. "Human sexual congress in order to be authentic, must involve intravaginal ejaculation by the husband and retention of the semen (or at least no deliberate effort at expulsion) by the wife."[31]

The exclusivity of the marriage contract forbids intercourse with a third person and/or the use of semen from a donor to effect artificial congress. Adultery, then, is considered adultery by the Church irrespective of the fact that a husband may consent to his wife's indulging in sexual "relations" with another man through artificial processes (AID).[32] The major point of emphasis is the invasion by a third party into an exclusive marriage contract. The unity of love and procreation must remain inviolate.

The normal way of obtaining semen is through masturbation. This very act is considered to be a "perversion of the sexual faculty;" it is non procreative.[33] Now, if semen were collected from a wife's husband (AIH) in a manner other than through auto-erotic techniques,[34] and then injected into the wife's reproductive tract, it has been submitted that Church teaching would allow this act as valid--since love and procreation are not really separated but, indeed, furthered by the act.[35] It is a physical disability which forces the husband to resort to AIH in the first instance. It is love which induces him to seek artificial means of impregnating his wife. The unity of love and procreation is, thus, strengthened.[36]

In contradistinction, fertilization by donor gametes in vivo or in vitro would be automatically rejected by the Church because--although no adulterous relation was evidenced--two different communities would be created: one procreative and the other loving. Although perhaps anonymous, the donor becomes a silent partner in an exclusive relationship which admits of no intruders.[37] Yet, the technological manipulation of a husband and wife's own gametes would appear to be inoffensive to and compatible with the principle of loving and procreation--since, there, the basic marital relationship remains intact.[38] Use of a woman's womb by another couple would be considered by the Church as analogous to allowing use of one's body solely for the sexual pleasure of another, and, thus immoral.[39]

The Roman Catholic view regarding sterilization--voluntary or involuntary--may be stated succinctly: The state exists for individuals--individuals do not exist for the state. Accordingly, the state should not compromise or invade the essential procreative nature of any human being unless it can be shown conclusively that it has the right to mutilate a citizen's body and, secondly, other punishment (in the case of ciminals) and/or restraint (in the case of mental defectives) cannot be imposed with equal if not greater success and effect the desired result than an act of sterilization.[40]

The Protestant View

The conservative Protestant Ethic maintains that some acts are specifically commanded in the Bible and must be followed by all. The literalist approach to the Bible has serious weaknesses as a basis for religious ethics primarily because often the moral precepts found within the Bible are both unclear and contradictory.[41] Under conservative protestantism, a monogamous marriage is--thus--the biblical expression of God's unalterable will. The only alternative to marriage is abstention from sexual intercourse.[42] The only inferences which may be drawn from this philosophy is that AID is morally objectionable as an invasion of a monogamous marriage unity and that genetic engineering qualifies as an offensive sexual relation.[43]

The liberal and more contemporary Protestant view is that since none of the biblical commandments--in their specific content--are unambiguous and, thus, clear expressions of God's will, there are no universal modes of conduct required of Christians.[44] In defining relationships between persons, the crucial determinant is whether love is present or absent. Therefore, the validity of one's actions, sanctified and legalized by a marriage contract, is of secondary importance. What is of central importance is whether coition is a truthful expression of a personal commitment to one another. Is it honest and carried out in such a manner so as not to exploit the other person?[45] So long as mutuality of love is expressed, then almost any procedure within the ambit of a practice of the New Biology would be tolerated.

Whether AID is considered adulterous is really but a question of semantics.[46] AID involves a far more responsible level of decision-making than the "normal" one-night stand act of adultery or the clandestine relationship. No infraction of the marriage vows is promoted by a consensual decision regarding the use of AID by a married couple. Moreover, when a husband allows his wife to be impregnated by a donor, it is this very consent and desire for offspring which assures that the subsequent child, itself, is of primary concern. There can be no allegation of broken faith in such a situation. In an adulterous relationship, the very essence of that relationship is ground in broken faith by one partner in the marriage contract.[47] In adultery, should a "careless mistake" be made and issue result as a consequence of the exultation of physical and emotional needs outside the bounds of the marriage, that "mistake" is the subject--usually in most cases--of concern, despair and non acceptance instead of love and acceptance as in a consensual act of AID.

According to the traditional Protestant ethic which is tied to respect and acceptance of the precedental value of law, compulsory sterilization may be justified and, thus, accepted without difficulty.[48]

## The Jewish View

Under Jewish law, a woman who participates in AID is not guilty of adultery. The child born of the artificial act is regarded as legitimate--regardless of whether its mother is married or single.[49] Only when it is established conclusively that a child has been born of an adulterous or incestuous relationship is the child regarded as illegitimate.[50] There is a strong presumption against adultery or incest. In fact, it is virtually impossible to prove any conception was adulterous or incestuous since the husband is always presumed to be the father of his wife's children.[51]

Interestingly, the donor in an act of heterologous insemination--although in no way stigmatized by his act--remains the natural father of the child and can never rid himself totally of this relationship. Yet, he may be relieved of liability for support of the issue and his estate removed from claims of inheritance by children whom he normally would never know of or see.[52] This strict rule of civil liability obviously does not preclude the development of a foster parent relationahip in addition to the natural relationship.[53]

The general position of Judaism on sterilization is unambiguous: it is prohibited.[54] Yet, sterilization designed to save life would be permissible.[55] Uncertainty surrounds the application of sterilization by the state to incompetents and criminals. Orthodox teaching would probably disallow such state action.[56]

What is clear and unambiguous in the theology of the New Biology is that belief in God and the perceptions of the divine will are not shared uniformly. Christians, Jews, Buddhists, Moslems and Hindus disagree within their own religious ranks regarding the use and application of the New Biology. Because of this variance within religious groups, perhaps it is better and wiser to only submit opinions about specific moral applications of genetic eningeering rather than zero in on positive condemnations of one development and its use as opposed to another.[57]

Religions should be careful not to be too terribly dogmatic in the area of the New Biology. A religious ethic rooted in human well-being should survive the pressures of the emerging, brave new world. Churches and religious teaching will either be molded to reflect the new ethics of the age or will die out.

> ". . . religion in the age of science cannot be sustained by the assumption of miraculous events abrogating the order of nature. Instead, we should see acts of God in events the natural causes of which we fully understand."[58]

Notes--Chapter 9

1. Hoagland, "Some Reflections on Science and Religion" (quoting the physicist P. W. Bridgman) in SCIENCE PONDERS RELIGION 17 at 18 (H. Shapley ed. 1966).

2. B. RUSSELL, THE SCIENTIFIC OUTLOOK 273 (1931).

3. B. RUSSELL, THE IMPACT OF SCIENCE ON SOCIETY 98 (1952).

4. Id. at 29.

The Greeks--with Archimedes being the exception--were interested only in the first function. The Arabs, however, were in quest of the elixir of life and the methods needed to transmute base metals into gold. Id.

5. Id. at 98.

During the past three centuries, the science which has been rated as successful has consisted "in a progressive mathematisation of the sensible order. . . ." Id.

The history of science reveals that it is based on creative leaps of imaginative vision. L. GILKEY, RELIGION AND THE SCIENTIFIC FUTURE 45 (1970).

See, S. MATTHEWS, CONTRIBUTIONS OF SCIENCE TO RELIGION (1924); J. MARITAIN, SCIENCE AND WISDOM (1940); H. MULLER, SCIENCE AND CRITICISM (1943); SCIENCE FOR A NEW WORLD (J. Crowther ed. 1934).

See also, H. HOVENKAMP, SCIENCE AND RELIGION IN AMERICA 1800-1860 (1978); F. TURNER, BETWEEN SCIENCE AND RELIGION: THE REACTION TO SCIENTIFIC NATURALISM IN LATE VICTORIAN ENGLAND (1974); Edsall, "Scientific Freedom and Responsibility," 188 SCIENCE 687 (1975).

6. Hoagland, "Some Reflections on Science and Religion," in SCIENCE PONDERS RELIGION 17 at 24 (H. Shapley ed. 1960). The examples used for support of this last statement are: the certainty that the earth is round, not flat and the realization that biological evolution--by natural selection--is no longer but a theory but is a high probability. Id.

In its fundamental phase, science is explanation by description using methods of observation and experiment. The fundamental assumptions which it makes are practical conclusions of common sense: namely, that the objects and the events constituting the material universe are in a necessary connection with one another and that man, by his decisions, can affect the order and events of the universe itself. W. SCHROEDER, SCIENCE, PHILOSOPHY AND RELIGION 44, 45, 58 (1933).

7. J. HUXLEY, SCIENCE, RELIGION AND HUMAN NATURE 20, 21 (1930).

8. Id. at 58, 21.

9. B. RUSSELL, RELIGION AND SCIENCE 3 (1935).

See Barbour, MYTHS, MODELS AND PARADIGMS--THE NATURE OF SCIENTIFIC AND RELIGIOUS LANGUAGE (1974); THE ENCOUNTER BETWEEN CHRISTIANITY AND SCIENCE (R. Bube ed. 1968); J. DRAPER, HISTORY OF THE CONFLICT BETWEEN RELIGION AND SCIENCE (1903).

10. B. RUSSELL, RELIGION AND SCIENCE 11 (1935).

Russell lists the fact that the historical religions have had a Church and a code of personal morals as a reason for further conflict. Id. at 4.

See generally, S. JAKI, THE ROAD OF SCIENCE AND THE WAYS OF GOD (1978).

11. Id. at 14.

See B. DeMARTINO, A. SWYHART, BIOETHICAL DECISION MAKING: RELEASING RELIGION FROM THE SPIRITUAL, Ch. 8 (1975).

See generally, Gustafson, "Theology Confronts Technology and The Life Sciences," COMMONWEAL 386 (June 16, 1978).

12. B. RUSSELL, THE IMPACT OF SCIENCE ON SOCIETY 16 (1952).

13. While religion seeks to explain the obvious in terms of mystery, science masters the simple and obvious and then witnesses, by the application of elemental principles, the dissolution of the complex. F. NORTHRUP, SCIENCE AND FIRST PRINCIPLES (1931).

See also, A. WHITEHEAD, SCIENCE AND THE MODERN WORLD, Ch. 13 (1926); J. MORTON, MAN, SCIENCE AND GOD (1972).

14. A. WHITEHEAD, THE INTERPRETATION OF SCIENCE 179 (A. Johnson, ed. 1961).

See also, L. GILKEY, RELIGION AND THE SCIENTIFIC FUTURE, Ch. 1 (1970).

See Briggs, "Theologian Weigh Risks to Scientists," N. Y. Times, July 15, 1979, at 19, col. 1, where a Conference on Science and Religion of the World Council of Churches recently found that while major conflicts have been largely overcome between science and religion, problems concerning evolution still exist.

See generally, Dobzhansky, "Evolution: Implications for Religion" in CHANGING MAN: THE THREAT AND THE PROMISE at 142 (K. Haselden, P. Hefner eds. 1968); P. QUINN, DIVINE COMMANDS AND MORAL REQUIREMENTS (1978).

15. Evans, "Rationalization, Superstition and Science" in SCIENCE, REASON AND RELIGION 43 at 45 (C. Macy ed. 1974).

See generally A. VanMELSEN, SCIENCE AND TECHNOLOGY (1961).

16. Hoagland, "Some Reflections on Science and Religion," in SCIENCE PONDERS RELIGION 17 (H. Shapley ed. 1960).

17. L. GILKEY, RELIGION AND THE SCIENTIFIC FUTURE 4 (1970).

18. WHITEHEAD, supra, note 14.

19. Id. at 176.

See also, GILKEY, supra, note 14.

20. C. COULSON, SCIENCE, TECHNOLOGY AND THE CRHISTIAN 48 (1960).

See generally, G. McLEAN, PHILOSOPHY IN A TECHNOLOGICAL CULTURE (1964).

21. GILKEY, supra, note 14 at 25.

Nonetheless, it is not so much the content of specific scientific theories but their methods of validation which trouble the thoughtful person with a religion perspective today. Barbour, "The Methods of Science and Religion," in SCIENCE PONDERS RELIGION 196 (H. Shapley ed. 1960).

See also, W. HOCKING, SCIENCE AND THE IDEA OF GOD 3 (1944).

22. GILKEY, supra, note 14 at 25.

23. W. SCHROEDER, SCIENCE, PHILOSOPHY AND RELIGION 61 (1933).

24. Id., at 62, 63.

25. Barbour, supra, note 21 at 214, 215.

26. Hocking, supra, note 21 at 5, 8; Murray, "Two Versions of Man," in SCIENCE PONDERS RELIGION 147 at 148 (H. Shapley, ed. 1960).

See generally, D. EDWARDS, RELIGION AND CHANGE (1969); J. SMURL, RELIGIOUS ETHICS (1972).

27. Burhoe, "Salvation in the Twentieth Century," in SCIENCE PONDERS RELIGION 65 at 77, 78 (H. Shapley, ed. 1960).

See also, T. FOWLER, THE RECONCILIATION OF RELIGION AND SCIENCE (1873); H. SCHILLING, THE NEW CONSCIOUSNESS IN SCIENCE AND RELIGION (1973).

28. Supra, note 23 at 60.

29. C. MILLER, A SCIENTIST'S APPROACH TO RELIGION 29 (1947).

30. Hasset, "Freedom and Order Before God: A Catholic View," 31 N.Y.U. REV. 1170, 1179 (1956).

31. H. SMITH, ETHICS AND THE NEW MEDICINE 64 (1970).

According to St. Augustine, a sexual act deprived of its procreative character was illegitimate. Thus--if, in the name of love--a couple choose to express themselves sexually, they should accordingly, perform the authentic sexual act not deprived of its procreative character. Love and procreation are inseparable. Smith, "Theological Reflections and the New Biology," 48 IND. L. J. 605, 619, 621 (1973).

See also, St. John-Stevas, "A Roman Catholic View of Population Control," 25 LAW & CONTEMP. PROB. 445, 446 passim (1960).

32. Supra, note 30.

33. Id., at 1180.

The concern of the Church, and more especially that of Pope Pius XII as enunciated in his address to the Fourth International Convention of Catholic Physicians, October, 1949, that the artificial means of obtaining semen was repulsive to the state of marriage and immoral is--today--no longer viewed by most moralists as a valid obstacle to AIH or homologous insemination. Donor insemination, or AID, is still regarded--however--as violative of Church dogma. HUMAN SEXUALITY--NEW DIRECTIONS IN AMERICAN CATHOLIC THOUGHT 137-139 (1977).

Central to any perception of the "problem" of artificial insemination is whether such acts destroy the idea of a family or merely enlarge and enrich the family role. Most Protestants and Jews favor the use of laboratory fertilization, while many traditional Catholics continue to condemn it because it is not viewed as a safe enough procedure, is a form of genetic manipulation and would produce troublesome "discards" when the process would fail from time to time. Hyer, "Theologians React Cautiously to Test-Tube Baby Process," Wash. Post, July 28, 1978, at A3, col. 1.

See generally, Smith, "A Close Encounter of The First Kind: Artificial Insemination and An Enlightened Judiciary," 17 J. FAM. L. 41 (1979) D. MARTIN, THE DILEMMAS OF CONTEMPORARY RELIGION (1978); K. HASELDEN, P. HEFNER, CHANGING MAN: THE THREAT AND THE PROMISE (1968).

34. See note 2, Chapter 5, supra.

35. Smith, "Theological Reflections," supra note 31 at 620.

36. One Catholic writer's position re artificial insemination has been interpreted as being that faith in God is final in the sense it over-rides any agreements for artificial insemination based on humanistic moral standards; that the Church's teaching has God's backing and God is final and that's that. J. FLETCHER, THE ETHICS OF GENETIC CONTROL 114 (1974).

37. Smith, supra, note 35 at 621.

38. Id., at 622.

39. Id., at 621.

40. Hassett, supra, note 30 at 1184. See also, Ramsey, "Freedom and Respectability in Medical and Sex Ethics: A Protestant View," 31 N.Y.U. L. REV. 1189, 1199 (1956); Note, "Elective Sterilization," 113 U. PA. L. REV. 415, 423-24 (1965); "Vatican Still Bars Birth Control Vasectomies," N.Y. Times, Aug. 8, 1977, at C3, col. 4.

Recently, the Catholic Bishops of the United States came out against contraceptive sterilization. If natural family planning methods do not work, then abstinence should be practiced and renewed emphasis placed on one's spiritual life. Hyer, "U.S. Bishops Bar Tubal Ligations--Catholics Tighten Sterilization Ban," Wash. Post, July 10, 1980, at 1, col. 2.

See generally, J. BOYLE, THE STERILIZATION CONTROVERSY: A NEW CRISIS FOR THE CATHOLIC HOSPITAL? (1977).

41. H. L. SMITH, ETHICS AND THE NEW MEDICINE 66, 67 (1970); Smith, "Theological Reflections and the New Biology," 48 IND. L. J. 605, 608 (1973).

42. H. L. SMITH, supra, at 67.

43. Id.

44. Id., at 68.

45. Id., at 69.

For a discussion of the prominent Protestant theologian, Helmut Thielicke's views of AID, see 70 passim in H. SMITH'S book, Id.

46. Ramsey, "Freedom and Responsibility in Medical and Sex Ethics: A Protestant View," 31 N.Y.U. L. REV. 1189, 1198 (1956).

47. For an argument regarding the compatibility of AID with the

Christian understanding of secularity, marriage and parenthood, see J. FLETCHER, MORALS & MEDICINE 118 (1960).

48. Supra, note 17 at 1199.

See also, Note, "Elective Sterilization," 113 U. PA. L. REV. 415, 423 (1965). See generally, Fagley, "A Protestant View of Population Control," 25 J. LAW & CONTEMP. PROBS. 470 (1960).

49. Rackman, "Morality in Medico-Legal Problems: A Jewish View," 31 N.Y.U. L. REV. 1203, 1208 (1956).

Although Jewish ethics would favor experiments and tests in order to discern possible genetic malfunctions which would result in congenital disease before the birth of a fetus, the artificiality of test tube babies and of cloning--for example--would be disregarded as tampering too much with the basic structures of Creation. Siegel, "The Ethical Dilemmas of Modern Medicine: A Jewish Approach," 3 THE KENNEDY INSTITUTE QUARTERLY REPORT 5 at 7 (No. 1, 1976-77).

50. Rackman, supra at 1210.

51. Id.

In 1958, the Chief Rabbi of Israel, Rabbi Nissim ruled that children born to parents as the result of artificial insemination will be recognized by the Jewish religion as legitimate. A. SCHEINFELD, YOUR HEREDITY AND ENVIRONMENT 665 (1965).

52. Rackman, supra note 49 at 1209.

53. Id. at 1209-10.

54. Id. at 1211.

55. Id. at 1212.

56. Id.

See generally, CHALLENGE--TORAH VIEWS ON SCIENCE AND ITS PROBLEMS (A. Carmell, C. Domb eds. 1976); C. LANCZOS, JUDAISM AND SCIENCE (1970); JUDAISM AND ETHICS (D. Silver ed. 1970).

57. J. FLETCHER, THE ETHICS OF GENETIC CONTROL 114, 115 (1974).

See generally, D. GOSLING, SCIENCE AND RELIGION IN INDIA (1976); R. WETTIMUNY, BUDDHISM AND ITS RELATION TO RELIGIOUS SCIENCE (1962). For an interesting perspective on atheist realism and marxist dialectics regarding the New Biology, see P. CHAUCHARD, SCIENCE AND RELIGION, Ch. 3 (1962).

It is far beyond the scope or purpose of this chapter to probe the various beliefs and reactions of the mainstream Protestant groups which would include: Evangelical Lutheran; Southern Baptists; Mormon; Lutheran-Church, Missouri Synod; Episcopal Church; United Presbyterian Church; United Church of Christ; United Methodist, and the Lutheran Church in America. While the most authoritative sources for discovery would be the various church statements published concerning individual issues of the New Biology, perhaps the best overview could be obtained from 3 ENCYCLOPEDIA OF BIOETHICS 901-1020, 1365-1378 (W. Reich ed. 1978). Other sources would include: D. GAINES, BELIEFS OF BAPTISTS (1952); A BAPTIST BIBLIOGRAPHY (E. Starr ed. 1947); NEW CATHOLIC ENCYCLOPEDIA (1967); ENCYCLOPEDIA OF RELIGION AND ETHICS (J. Hastings ed. 1927); THEOLOGICAL DICTIONARY OF THE OLD TESTAMENT (G. Botterweck,

H. Ringgsen eds. 1978); THE INTERPRETERS DICTIONARY OF THE BIBLE (1962).

Since marked development and, indeed, progress has been recorded as to what has been called the "Genetic Revolution" only within the last five or so years, many of the religious denominations have yet to evolve a statement of position here. It would be practical, for example, to conceive of the Southern Baptists taking a different point of view from their brothers on the East Coast. In such cases, one would perhaps try to reconcile the two views by considering the fundamental tenets of the Baptist Church. Or, one could study the basic precepts of the Baptist faith and predict what direction one group would take as opposed to another group geographically located elsewhere.

58. FLETCHER, supra, at 127.

The biological revolution of today must be regarded by religion as neither a threat nor an annoyance--but, rather, as an integral and extremely important part of God's gift and his continuing revelation to all to become more effective tools and worthier stewards. Thus appreciated, the revolution will lead to what might be considered an ultimate ecumenism or final rapproachment between science and religion. R. ETTINGER, MAN INTO SUPERMAN 216, 219 (1972).

CONCLUSIONS

Man currently possesses knowledge that permits him to disregard governance by certain biological facts of life. He can alter those facts and can even cite an ethical mandate for such acts:

> "We cannot accept the 'invisible hand' of blind, natural chance or random nature in genetics. . . . To be men we must be in control. That is the first and last ethical word. For, when there is no choice, there is no possibility of ethical action. Whatever we are compelled to do is amoral."[1]

The scientific method through which knowledge is revealed is less a set of codified principles than a habit of workmanship that skilled investigators possess.[2] The cultivation and development of an intellectual temper that reveals both the nature of scientific reason and the grounds for continued confidence in it are fundamental conditions for every liberal civilization to achieve.[3]

Interestingly, law--itself--has often been viewed as little more than a mask used in order to invoke a higher authority, principle or ethic or in order to channel emotions and events into what might well be considered fixed styles of reasoning which are--regardless of their intrinsic truth--not only of aesthetic value, but considered persuasive to those who enforce and implement it and to the community in which it is applied.[4] Legal reasoning which is employed in construing that which has already been mandated as law by the courts, administrative bodies or the legislatures--or, as a given in initially shaping or structuring a jurisprudential response to that which is perceived as a social, economic or ethical conundrum--must always be applied pragmatically not as a mask which would dehumanize those persons to whom it would rightly apply but as a construct embodying and reaching an equitable balance between the gravity of the harm and the utility of the good in any given situation which legal control is deemed necessary and proper. And, similarly, in applying legal reasoning to fresh frontier issues yet to be resolved by either legislative, judicial, administrative or executive direction, the same balancing point mechanism in the reasoning process must be utilized.

Controlled breeding through genetic manipulation is not far behind the legalization of artificial insemination.[5] Once public acceptance of AID is achieved, rapid progress will be made in achieving similar recognition of other new techniques. The law, then, will be in a stronger position to begin to chart a course of action and keep pace with science instead of remaining behind in grappling with the scientific, legal, ethical, and social issues of the brave new world. Although some assert that eugenic control or controlled breeding is dangerous, foolhardy, destructive of the integrity of the family, and violative of the human right to determine the size of the family unit, the unalterable fact is that population forecasts indicate that the world soon will be overpopulated if appropriate actions are not taken.[6] Genetic planning and eugenic programming are more rational and humane alternatives to population regulation than death by famine and war. Efforts at fetal experimentation designed to master the genetic code and thus improve the overall quality of life are to be encouraged and, indeed, valued to an extent beyond the sanctity of creation.

If we approach mastery of the genetic code with careful resolve to minimize human suffering and maximize the social good, we will approach the future with assurance that the night will no longer be ominous and foreboding. Herman J. Muller, the Nobel Laureate of Indiana University, summed it all up very eloquently when he said:

> "The fact that man is already a creature of thought, of

aspiration, of struggle, of the will-to-power is a sine qua non in the successful issue of his passage from the present darkness to the light that is dawning. . . . The mind of man must more and more become the master, not only of the outer material world, and so too of his social world, but also of the genetic thread of life within him. Thus, there will come an even greater freedom. . . . Intelligence. . . has of late grown astonishingly; but without a corresponding growth in social motivation and in the means of carrying it out, man's great new tools so much more dangerous and more easily misdirected on a large scale than were the primitive instruments of the past--may work only misery and even destruction. Love must balance knowledge or we fail."[7]

Notes--Conclusions

1. Rorvik, "The Embryo Sweepstakes", N. Y. Times Mag., Sept. 15, 1974 at 62 (quoting Dr. Joseph Fletcher).

2. E. NAGEL, SOVERIGN REASON 300 (1954).

3. Id. at 308.

   In considering the myriad problems of the New Biology, what is correct or incorrect cannot be settled dogmatically in advance of an evaluation of the facts of each problem. Thus, abortion may be proper in some cases, while in others wrong; sometimes egg transfers may be valid, other times not. The same approach must be taken as to cloning, artificial insemination, in vitro fertilization, etc. The principles which should be of guiding importance here may be listed as: compassion, a consideration of consequences, proportionate good, the priority of actual needs over the ideal or the potential, a desire to enlarge choice and cut down on chance, and, foremost a courageous acceptance of responsibility to make decisions. If followed, these principles will enable some reasonable conclusions about this morality of human reproduction and genetic engineering to be made. J. FLETCHER, THE ETHICS OF GENETIC CONTROL 147 (1974).

4. Weyrauch, "Law as Mask--Legal Ritual and Reliance", 66 CAL. L. REV. 699, 725, 726 (1978).

   See generally, J. NOONAN, PERSONS AND MASKS OF THE LAW--CARDOZO, HOLMES, JEFFERSON AND WYTHE AS MAKERS OF THE MASKS (1976); Dworkin, "The Model of Rules", 35 U. CHI. L. REV. 14 (1967); Pound, "Juristic Science and the Law", 31 HARV. L. REV. 1047 (1918).

5. Smith, "Through A Test-Tube Darkly: Artificial Insemination and The Law", 67 MICH. L. REV. 127 at 149 (1968).

6. Id.

7. H. J. MULLER, OUT OF THE NIGHT 43, 158 (1935).

   Bertrand Russell phrased the problem man faces in developing science to an even fuller degree in this manner: "Man has been disciplined hitherto by his subjection to nature. Having emancipated himself from this subjection, he is showing something of the defects of a slave turned master. A new moral outlook is called for in which submission to the powers of nature is replaced by respect for what is best in man. It is where this respect is lacking that scientific technique is dangerous." B. RUSSELL, THE SCIENTIFIC OUTLOOK 379 (1931).

   See generally, R. NIEBUHR, MORAL MAN AND IMMORAL SOCIETY (1932).

APPENDIX A

Model Informed Consent Law

Prepared by: The American Medical Association, Office of the General Counsel, Legal Research Department, June 14, 1976.

Section XXX. Consent to Treatment.

A consent in writing to any medical or surgical procedure or course of procedures which (a) sets forth in general terms the nature and purpose of the procedure or procedures, together with the known risks, if any, of death, brain damage, quadriplegia, paraplegia, the loss or loss of function of any organ or limb, or disfiguring scars associated with such procedure or procedures, with the probability of each such risk if reasonably determinable, (b) acknowledges that such disclosure of information has been made and that all questions asked about the procedure or procedures have been answered in a satisfactory manner, and (c) is signed by the patient for whom the procedure is to be performed, or if the patient for any reason lacks legal capacity to consent by a person who has legal authority to consent on behalf of such patient in such circumstances, such consent shall be conclusively presumed to be valid and effective, in the absence of clear proof that execution of the consent was induced by fraudulent misrepresentation of material facts. Except as herein provided, no evidence shall be admissible to impeach, modify or limit the authorization for performance of the procedure or procedures set forth in such written consent.

For legal forms of consent for various medical procedures, see 15 AM. JUR. LEGAL FORMS 2d Physician and Surgeons, Secs. 202:84--154 (1973).

## APPENDIX B

### Proposed Voluntary Sterilization Act*

*Champlin & Winslow, "Elective Sterilization,"
113 PA. L. REV. 415, 442, 443 (1965)

Sec. 1.  Sterilization operation upon person twenty-one years of age or older.

It shall be lawful for any physician or surgeon licensed by this State, when so requested by any person twenty-one years of age or over, to perform upon such person a surgical interruption of vas deferens or fallopian tubes provided a request in writing is made by such person at least thirty (30) days prior to the performance of such surgical operation; and provided, further, that prior to or at the time of such request a full and reasonable medical explanation is given by such physician or surgeon to such person as to the nature and consequences of such operation; and provided, further, that a request in writing is also made at least thirty (30) days prior to the performance of the operation by the spouse of such person, if there be one, unless the spouse has been declared mentally incompetent, or unless a separation agreement has been entered into between the spouse and the person to be operated upon, or unless the spouse and the person to be operated upon have been divorced from bed and board or have been divorced absolutely.

Sec. 2.  Sterilization operation upon person under twenty-one.

Any such physician or surgeon may perform a surgical interruption of vas deferens or fallopian tubes upon any person under the age of twenty-one years when so requested in writing by such minor and in accordance with the conditions and requirements set forth in Section 1 of this act, provided that the juvenile court of the county wherein such minor resides, upon petition of the parent or parents, if they be living, or the guardian or next firned of such minor shall determine that the operation is in the best interest of such minor and shall enter an order authorizing the physician or surgeon to perform such operation.

Sec. 3.  Thirty-day waiting period.

No operation shall be performed pursuant to the provisions of this Act prior to thirty (30) days from the date of consent or request therefor, or in the case of a minor, from the date of the order of the court authorizing the same, and in neither event if the consent for such operation is withdrawn prior to its commencement.

Sec. 4.  No liability for non-negligent performance of operation.

Subject to the rules of law applicable generally to negligence, no physician or surgeon licensed by this State shall be liable either civilly or criminally by reason of having performed a surgical interruption of vas deferens or fallopian tubes authorized by the provisions of this Act upon any person in this State.

Sec. 5.  Prohibition of sterilization operations not performed in accordance with this Act.

Any person who shall perform a surgical interruption of vas deferens or fallopian tubes upon any person in this State, except as authorized by this Act (or by a eugenic sterilization statute) shall be guilty of:

(a) a misdemeanor where the person performing the operation believes that the person upon whom such operation is performed consents to its performance;

(b) a felony where the person performing the operation does not believe that the person upon whom such operation is performed consents to its performance.

Sec. 6. Therapeutic and eugenic sterilizations excepted.
Nothing in this act shall restrict the performance of a surgical interruption of vas deferens or fallopian tubes for sound therapeutic reasons (or affect the provisions of a state eugenic sterilization statute).

This proposed statute does not contain provisions, such as the requirement of consultation, which serve no valid purpose in cases of elective sterilization. Other principal innovations of the suggested statute are the provisions for criminal liability for failure to conform to the prescribed procedures and for the waiver of the required procedures in cases of therapeutic indications for the operation.

It has been argued previously that the caluses in eugenic sterilization statutes which waive the procedural requirements in cases of a medical indication should not be construed to preclude lawful sterilizations for other than therapeutic purposes. This argument was based upon the rationale that it would be unwarranted to conclude that the legislature had considered and disposed of the controversial subject of elective sterilization in the context of a eugenic statute. However, a statute such as the one presently proposed would demand full consideration of the entire problem. In this context, a legislature should properly devise procedures for elective sterilization which are intended to be exclusive--and they should express this purpose in the statue so that the courts are not faced with the dilemma of searching for their intent or the force of public policy whenever the procedures are not followed.

# APPENDIX C

## Model Statute for Artificial Insemination*

*Klayman, "Therapeutic Impregnation: Prognosis of A Lawyer--
Diagnosis of a Legislature,"
39 CINN. L. REV. 291, 328-330 (1970)

Sec. 1. Therpaeutic impregnation is the introduction of human semen into the female genital tract by means other than coitus for the purpose of causing pregnancy, where there exists a matrimonial inability or inadvisability to conceive by normal intercourse.

Sec. 2. Homologous therapeutic impregnation is therapeutic impregnation where only the husband's semen is used.

Sec. 3. Heterologous therapeutic impregnation is therapeutic impregnation where semen other than the husband's is used.

Sec. 4. Donor is a male who donates his semen with or without consideration for use in therapeutic impregnation.

Sec. 5. The technique of therapeutic impregnation is permissible when performed:
- (a) on a married woman,
- (b) by a physician licensed to practice medicine in this state or under the supervision or advice of such physician,
- (c) at the written request of the husband and wife desiring therapeutic impregnation,
- (d) after a complete medical examination of husband and wife, with particular attention to the reproductive organs, if the physician has reason to believe that the technique will cause pregnancy.

Sec. 6. A child born to a married woman by means of therapeutic impregnation with the express or implied consent of her husband shall be deemed the legitimate biological child of such wife and husband.

Sec. 7. A child born to a married woman by means of therapeutic impregnation without the express or implied consent of her husband shall not be deemed the legitimate biological child of such husband, unless he legally adopts or acknowledges the child.

Sec. 8. Submission by a married woman to heterologous therapeutic impregnation without the express or implied consent of her husband, regardless of whether such submission results in conception, shall be grounds for divorce.

Sec. 9. No physician shall extract or supervise the extraction of semen for purposes of therapeutic impregnation unless he ascertains by standard serological tests, written questionnaires and physician examination that the donor is not suffering from veneral disease, defects or diseases that may be transmissible by the genes, or any other disorder that is known to by communicable by semen transfer.

Sec. 10. No physician shall use semen for the purpose of therapeutic impregnation unless he:
- (a) extracts or supervises the extraction of the semen in compliance with Section 9 or obtains the semen from a semen bank certified by the Sanitation Divison of the Board of Health, and
- (b) ascertains that the proposed donor's and proposed recipient's Rh and ABO blood factors are compatible.

Sec. 11. A physician performing therapeutic impregnation shall keep a record in confidential files containing:
    (a) the name and address of the donor.
    (b) the name and address of the recipient and her husband.
    (c) the written form signed by the husband and wife requesting therapeutic impregnation.
    (d) the date(s) of the therapeutic impregnations(s) and the method used.
    (e) the results of such therapeutic impregnation, if known.

Sec. 12. No donor other than the recipient's husband shall be deemed to be the lawful or biologic father or any offspring produced as a result of the use of his semen.

Sec. 13. A donor shall answer truthfully the questionnaire form provided by the physician intending to extract or supervise the extraction of semen.

Sec. 14. Any person who knowingly performs, attempts to perform or aid or abets in the performance of therapeutic impregnation in violation of Section 5, or who violates Sections 9, 10 or 11 shall be fined not more than $1,000.00 or be imprisoned not more than one year or both.

Sec. 15. Any person who violates Section 13 shall be fined nor more than $5,000.00 or be imprisoned not more than one year or both.

APPENDIX D

Guide to Clinical Research--The Declaration of Helsinki*

It is the mission of the doctor to safeguard the health of the people. His knowledge and conscience are dedicated to the fulfillment of this mission.

The Declaration of Geneva of the World Medical Association binds the doctor with the words: "The health of my patient will be my first consideration" and the International Code of Medical Ethics declares that, "Any act or advice which could weaken physical or mental resistance of a human being may be used only in his interest."

Because it is essential that the results of laboratory experiments be applied to human beings to further scientific knowledge and to help suffering humanity, the World Medical Association has prepared the following recommendations as a guide to each doctor in clinical research. It must be stressed that the standards as drafted are only a guide to physicians all over the world. Doctors are not relieved from criminal, civil and ethical responsibilities under the laws of their own countries.

In the field of clinical research, a fundamental distinction must be recognized between clinical research in which the aim is essentially therapeutic for a patient, and clinical research, the essential object of which is purely scientific and without therapeutic value to the person subjected to the research.

I. Basic Principles

Clinical research must conform to the moral and scientific principles that justify medical research and should be based on laboratory and animal experiments or other scientifically established facts.

Clinical research should be conducted only by scientifically qualified persons and under the supervision of a qualified medical man.

Clinical research cannot legitimately be carried out unless the importance of the objective is in proportion to the inherent risk to the subject.

Every clinical research project should be preceded by careful assessment of inherent risks in comparison to foreseeable benefits to the subject or to others.

Special caution should be exercised by the doctor in performing clinical research in which the personality of the subject is liable to be altered by drugs or experimental procedure.

II. Clinical Research Combined with Professional Care

In the treatment of the sick person, the doctor must be free to use a new therapeutic measure, if in his judgment it offers hope of saving life, re-establishing health, or alleviating suffering.

If at all possible, consistent with patient psychology, the doctor should obtain the patient's freely given consent after the patient has been given a full explanation. In case of legal incapacity, consent should also be procured from the legal guardian; in case of physical incapacity, the permission of the legal guardian replaces that of the patient.

### III. Nontherapeutic Clinical Research

In the purely scientific application of clinical research carried out on a human being, it is the duty of the doctor to remain the protector of the life and health of that person on whom clinical research is being carried out.

The nature, the purposes and the risk of clinical research must be explained to the subject by the doctor.

Clinical research on a human being cannot be undertaken without his free consent after he has been informed; if he is legally incompetent, the consent of the legal guardian should be procured.

The subject of clinical research should be in such a mental, physical and legal state as to be able to exercise fully his power of choice.

Consent should, as a rule, be obtained in writing. However, the responsibility for clinical research always remains with the research worker; it never falls on the subject even after consent is obtained.

The investigator must respect the right of each individual to safeguard his personal integrity, especially if the subject is in a dependent relationship to the investigator.

At any time during the course of clinical research the subject or his guardian should be free to withdraw permission for research to be continued.

The investigator or the investigating team should discontinue the research if, in his or their judgment, it may, if continued, be harmful to the individual.

---

*Adopted by the World Association in June, 1964, at the Eighteenth World Medical Assembly, as a guide to doctors engaged in clinical research.

## APPENDIX E

## The Nuremberg Code of Ethics in Medical Research*

1. The voluntary consent of the human subject is absolutely essential. This means that the person involved should have legal capacity to give consent; should be so situated as to be able to exercise free power of choice without the intervention of any element of force, fraud, deceit, duress, overreaching, or other ulterior form of constraint or coercion; and should have sufficient knowledge and comprehension of the elements of the subject matter involved as to enable him to make an understanding and enlightened decision. This latter element requires that before the acceptance of an affirmative decision by the experimental subject there should be made known to him the nature, duration, and purpose of the experiment; the method and means by which it is to be conducted; all inconveniences and hazards reasonably to be expected; and the effects upon his health or person which may possibly come from his participation in the experiment.

The duty and responsibility for ascertaining the quality of the consent rests upon each individual who initiates, directs, or engages in the experiment. It is a personal duty and responsibility which may not be delegated to another with impunity.

2. The experiment should be such as to yield fruitful results for the good of society, unprocurable by other methods or means of study, and not random and unnecessary in nature.

3. The experiment should be so designed and based on the results of animal experimentation and a knowledge of the natural history of the disease or other problem under study that the anticipated results will justify the performance of the experiment.

4. The experiment should be so conducted as to avoid all unnecessary physical and mental suffering and injury.

5. No experiment should be conducted where there is an a priori reason to believe that death or disabling injury will occur; except, perhaps, in those experiments where the experimental physicians also serve as subjects.

6. The degree of risk to be taken should never exceed that determined by the humanitarian importance of the problem to be solved by the experiment.

7. Proper preparations should be made and adequate facilities provided to protect the experimental subject against even remote possibilities of injury, disability, or death.

8. The experiment should be conducted only by scientifically qualified persons. The highest degree of skill and care should be required through all stages of the experiment of those who conduct or engage in the experiment.

9. During the course of the experiment the human subject should be at liberty to bring the experiment to an end if he has reached the physical or mental state where continuation of the experiment seems to him to be impossible.

10. During the course of the experiment the scientist in charge must be prepared to terminate the experiment at any stage, if he has probable cause to believe, in the exercise of the good faith, superior skill, and careful judgment required of him, that a continuation of the experiment is likely to result in injury, disability, or death to the experimental subject.

---

*Report of the National Conference on the Legal Environment of Medical Science--sponsored by the National Society for Medical Research and the University of Chicago, May 27-28, 1977, at 91-92.

APPENDIX F

Research Involving Prisoners: Report and Recommendations
of the National Commission for the Protection of Human
Subjects of Biomedical and Behavioral Research--1976

Preface

The National Commission for the Protection of Human Subjects of Biomedical and Behavioral Research was established under the National Research Act (P.L. 93-348) to develop ethical guidelines for the conduct of research involving human subjects and to make recommendations for the application of such guidelines to research conducted or supported by the Department of Health, Education and Welfare (DHEW). The legislative mandate also directs the Commission to make recommendations to Congress regarding the protection of human subjects in research not subject to regulation by DHEW. Particular classes of subjects that must receive the Commission's attention include children, prisoners and the institutionalized mentally infirm.

The duties of the Commission with regard to research involving prisoners are specifically set forth in Section 202(a) (2) of the National Research Act, as follows:

> The Commission shall identify the requirements for informed consent to participation in biomedical research by...prisoners.... The Commission shall investigate and study biomedical and behavioral research conducted or supported under programs administered by the Secretary (DHEW) and involving...prisoners....to determine the nature of the consent obtained from such persons or their legal representatives before such persons were involved in such research; the adequacy of the information given them respecting the nature and purpose of the research, procedures to be used, risks and discomforts, anticipated benefits from the research, and other matters necessary for informed consent; and the competence and the freedom of the persons to make a choice for or against involvement in such research. On the basis of such investigation and study the Commission shall make such recommendations to the Secretary as it determines appropriate to assure that biomedical and behavioral research conducted or supported under programs administered by him meet the requirements respecting informed consent identified by the Commission.

This responsibility is broadened by the provision (Section 202(a) (3) that the Commission make recommendations to Congress regarding the protection of subjects involved in research not subject to regulation by DHEW, such as research involving prisoners that is conducted or supported by other federal departments or agencies, as well as research conducted in federal prisons or involving inmates from such prisons.

To carry out its mandate, the Commission studied the nature and extent of research involving prisoners, and the conditions under which such research is conducted, and the possible grounds for continuation, restriction or termination of such research. Commission members and staff made site visits to four prisons and two research facilities outside prisons that use prisoners, in order to obtain first-hand information on the conduct of biomedical research and the operation of behvioral programs in these settings. During the visits, interviews were conducted with many inmates who have participated in research or behavioral programs as well as with non-participants.

The Commission held a public hearing at which research scientists, prisoner advocates and providers of legal services to prisoners, representatives of the pharmaceutical industry, and members of the public presented their views on research involving prisoners. This hearing was duly announced, and no request to testify was denied. The National Minority Conference on Human Experimentation, which was convened by the Commission in order to assure that viewpoints of minorities would be expressed, made recommendations to the Commission on research in prisons. In addition to papers, surveys and other materials prepared by the Commission staff, studies on the following topics were prepared under contract: (1) alternatives to the involvement of prisoners; (2) foreign practices with respect to drug testing; (3) philosophical, sociological and legal perspectives on the involvement of prisoners in research; (4) behavioral research involving prisoners; and (5) a survey of research review procedures, investigators and prisoners at five prisons. Finally, at public meetings commencing in January, 1976, the Commission conducted extensive deliberations and developed its recommendations on the involvement of prisoners in research.

Part I of this Report contains the recommendations as well as the deliberations and conclusions of the Commission and a summary of background materials. The nature and extent of research involving prisoners are described in Part II. The activities of the Commission and reports that were prepared for it are summarized in Parts III and IV, respectively. An appendix to this Report contains papers, surveys, reports and other materials that were prepared or collected for the Commission on various topics related to research involving prisoners ....Most of such materials are summarized in Part IV of this Report.

. . . .

Glossary of Terms Used in this Report.

"Phases of drug testing."
FDA regulations require three phases for the testing of new drugs. Phase 1 is the first introduction of a new drug into humans (using normal volunteers), with the purpose of determining human toxicity, metabolism, absorption, elimination and other pharmacological action, preferred route of administration, and safe dosage range. Phase 2 covers the initial trials on a limited number of patients for specific disease control or prophylaxis purposes. Phase 3 involves extended clinical trials, providing assessment of the drug's safety and effectiveness and optimum dosage schedules in the diagnosis, treatment or prophylaxis or groups of subjects involving a given disease or condition. (Source: 21 C.F.R. 312.1).

"Prisoner."
Any individual involuntarily confined in a prison.

"Therapeutic research, nontherapeutic research."
The Commission recognizes problems with employing the terms "therapeutic" and "nontherapeutic" research, notwithstanding their common usage, because they convey a misleading impression. Research refers to a class of activities designed to develop generalizable new knowledge. Such activities are often engaged in to learn something about practices designed for the therapy of the individual. Such research is often called "therapeutic" research; however, the research is not solely for the therapy of the individual. In order to do research, additional interventions over and above those necessary for therapy may need to be done, e.g., randomization, blood drawing, catherterization; these interventions are not "therapeutic" for the

individual. Some of these interventions may present risk to the individual--risk clearly unrelated to the therapy of the subject. The Commission has employed the term "research" on practices which have the intent and reasonable probability of improving the health or well-being of the subject or variants of this term. Since the reports prepared for the Commission by outside contractors or consultants generally employ the terms in common usage, those terms have been retained in the summaries of those reports.

Part I. Deliberations, Conclusions and Recommendations

Chapter 1. Deliberations and Conclusions.

Introduction.

Prior to 1940, prisoners in the United States seldom participated in biomedical research that had no reasonable expectation of improving the health or well-being of the research subjects. During World War II, however, large numbers of prisoners participated in voluntary research programs to develop treatment for infectious diseases that afflicted our armed forces. This involvement of prisoners was considered to be not only acceptable, but praiseworthy. Following the war, the growth of biomedical research and the imposition of requirements for testing drugs as to safety led to the increased use of prisoners. Their participation in biomedical research not related to their health or well-being has continued in this country to the present time. This participation is now primarily in phase 1 drug and cosmetic testing, which is conducted or supported by pharmaceutical manufacturers in connection with applications to the Food and Drug Administration for licensing new drugs. Other research of this sort in which prisoners participate, or have participated, includes studies of normal metabolism and physiology, conducted by the Public Health Service (PHS); studies of the prevention or treatment of infectious diseases, conducted or supported by the PHS and the Department of Defense; a study of the effects of irradiation on the male reproductive function, supported by the Atomic Energy Commission; and testing of the addictive properties of new analgesics by giving them to prisoners with a history of narcotic abuse, conducted at the Addiction Research Center in Lexington, Kentucky. (The involvement of federal prisoners in the Lexington program is scheduled to be phased out).

Prisoners also participate in research on practices that have the intent and reasonable probability of improving their health or well-being. This research includes, for example, studies (supported by various components of DHEW and the Federal Bureau of Prisons) to develop methods to reduce the spread of infections, improve dental care help the subjects stop smoking, and remove tatoos. A major focus of this sort of research involving federal prisoners has been the development of new treatments for narcotic addiction.

A third type of research in which prisoners participate includes sutides of the possible causes, effects and process of incarceration, and studies of prisons as institutional structures or of prisoners as incarcerated persons. Components of DHEW have undertaken research of this sort for purposes as learning the etiology of drug addiction and deviant or self-destructive behavior, and the factors relating to parole performance and recidivism.

Research is also conducted on the methods of treatment or "rehabilitation" of prisoners. The National Institute of Mental Health, the Federal Bureau of Prisons, and the Law Enforcement Assistance Administration have supported research on the experimental treatment of aggressive behavior with drugs and aversive conditioning techniques, as well as behavior modification based upon depriving inmates of basic amenities which they must then earn back as privileges. Rehabilitative practices have not always been based upon prior scientific design and evaluation, however, despite the fact that there are few, if any, approaches to the treatment or rehabilitation of prisoners for which effectiveness has been clearly demonstrated.

Outside the United States, prisoners do not generally participate in biomedical research. This exclusion may be ascribed in part to continuing concern over experiments that were conducted on prisoners in Nazi concentration camps. Revelations of those experiments led to the enunciation of the Nuremberg Code (1946-1949), which required that human subjects of research "be so situated as to be able to exercise free power of choice" but did not expressly prohibit research involving civil prisoners. The Declaration of Helsinki, adopted by the World Medical Association in 1964 and endorsed by the American Medical Association in 1966, contained similar language that was subsequently deleted in 1975. Although little if any drug testing is conducted in foreign prisons, other kinds of research have been conducted in prisons throughout the world, such as studies dealing with the incidence and implications of chromosome abnormalities.

Since the 1960's, the ethical propriety of participation by prisoners in research has increasingly been questioned in this country. Among the events that have focused public attention on this issue was the publication of Jessica Mitford's book, KIND AND USUAL PUNISHMENT, in 1973. Eight states and the Federal Bureau of Prisons have formally moved to abandon research in prisons. The Health Subcommittee of the Senate Committee of Labor and Public Welfare held hearings (Quality of Health Care--Human Experimentation, 1973) on research involving prisoners. Those speaking against the use of prisoners cited exploitation, secrecy, danger, and the impossibility of obtaining informed consent as reasons to impose a prohbition or moratorium on the conduct of research in prisons. The advantages of using prisoners in research (e.g., opportunity for close mointoring and controlled environment) and the procedures that are employed to protect prisoner participants were also described in the hearings. The Health Subcommittee held extensive hearings on other areas of human experimentation as well, and reported the bill establishing the Commission with a mandate that included a directive to study and make recommendations concerning the involvement of prisoners in research.

More recently, the House Subcommittee on Courts, Civil Liberties, and the Administration of Justice held hearings (Prison Inmates in Medical Research, 1975) on a bill (H.R. 3603) to prohibit "medical research" in federal prisons and prisons of states that receive certain federal support. Following these hearings, the Director of the Federal Bureau of Prisons determined that "continued use of prisoners in any medical experimentation should not be permitted," and he ordered that such participation by prisoners under federal jurisdiction be phased out.

Some of the more extreme behavioral programs have also raised questions. In her 1973 book, Jessica Mitford expressed concern about new approaches to "treatment" for offenders. Concurrently, others raised questions about the use of psychosurgery in prisons. In the early 1970's, the first challenges to behavior modification and aversive conditioning programs in prisons were argued in the courts, with mixed results. Most of the cases involved the right to refuse to participate in such programs, although prisoners have also petitioned for the right to be included in programs designed to alter sexually aggressive behavior.

Concern over behavior modification programs in prisons was expressed in a study, Individual Rights And The Federal Role In Behavior

Modification (1974), prepared by the staff of the Constitutional Rights Subcommittee of the Senate Judiciary Committee. The study contained information on a number of such programs and suggested that this Commission make use of the information in attempting to resolve the issues that they raised. It should be noted that a number of the "treatment" programs mentioned in the study are reported to have been discontinued.

General Concerns.

In conducting its investigations and studies, the Commission has noted and cannot ignore serious deficiencies in living conditions and health care that generally prevail in prisons. Nor can the Commission ignore the potential for arbitary exercise of authority by prison officials and for unreasonable restriction of communication to and from prisoners. The Commission, although acknowledging that it has neither the expertise nor the mandate for prison reform, nevertheless urges that unjust and inhumane conditions be eliminated from all prisons, whether or not research activities are conducted or contemplated.

Ethical Considerations About Using Prisoners As Research Subjects

There are two basic ethical dilemmas concerning the use of prisoners as research subjects: (1) whether prisoners bear a fair share of the burdens and receive a fair share of the benefits of research; (2) whether prisoners are, in the words of the Nuremberg Code, "so situated as to be able to exercise free power of choice"--that is, whether prisoners can give truly voluntary consent to participate in research.

These two dilemmas relate to two basic ethical principles: the principle of justice, which requires that persons and groups be treated fairly, and the principle of respect for persons, which requires that the autonomy of persons be promoted and protected. Disproportionate use of prisoners in certain kinds of research (e.g., phase 1 drug testing) would constitute a violation of the first principle; closed and coercive prison environments would compromise the second principle. It is within the context of a concern to implement these principles that the Commission has deliberated the question of use of prisoners as research subjects.

The Commission recognizes, however, that the application of these principles to the problem is not unambiguous. To respect a person is to allow that person to live in accord with his or her deliberate choices. Since the choices of prisoners in all matters except those explicitly withdrawn by law should be respected, as courts increasingly affirm, it seems at first glance that the principle of respect for persons requires that prisoners not be deprived of the opportunity to volunteer for research. Indeed, systematic deprivation of this freedom would also violate the principle of justice, since it would arbitrarily deprive one class of persons of benefits available to others--namely, the benefits of participation in research.

However, the application of the principles of respect and justice allows another intepretation, which the Commission favors. When persons seem regularly to engage in activities which, were they stronger or in better circumstances, they would avoid, respect dictates that they be protected against those forces that appear to compel their choices. It has become evident to the Commission that, although prisoners who participate in research affirm that they do so freely, the conditions of social and economic deprivation in which they live compromise their freedom. The Commission believes, therefore, that the appropriate expression of respect consists in protection from exploitation. Hence it calls for certain safeguards intended to reduce the elements of constraint under which prisoners give consent and suggests that certain kinds of research would not be permitted where such safeguards cannot be assured.

Further, a concern for justice raises the question whether social institutions are so arranged that particular persons or groups are burdened with marked disadvantages or deprived of certain benefits for reasons unrelated to their merit, contribution, deserts or need. While this principle can be interpreted, as above, to require that prisoners not be unjustly excluded from participation in research, it also requires attention to the possibility that prisoners as a group bear a disproportionate share of the burdens of research or bear those burdens without receiving a commensurate share of the benefits that ultimately derive from research. To the extent that participation in research may be a burden, the Commission is concerned to ensure that this burden not be unduly visited upon prisoners simply because of their captive status and administrative availability. Thus it specifies some conditions for the selection of prisoners as a subject pool for certain kinds of research. In so doing, the Commission is not primarily intending to protect prisoners from the risks of research; indeed, the commission notes that the risks of research, as compared with other kinds of occupations, may be rather small. The Commission's concern, rather, is to ensure the equitable distribution of the burdens of research no matter how large or small those burdens may be. The Commission is concerned that the status of being a prisoner makes possible the perpetration of certain systematic injustices. For example, the availability of a population living in conditions of social and economic deprivation makes it possible for researchers to bring to these populations types of research which persons better situated would ordinarily refuse. It also establishes an enterprise whose fair administration can be readily corrupted by prisoner control or arbitrarily manipulated by prison authorities. And finally, it allows an inequitable distribution of burdens and benefits, in that those social classes from which prisoners often come are seldom full beneficiaries of improvements in medical care and other benefits accruing to society from the research enterprise.

Reflection upon these principles and upon the actual conditions of imprisonment in our society has lead the Commission to believe that prisoners are, as a consequence of being prisoners, more subject to coerced choice and more readily available for the imposition of burdens which others will not willingly bear. Thus, it has inclined toward protection as the most appropriate expression of respect for prisoners as persons and toward redistribution of those burdens of risk and inconvenience which are presently concentrated upon prisoners. At the same time, it admits that, should coercions be lessened and more equitable systems for the sharing of burdens and benefits be devised, respect for persons and concern for justice would suggest that prisoners not be deprived of the opportunity to participate in research. Concern for principles of respect and justice leads the Commission to encourage those forms of inquiry that could form a basis for improvement of current prison conditions and practices, such as studies of the effects of incarceration, or prisons as institutions and of prisoners as prisoners, and also to allow research on practices clearly intended to improve the health or well-being of individual prisoners.

The Commission has noted the concern, expressed by participants at the National Minority Conference and by others, that minorities bear a disproportionate share of the risks of research conducted in prisons. This concern is fostered, in part, by eivdence that prison populations are disproportionately non-white. Evidence presented to the Commission indicates that where research is done in prison, those prisoners who participate tend to be predominantly white, even in institutions where the population as a whole is predominantly non-white; further, those who participate in research tend to be better educated and more frequently employed at better jobs than the prison population as a whole. This evidence suggests that non-whites and poor or less educated persons in prison do not carry a greater share of the burdens of research. However, the evidence is inconclusive for two reasons: first, because it does not fully satisy questions related to the risks of research; and second, because it raises questions of justice with respect to the equitable distribution of benefits (as well as burdens) of research.

With respect to risks, the Commission notes that different research projects carry different risks; it is possible, though the Commission has no evidence to this effect, that one race or another may participate in more research of higher risk. And of course, the ratio of non-whites to whites participating in research and hence bearing the burdens of research may still be disproportionate when compared to the ratio of the populations as a whole.

But the Commission also notes that those who participate in research consider the benefits sufficient to outweigh the burdens. Thus, the greater participation of whites may mean that there is an inequitable distribution of benefits between racial groups. Hence the greater participation by whites does not necessarily resolve the issue of distributive justice.

Similarly, the Commission notes that less research is conducted in women's prisons. While the reasons for this may well be the same reasons that women in general are used less frequently than men as research subjects (e.g., the possibility of pregnancy), questions of distributive justice, similar to those raised above, may still need to be addressed with respect to participation in research by women prisoners.

Discussion.

Among the issues discussed by the Commission are two on which no specific recommendations are made, but concerning which the considerations of the Commission should be expressed: (1) remuneration, and (2) alternatives to conducting research in prisons. (1) Remuneration is a subject that should be analyzed by human subjects review committees, in consultation with prison grievance committees and prison authorities. There are at least two considerations that must be balanced in the determination of appropriate rates for participation in research not related to the subjects' health or well-being. On the one hand, the pay offered to prisoners should not be so high, compared to other opportunities for employment within the facility, as to constitute undue inducement to participate. On the other hand, those who sponsor the research should not take economic advantage of captive populations by paying significantly less than would be necessary if non-prisoner volunteers were recruited. Fair solutions to this problem are difficult to achieve.

One suggestion is that those who sponsor research pay the same rate for prisoners as they pay other volunteers, but that the amount actually going to the research subjects be comparable to the rates of pay otherwise available within the facility. The difference between the two amounts could be paid into a general fund, either to subsidize the wages for all inmates within the prison, or for other purposes that benefit the prisoners or their families. Prisoners should participate in managing such a fund and in determining allocation of monies. Another suggestion is that the difference be held in escrow and paid to each participant at the time of release or, alternatively, that it be paid directly to the prisoner's family.

A requirement related to the question of appropriate remuneration for participation in research is that prisoners should be able to obtain an adequate diet, the necessities of personal hygiene, medical attention and income without recourse to participation in research.

Some of the Commission members endorse the alternative of permitting prisoners to participate in research provided it is conducted in a clinic or hospital outside the prison grounds, and provided also that non-prisoners participate in the same projects for the same wages. Other members of the Commission believe that such a mechanism would serve only to increase the disparity between the conditions within the prison and those within the research unit, thereby heightening the inducement to participate in research in order to escape from the constraints of the prison setting. All of the members of the Commission endorse the suggestion that the use of alternative populations be explored and utilized more fully than is presently the case. This may be especially important to permit drugs to continue to be tested, as required by current law and regulations of the FDA, during any period in which prisons have not satisfied the conditions that are recommended for the conduct of such research. Increased utilization of alternative populations would have the added benefit of providing non-prisoner populations to participate in research projects along with prisoners, or in parallel with similar projects within prisons, in order to satisfy the general concern that prisoners not participate in experiments that non-prisoners would find unacceptable. The Commission also suggests that Congress and the FDA consider the advisability of undertaking a study and evaluation to determine whether present requirements for phase 1 drug testing in normal volunteers should be modified.

Conclusions.

In the course of its investigations and review of evidence presented to it, the Commission did not find in prisons the conditions requisite for a sufficiently high degree of voluntariness and openness, notwithstanding that prisoners currently participating in research consider, in nearly all instances, that they do so voluntarily and want the research to continue. The Commission recognizes the role that research involving prisoners has played. It does not consider, however, that administrative convenience or availability of subjects is, in itself, sufficient justification for selecting prisoners as subjects.

Throughout lengthy deliberations, the strong evidence of poor conditions generally prevailing in prisons and the paucity of evidence of any necessity to conduct research in prisons have been significant considerations of the Commission. An equally important consideration has been the closed nature of prisons, with the resulting potential for abuse of authority. Some of the Commission members, who are opposed to research not related to the health or well-being of prisoner-participants, have, however, agreed to permit it to be conducted, but only under the following standards: adequate living conditions, separation of research participation from any appearance of parole consideration, effective grievance procedures and public scrutiny at the prison where research will be conducted or from which prospective subjects will be taken; importance of the research; compelling reasons to involve prisoners; and fairness of such involvement. Compliance with these requirements must be certified by the highest responsible federal official, assisted by a national ethical review body. The Commission has concluded that the burden of proof that all the requirements are satisfied should be on those who wish to conduct the research.

Chapter 2. Recommendations.

The National Commission for the Protection of Human Subjects of Biomedical and Behavioral Research makes the following recommendations on research involving prisoners to:

(i) the Secretary, DHEW, with respect to research subject to his regulation, i.e., research conducted or supported under programs administered by him and research reported to him in fulfillment of regulatory requirements; and

(ii) the Congress, except as otherwise noted, with respect to research not subject to regulation by the Secretary, DHEW.

Recommendation (1): STUDIES OF THE POSSIBLE CAUSES, EFFECTS AND PROCESSES OF INCARCERATION AND STUDIES OF PRISONS AS INSTITUTIONAL STRUCTURES OR OF PRISONERS AS INCARCERATED PERSONS MAY BE CONDUCTED OR SUPPORTED, PROVIDED THAT (A) THEY PRESENT MINIMAL OR NO RISK AND NO MORE THAN MERE INCONVENIENCE TO THE SUBJECTS, AND (B) THE REQUIREMENTS UNDER RECOMMENDATION (4) ARE FULFILLED.

Comment. The Commission encourages the conduct of studies of prisons as institutions and prisoners as incarcerated persons. Because the inadequacies of the prisons may themselves be the object of such studies, the Commission has not set any conditions for the conduct of such research other than a limitation of this category to research that presents minimal or no risk and no more than mere inconvenience, and the requirements of Recommendation (4).

Studies of prisoners consisting of questionnaires, surveys, analyses of census and demographic data, psychological tests, personality inventories and the like rarely involve risk and are essential for proper understanding of prisons and the effects of their practices. Research designed to determine the effects on general health of institutional diets and restircted activity, and similar studies that do not manipulate bodily conditions (except innocuously, e.g., obtaining blood samples) but merely monitor or analyze such conditions, also present little physical risk and are necessary to gain some knowledge of the effects of imprisonment. Such research is a necessary step toward understanding prison practices and alternatives, without which there can be no improvement.

RECOMMENDATION (2): RESEARCH ON PRACTICES, BOTH INNOVATIVE AND ACCEPTED, WHICH HAVE THE INTENT AND REASONABLE PROBABILITY OF IMPROVING THE HEALTH OR WELL-BEING OF THE INDIVIDUAL PRISONERS MAY BE CONDUCTED OR SUPPORTED, PROVIDED THE REQUIREMENTS UNDER RECOMMENDATION (4) ARE FULFILLED.

Comment. Research would fall under this recommendation if the practices under study are designed solely to improve the health or well-being of the research subject by prophylactic, diagnostic or treatment methods that may depart from standard practice but hold out a reasonable expectation of success. The Commission intends that prisoners not be discriminated against with respect to research protocols in which a therapeutic result might be realized for the individual subject. The committees that review all research involving prisoners should analyze carefully any claims that research projects are designed to improve the health or well-being of subjects and should be particularly cautious with regard to research in which the principal purpose of the practice under study is to enforce conformity with behavioral norms established by prison officials or even by society. Such conformity cannot be assumed to improve the condition of the individual prisoner. If the review committee does not consider such claims to be sufficiently substantiated, the research should not be conducted unless it conforms to the requirements of Recommendation (3).

RECOMMENDATION (3): EXCEPT AS PROVIDED IN RECOMMENDATIONS (1) AND (2), RESEARCH INVOLVING PRISONERS SHOULD NOT BE CONDUCTED OR SUPPORTED, AND REPORTS OF SUCH RESEARCH SHOULD NOT BE ACCEPTED BY THE SECRETARY, DHEW, IN FULFILLMENT OF REGULATORY REQUIREMENTS, UNLESS THE REQUIREMENTS UNDER RECOMMENDATION (4) ARE FULFILLED AND THE HEAD OF THE RESPONSIBLE FEDERAL DEPARTMENT OR AGENCY HAS CERTIFIED, AFTER CONSULTATION WITH THE NATIONAL ETHICAL REVIEW BODY, THAT THE FOLLOWING THREE REQUIREMENTS ARE SATISFIED:

(A) THE TYPE OF RESEARCH FULFILLS AN IMPORTANT SOCIAL AND SCIENTIFIC NEED, AND THE REASONS FOR INVOLVING PRISONERS IN THE TYPE OF RESEARCH ARE COMPELLING;

(B) THE INVOLVEMENT OF PRISONERS IN THE TYPE OF RESEARCH SATISFIES CONDITIONS OF EQUITY; AND

(C) A HIGH DEGREE OF VOLUNTARINESS ON THE PART OF THE PROSPECTIVE PARTICIPANTS AND OF OPENNESS ON THE PART OF THE INSTITUTIONS(S) TO BE INVOLVED WOULD CHARACTERIZE THE CONDUCT OF THE RESEARCH; MINIMUM REQUIREMENTS FOR SUCH VOLUNTARINESS AND OPENNESS INCLUDE ADEQUATE LIVING CONDITIONS, PROVISIONS FOR EFFECTIVE REDRESS OF GRIEVANCES, SEPARATION OF RESEARCH PARTICIPATION FROM PAROLE CONSIDERATIONS, AND PUBLIC SCRUTINY.

Comment. Detailed standards expressing the intent of the Commission with respect to paragraph (C) of this recommendation are as follows:

(i) Public scrutiny. Prisoners should be able to communicate, without censorship, with persons outside the prison and, on a privileged, confidential basis, with attorneys, legal organizations which assist prisoners, the accrediting office which assists the certifying federal official or national ethical review body, the grievance committee referred to in paragraph (ii) below, and the human subjects review committee or institutional review board referred to in Recommendation (4). Each of such persons or organizations with whom prisoners should be able to communicate on a privileged, confidential basis should be able to conduct private interviews with any prisoner who so desires. The accrediting office, grievance committee and human subjects review committee or institutional review board should be allowed free access to the prison.

(ii) Grievance procedures. There should exist a grievance committee composed of elected prisoner representatives, prisoner advocates and representatives of the community. The committee should enable prisoners to obtain effective redress of their grievances and should facilitate inspections and monitoring by the accrediting office to assure continuing compliance with requirement (C).

(iii) Standard of living. Living conditions in the prison in which research will be conducted or from which subjects will be recruited should be adequate, as evidenced by compliance with all of the following standards:

(1) The prison population does not exceed designed capacity, and each prisoner has an adequate amount of living space;

(2) There are single occupancy cells available for those who desire them;

(3) There is segregation of offenders by age, degree of violence, prior criminal record, and physical and mental health requirements;

(4) There are operable cell doors, emergency exits and fire extinguishers, and compliance with state and local fire and safety codes is certified;

(5) There are operable toilets and wash basins in cells;

(6) There is regular access to clean and working showers;

(7) Articles of personal care and clean linen are regularly issued;

(8) There are adequate recreation facilities, and each prisoner is allowed an adequate amount of recreation;

(9) There are good quality medical facilities in the prison, adequately staffed and equipped, and approved by an outside medical accrediting organization such as the Joint Commission on Accreditation of Hospitals or a state medical society;

(10) There are adequate mental health services and professional staff;

(11) There is adequate opportunity for prisoners who so desire to work for remuneration comparable to that received for participation in research;

(12) There is adequate opportunity for prisoners who so desire to receive education and vocational training;

(13) Prisoners are afforded opportunity to communicate privately with their visitors, and are permitted frequent visits;

(14) There is a sufficiently large and well-trained staff to provide assurance of prisoners' safety;

(15) The racial composition of the staff is reasonably concordant with that of the prisoners;

(16) To the extent that it is consistent with the security needs of the prison, there should be an opportunity for inmates to lock their own cells; and

(17) Conditions in the prison satisfy basic institutional environmental health, food service and nutritional standards.

(iv) Parole. There should be effective procedures assuring that parole boards cannot take into account prisoners' participation in research and that prisoners are clearly informed that there is absolutely no relationship between research participation and determinations by their parole boards.

If an investigator wishes to present evidence of the importance and fairness of conducting a type of research on a prison population (requirements (A) and (B) propose that the conditions of voluntariness and openess would be satisfied at a particular prison (requirement (C) ), the case should be presented to the Secretary, DHEW (or the head or any other department or agency under whose authority the research would be conducted). Such official should seek the advice of an existing or newly created advisory body (such as the Ethical Advisory Board established within the Public Health Service) in determining whether to approve the type of research at the specific institution. Such official or advisory body should be assisted by an accrediting office, which makes inspections, certifies compliance with requirements (C), and monitors continuing compliance of any prison involved in research. In determining such compliance, the accrediting office should be guided by the above description of the Commission's intent in recommending requirement (C).

RECOMMENDATION (4): (A) THE HEAD OF THE RESPONSIBLE FEDERAL DEPARTMENT OR AGENCY SHOULD DETERMINE THAT THE COMPETENCE OF THE INVESTIGATORS AND THE ADEQUACY OF THE RESEARCH FACILITIES INVOLVED ARE SUFFICIENT FOR THE CONDUCT OF ANY RESEARCH PROGRAM IN WHICH PRISONERS ARE TO BE INVOLVED.

(B) ALL RESEARCH INVOLVING PRISONERS SHOULD BE REVIEWED BY AT LEAST ONE HUMAN SUBJECTS REVIEW COMMITTEE OR INSTITUTIONAL REVIEW BOARD COMPRISED OF MEN AND WOMEN OF DIVERSE RACIAL AND CULTURAL BACKGROUNDS THAT INCLUDE AMONG ITS MEMBERS PRISONERS OR PRISONER ADVOCATES AND SUCH OTHER PERSONS AS COMMUNITY REPRESENTATIVES, CLERGY, BEHAVIORAL SCIENTISTS AND MEDICAL PERSONNEL NOT ASSOCIATED WITH THE CONDUCT OF THE RESEARCH OR THE PENAL INSTITUTION. IN REVIEWING PROPOSED RESEARCH, THE COMMITTEE OR BOARD SHOULD CONSIDER AT LEAST THE FOLLOWING: THE RISKS INVOLVED, PROVISIONS FOR OBTAINING INFORMED CONSENT, SAFEGUARDS TO PROTECT INDIVIDUAL DIGNITY AND CONFIDENTIALITY, PROCEDURES FOR THE SELECTION OF SUBJECTS, AND PROVISIONS FOR PROVIDING COMPENSATION FOR RESEARCH-RELATED INJURY.

Comment. The risks involved in research involving prisoners should be commensurate with risks that would be accepted by non-prisoner volunteers. If it is questionable whether a particular project is offered to prisoners because of the risk involved, the review committee might require that non-prisoners be included in the same project.

In negotiations regarding consent, it should be determined that the written or verbal comprehensibility of the information presented is appropriate to the subject population.

Procedures for the selection of subjects within the prison should be fair and immune from arbitrary intervention by authorities or prisoners.

Compensation and treatment for research-related injury should be provided, and the procedures for requesting such compensation and treatment should be described fully on consent forms retained by the subjects.

Prisoners who are minors, mentally disabled or retarded should not be included as subjects unless the research is related to their particular condition and complies with the standards for research involving those groups as well as those for prisoners. (Recommendations concerning research participation of children and the institutionalized mentally infirm will hereafter be made by the Commission).

There should be effective procedures assuring that parole boards cannot take into account prisoners' participation in research, and that prisoners are made certain that there is absolutely no relationship between research participation and determinations by their parole boards.

RECOMMENDATION (5): IN THE ABSENCE OF CERTIFICATION THAT THE REQUIREMENTS UNDER RECOMMENDATION (3) ARE SATISFIED, RESEARCH PROJECTS COVERED BY THAT RECOMMENDATION THAT ARE SUBJECT TO REGULATION BY THE SECRETARY, DHEW, AND ARE CURRENTLY IN PROGRESS SHOULD BE PERMITTED TO CONTINUE NOT LONGER THAN ONE YEAR FROM THE DATE OF PUBLICATION OF THESE RECOMMENDATIONS IN THE FEDERAL REGISTER OR UNTIL COMPLETED, WHICHEVER IS EARLIER.

Part II.  Background

Chapter 3.  Nature of Research Involving Prisoners

Research activities involving prisoners may be divided into four broad categories: biomedical research not realted to the health or well-being of the subject, social research, and behavioral research on practices intended to improve the health or well-being of the subject. The first category of research using prisoners mainly involves phase 1 testing of new drugs and testing of vaccines as to efficacy. Biomedical and behavioral research related to the health or well-being of the prisoner-participants generally involves the study of conditions associated with prisoners or prisons. In addition, innovative practices in prisons, intended to rehabilitate or treat prisoners, often have many attributes of behavioral research but are seldom introduced as such. The major controversy over participation of prisoners surrounds their use as subjects of biomedical research not related to their health or well-being and their unwilling involvement in experimental treatment or rehabilitative programs.

Biomedical research unrelated to the health or well-being of prisoner-participants was conducted in the United States only in isolated instances prior to the establishment in 1934 of a program at Leavenworth Prison to assess the abuse potential of narcotic analgesics; such research is now conducted at the Addiction Research Center in Lexington, Kentucky, although it was announced recently that the program will be terminated by the end of 1976. The current involvement of prisoners in biomedical research unrelated to their health or well-being can be traced to three sources. First, during World War II, prisoners volunteered in large numbers for studies, such as those to develop effective anti-malarial drugs, which were viewed as contributing to the national interest. Reviews of these prison research activites by several state commissions resulted in their endorsement. In fact, prisoner participation in research was felt to be such a salutary experience that the American Medical Association formally opposed allowing persons convicted or participating in scientific experiments. Second, the enthusiastic support of biomedical research by the government and the public following the war brought an enormous growth to research enterprises, and prisoners served as subjects in many of these new endeavors. Third, the thalidomide experience was followed by passage in 1962 of the Keauver-Harris Amendments to the Food and Drug Act, which established additional requirements for testing the safety and efficacy of all drugs to be sold in interstate commerce and thereby encouraged the continued use of prisoners in research. The phase 1 testing requirements established under these amendments required evaluation of the safety of new drugs in normal volunteers under controlled conditions, and prisoners became the population on which much of this testing was performed.

Innovative prison practices are often difficult to distinguish from what might be termed behavioral research on practices intended to improve the health or well-being of prisoner-participants. Since the early 1900's, innovations such as flexible sentences, indeterminate sentences, behavioral therapies during imprisonment, and parole and probation based on evidence of rehabilitation have been introduced into the prison system. These innovations have not generally included provisions for design, review and evaluation as research. Frequently, though, the behavioral programs have had many characteristics of behavior modification research. Examples range from use of "therapeutic community" and reinforcement techniques in prison, to use of adversive conditioning (employing electric shock or drugs with unpleasant effects) in treating sex offenders or uncontrollably violent prisoners, to use of a structured tier system (token economy) in which a prisoner progresses from living conditions of severe deprivation to relative freedom and comfort as a reward for socially acceptable behavior. At the extreme of research or treatment designed to change behavior were castration for sexual offenders and psychosurgery for uncontrollable violence.

The peak of enthusiasm for the application of behavior modification techniques in the prison system was marked by the establishment of the Special Treatment and Rehabilitation Training (START) program in the Federal Bureau of Prisons, and the planning of a new federal prison at Butner, North Carolina, with research in applying behavioral modification throughout a prison as its primary purpose. The START program was abandoned, after one and one-half years of operation, under considerable criticism and after some challenges in court. Similar activities led to a re-evaluation of the programs planned for Butner, which opened in May 1976. It now offers a variety of vocation and academic courses as well as general counseling. Participation in these programs is voluntary, and changes in the program content will be introduced only with the approval of both the inmates and the staff.

Social research and psychological testing are also conducted in prisons. Projects include studies of the factors which may contribute to criminal behavior (such as cytogenetic anomalies or socio-economic and psychological stress), comparison of effectiveness of various rehabilitative programs in reducing recidivism, psychological assessment of criminals as compared with non-criminal counterparts, tracking the outcome of judgments concerning "dangerousness," and evaluating standards for determining competency to stand trial.

Examples of biomedical research on practices intended to improve the health or well-being of subjects in prisons are studies to reduce the spread of infections in crowded environments or to develop new methods of treating drug addiction. Other research, which may or may not be intended to benefit subjects, includes investigations to increase understanding of the nature and causes of narcotic or alcohol abuse and addiction.

Research conducted or supported by DHEW. Information was made available to the Commission by the Public Health Service (PHS) regarding all biomedical research projects involving prisoners that were conducted or supported since January 1, 1970. In addition, the National Institute of Mental Health (NIMH) provided information on all behavioral research with prisoners that was conducted or supported since July 1, 1971. A summary of this information follows.

Biomedical research with prisoners was conducted or supported by five of the six PHS agencies, the exception being the Health Resources Administration. The Alcohol, Drug Abuse, and Mental Health Administration (ADAMHA) reported conducting over 40 intramural research projects in its testing facility at the Addiction Research Center in Lexington, Kentucky. These sudies involved a wide range of activities, such as developing methods for detecting drugs of abuse through urinalysis, studies of various properties of morphine and other narcotics, evaluations of methadone, studies of the effects of amphetamines, analysis of interactions of various drugs with narcotics, and assessment of the addictive or abuse potential and psychoactive effects of new drugs. ADAMHA also supported nine extramural studies involving prisoners, including studies of the XYY chromosome anomaly, assessment of clinical methods to predict episodic violence, study of the use of narcotic antagonists to treat addict inmates in a prison and in a work release program, and study of behavioral and biological correlates of alcoholism.

The Center for Disease Control reported three studies with prisoners; these involved vaccines and skin test studies for a parasitic disease. FDA conducted five studies with prisoners, all of which involved oral administration of a standard dose of a commercially available antibiotic (Penicillin or Tetracycline). FDA also supported three studies with prisoners (two evaluating skin sensitization by irritants and one studying cyclamates). In the Health Services Administration, research involving prisoners was conducted by physicians at one PHS hospital (13 studies of metabolic responses to prolonged bed rest) and by physicians and behavioral scientists at the Research Division, Bureau of Prisons (33 studies involving a wide range of activities, such as dental care, weight reduction and tattoo removal; many were behavioral and rehabilitative rather than biomedical in focus). Seven Institutes of the National Institutes of Health reported support of a total of 19 research programs involving prisoners. This research included studies of vaccines (rubella, rubeola, cholera toxoid, influenza and other respiratory viruses, streptococcus), testicular cell function, treatment of sun-induced skin conditions, responses to infectious diseases (colds, cholers), pathogenesis of acne, and the effect of diet on blood pressure and lipids.

Behavioral research with prisoners conducted or supported by NIMH included psychological and social research studies of crime and delinquency, individual violecne, institutionalization, and law-mental health interactions. Participation of prisoners as subjects in these studies was essential due to the nature of the inquiries. A small number of intramural studies conducted at St. Elizabeth's Hospital were related to analysis of procedures used to determine competency to stand trial or assess dangerousness of criminally insane patients. Support was provided for 19 extramural studies, some of which had biomedical as well as behavioral components. This research included studies: (1) to identify sources and patterns of criminal and delinquent behavior (the XYY syndrome, attitudes toward criminal behavior); (2) to develop, test or evaluate models for the prevention, treatment or remediation of criminal behaviors (prediction of violence, lithium treatment for aggressive behavior, impact of imprisonment on the families of black prisoners, perceptions of the minority prison community, effects of prison environment stress on physical and mental health of inmates and staff); and (3) to define and analyze critical issues in law and mental health interactions (due process in determination of criminal insanity, assessment of adequacy of treatment for affenders commited to mental institutions, release of dangerous mental patients, the impact of a "dangerousness" standard as the sole criterion for involuntary commitment). In addition, NIMH has been directed by Congress to study contributing to homosexual rape in prisons.

Chapter 4.  Extent of Research Involving Prisoners.

The Commission obtained information from all fifty states and the Federal Bureau of Prisons on the policies of each toward research involving prisoners and whether or not research, if permitted, is being conducted. Also, the Pharmaceutical Manufacturers Association surveyed its members to assess the extent of pharmaceutical research involving prisoners. These surveys do not document what is generally considered to be a significant amount of social and behavioral research conducted by scholars and by the prison system itself.

Research in state and federal prisons. To ascertain the status of state laws, regulations and policies governing research involving prisoners, and to determine where such research is being conducted, state correctional agencies and the Federal Bureau of Prisons were surveyed during the summer of 1975. The following information is based on the reports received at the time from the state-wide agencies and the Bureau of Prisons. It should be noted that the policies and research activities of county and municipal jails were not surveyed.

1. Of the twenty-one states that permit biomedical research and the twenty-three states that permit behavioral research in prisons, studies are being conducted in the state prisons of only seven and five states, respectively.

2. Of the seven states in which biomedical research is conducted, all of the programs are unrelated to the health or well-being of the subjects and primarily involve drug and cosmetic testing.

3. Of the five states in which behavioral research is conducted, all of the programs are characterized as therapeutic in four states, and both therapeutic and non-therapeutic research programs involving behavior modification.

4. Eight states prohibit biomedical research: one by legislation, three by departmental policy, and one by moratorium; twenty-three have no specific policy.

5. Five states prohibit behavioral presearch: one by legislation, three by departmental policy, and one by moratorium; twenty-three have no specific policy,

6. Research is being conducted only in states that have specific legislation or departmental policies permitting and regulating it.

7. Information provided by the Federal Bureau of Prisons indicated that both biomedical and behavioral research are permitted by deparmtnetal policy. Biomedical research (limited to addiction research at Lexington) and behavioral research projects are being conducted. In March, 1976, the Director of the Federal Bureau of Prisons announced that all biomedical research in federal prisons would be discontinued.

Participation of prisoners in pharmaceutical testing. The Pharmaceutical Manufacturers Association conducted a survey of its members to ascertain the extent to which they used prisoner volunteers as subjects for drug testing in 1975, with the focus primarily on phase 1 studies. Fifty-one companies, representing three-fourths of the members' annual expenditures for research and development, responded to the survey. Sixteen of the fifty-one used prisoners as subjects.

Of these sixteen companies, fourteen conducted phase 1 drug research with prisoners, employing a total of nearly three thousand six hundred prisoners in one hundred protocols studying seventy-one substances. For nine companies, phase 1 testing represented their only use of prisoners as subjects. The percentage of phase 1 testing subjects who were prisoners rangaed from one hundred per-cent (one comapny) to two per-cent, with a median of fifty per-cent (an average could not be calculated from the data given). The companies listed a total of eight state and six county or municipal prisons as research sites. Ten companies used only minimum security prisons. No companies used detainees in their research. Other categories of volunteer subjects which the companies reported using in phase 1 studies included college students, medical students, company employees, residents of foreign countries, military personnel, members of fraternal organizations, medical personnel, and the general population.

Thirty-three of the fifty-one companies indicated that they had insurance policies or other mechanisms for compensating subjects who might be injured in research. (There was no determination of the extent to which such policies or other mechanisms would provide compensation in the absence of legal liability).

# APPENDIX G

Decision Of The Director, National Institutes Of Health, To Release Guidelines For Research On Recombinant DNA Molecules--June 23, 1976

## Introduction

Today, with the concurrence of the Secretary of Health, Education, and Welfare and the Assistant Secretary for Health, I am releasing guidelines that will govern the conduct of NIH-supported research on recombinant DNA molecules (molecules resulting from the recombination in cell-free systems of segments of deoxyribonucleic acid, the material that determines the hereditary characteristics of all known cells). These guidelines establish carefully controlled conditions for the conduct of experiments involving the insertion of such recombinant genes into organisms, such as bacteria. The chronology leading to the present guidelines and the decision to release them are outlined in this introduction.

In addition to developing these guidelines, NIH has undertaken an environment impact assessment of these guidelines for recombinant DNA research in accordance with the National Environmental Policy Act of 1969 (NEPA). The guidelines are being released prior to completion of this assessment. They will replace the current Asilomar guidelines, discussed below, which in many instances allow research to proceed under less strict conditions. Because the NIH guidelines will afford a greater degree of scrutiny and protection, they are being released today, and will be effective while the environmental impact assessment is under way.

Recombinant DNA research brings to the fore certain problems in assessing the potential impact of basic science on society as a whole, including the manner of providing public participation in those assessments. The field of research involved is a rapidly moving one, at the leading edge of biological science. The experiments are extremely technical and complex. Molecular biologists active in this research have means of keeping informed, but even they may fail to keep abreast of the newest developments. It is not surprising that scientists in other fields and the general public have difficulty in understanding advances in recombinant DNA research. Yet public awareness and understanding of this line of investigation is vital.

It was the scientists engaged in recombinant DNA research who called for a moratorium on certain kinds of experiments in order to assess the risks and devise appropriate guidelines. The capability to perform DNA recombinations, and the potential hazards, had become apparent at the Gordon Research Conference on Nucleic Acids in July 1973. Those in attendance voted to send an open letter to Dr. Philip Handler, President of the National Academy of Sciences, and to Dr. John R. Hogness, President of the Institute of Medicine, NAS. The letter, appearing in SCIENCE 181, 1114, (1973), suggested "that the Academies [sic] establish a study committee to consider this problem and to recommend specific actions or guidelines, should that seem appropriate."

In response, NAS formed a committee, and its members published another letter in SCIENCE 185, 303, (1974). Entitled "Potential Biohazards of Recombinant DNA Molecules," the letter proposed:

> First, and most important, that until the potential hazards of such recombinant DNA molecules have been better evaluated or until adequate methods are developed for preventing their spread, scientists throughout the world join with the members of this committee in voluntarily deferring . . . [certain] experiments . . . .

> Second, plans to link fragments of animal DNAs to bacterial plasmid DNA or bacteriophage DNA should be carefully wrighed . . . .
>
> Third, the Director of the National Institutes of Health is requested to give immediate consideration to establishing an advisory committee charged with (i) overseeing an experimental program to evaluate the potential biological and ecological hazards of the above types of recombinant DNA molecules; (ii) developing procedures which will minimize the spread of such molecules within human and other populations; and (iii) devising guidelines to be followed by investigators working with potentially hazardous recombinant DNA molecules.
>
> Fourth, an international meeting of involved scientists from all over the world should be convened early in the coming year to review scientific progress in this area and to further discuss appropriate ways to deal with the potential biohazards of recombinant DNA molecules.

On October 7, 1974, the NIH Recombinant DNA Molecule Program Advisory Committee (hereafter "Recombinant Advisory Committee") was established to advise the Secretary, HEW, the Assistant Secretary for Health, and the Director, NIH, "concerning a program for developing procedures which will minimize the spread of such molecules within human and other populations, and for devising guidelines to be followed by investigators working with potentially hazardous recombinants."

The international meeting proposed in the SCIENCE article (185, 303, 1974) was held in February 1975 at the Asilomar Conference Center, Pacific Grove, California. It was sponsored by the National Academy of Sciences and supported by the National Institutes of Health and the National Science Foundation. One hundred and fifty people attended, including 52 foreign scientists from 15 countries, 16 representatives of the press, and 4 attorneys.

The conference reviewed progress in research on recombinant DNA molecules and discussed ways to deal with the potential bio-hazards of the work. Participants felt that experiments on construction of recombinant DNA molecules should proceed, provided that appropriate biological and physical containment is utilized. The conference made recommendations for matching levels of containment with levels of possible hazard for various types of experiments. Certain experiments were judged to pose such serious potential dangers that the conference recommended against their being conducted at the present time.

A report on the conference was submitted to the Assembly of Life Sciences, National Research Council, NAS, and approved by its Executive Committee on May 20, 1975. A summary statement of the report was published in SCIENCE 188, 991 (1975), NATURE 225, 442, (1975), and the PROCEEDINGS OF THE NATIONAL ACADEMY OF SCIENCES 72, 1981, (1975). The report noted that "in many countries steps are already being taken by national bodies to formulate codes of practice for the conduct of experiments with known or potential biohazard. Until these are established, we urge individual scientists to use the proposals in this document as a guide."

The NIH Recombinant Advisory Committee held its first meeting in San Francisco immediately after the Asilomar conference. It proposed that NIH use the recommendations of the Asilomar conference as guidelines for research until the committee had an opportunity to elaborate more specific guidelines, and that NIH establish a newsletter for informal distribution of information. NIH accepted these recommendations.

At the second meeting, held on May 12-13, 1975, in Bethesda, Maryland, the committee received a report on biohazard-containment facilities in the United States and reviewed a proposed NIH contract program for the construction and testing of micro-organisms that would have very limited ability to survive in natural environments and would thereby limit the potential hazards. A subcommittee chaired by Dr. David Hogness was appointed to draft guidelines for research involving recombinant DNA molecules, to be discussed at the next meeting.

The NIH committee, beginning with the draft guidelines prepared by the Hogness subcommittee, prepared proposed guidelines for research with recombinant DNA molecules at its third meeting, held on July 18-19, 1975, in Woods Hole, Massachusetts.

Following this meeting, many letters were received which were critical of the guidelines. The majority of critics felt that they were too lax, others that they were too strict. All letters were reviewed by the committee, and a new subcommittee, chaired by Dr. Elizabeth Kutter, was appointed to revise the guidelines.

A fourth committee meeting was held on December 4-5, 1975, in La Jolla, California. For this meeting a "variorum edition" had been prepared, comparing line-for-line the Hogness, Woods Hole, and Kutter guidelines. The committee reviewed these, voting item-by-item for their preference among the three variations and, in many cases, adding new material. The result was the "Proposed Guidelines for Research Involving Recombinant DNA Molecules," which were referred to the Director, NIH, for a final decision in December 1975.

As Director of the National Institutes of Health, I called a special meeting of the Advisory Committee to the Director to review these porposed guidelines. The meeting was held at NIH, Bethesda, on February 9-10, 1976. The Advisory Committee is charged to advise the Director, NIH, on matters relating to the broad setting--scientific, technological, and socioeconomic--in which the continuing development of the biomedical sciences, education for the health professions, and biomedical communications must take place, and to advise on their implications for NIH policy, program development, resource allocation, and administration. The members of the committee are knowledgeable in the fields of basic and clinical biomedical sciences, the social sciences, physical sciences, research, education, and communications. In addition to current members of the committee, I invited a number of former committee members as well as other scientific and public representatives to participate in the special February session.

The purpose of the meeting was to seek the committee's advice on the guidelines proposed by the Recombinant Advisory Committee. The Advisory Committee to the Director was asked to determine whether, in their judgment, the guidelines balanced scientific responsibility to the public with scientific freedom to pursue new knowledge.

Public responsibility weighs heavily in this genetic research area. The scientific community must have the public's confidence that the goals of this profoundly important research accord respect to important ethical, legal, and social values of our society. A key element in achieving and maintaining this public trust is for the scientific community to ensure an openness and candor in its proceedings. The meerings of the Director's Advisory Committee, the Asilomar group, and the Recombinant Advisory Committee have reflected the intent of science to be an open community in considering the conduct of recombinant DNA experiments. At the Director's Advisory Committee meeting, there was ample opportunity for comment and an airing of the issues, not only by the committee members but by public witnesses as well. All major points of view were broadly represented.

I have been reviewing the guidelines in light of the comments and suggestions made by participants at that meeting, as well as the written comments received afterward. As part of that review, I asked the Recombinant Advisory Committee to consider at their meeting of April 1-2, 1976, a number of selected issues raised by the commentators. I have taken those issues and the response of the Recombinant Advisory Committee into account in arriving at my decision on the guidelines. An analysis of the issues and the basis for my decision follow.

I. General Policy Considerations.

A word of explanation might be interjected at this point as to the nature of the studies in question. Within the past decade, enzymes capable of breaking DNA strands at specific sites and of coupling the broken fragments in new combinations were discovered, thus making possible the insertion of foreign genes into viruses or certain cell particles (plasmids). These, in turn, can be used as vectors to introduce the foreign genes into bacteria or into cells of plants or animals in test tubes. Thus transplanted, the genes may impart their hereditary properties to new hosts. These cells can be isolated and cloned--that is, bred into a genetically homogeneous culture. In general, there are two potential uses for the clones so produced: as a tool for studying the transferred genes, and as a new useful agent, say for the production of a scarce hormone.

Recombinant DNA research offers great promise, particularly for improving the understanding and possibly the treatment of various diseases. There is also a potential risk--that micro-organisms with transplanted genes may prove hazardous to man or other forms of life. Thus special provisions are necessary for their containment.

All commentators acknowledged the exemplary responsibility of the scientific community in dealing publicly with the potential risks in DNA recombinant research and in calling for a self-imposed moratorium on certain experiments in order to assess potential hazards and devise appropriate guidelines. Most commentators agreed that the process leading to the formulation of the proposed guidelines was a most responsible and responsive one. Suggestions by the commentators on broad policy aspects of the guidelines, the implementation of the guidelines for NIH grantees and contractors, and the scope and impact of the guidelines nationally and internationally.

A. Science Policy Considerations. Commentators were divided on how best to steer a course between stifling research through excessive regulation and allowing it to continue with sufficient controls. Several emphasized that the public must have assurance that the controls afford adequate protection against potential hazards. In the views of these commentators, the burden is on the scientific community to show that the danger is minimal and that the benefits are substantial and far outweigh the risks.

Opinion differed on whether the proposed guidelines were an appropriate response to the potential benefits and hazards. Several found the guidelines to so exaggerate safety procedures that inquiry would be unnecessarily retarded, while others found the guidelines weighted toward promoting research. The issue was how to strike a reasonable balance--in fact, a proper policy "bias"--between concerns to "to slow" and those to progress rapidly.

There was strong disagreement about the nature and level of the possible hazards of recombinant DNA research. Several commentators believed that the hazards posed were unique. In their view, the occurrence of an accident or the escape of a vector could initiate an irreversible process, with a potential for creating problems many times greater than those arising from the multitude of genetic recombinations that occur spontaneously in nature. These commentators stress the moral obligation on the part of the scientific community to do no harm.

Other commentators, however, found the guidelines to be adequate to the hazards posed. In their view, the guidelines struck an appropriate balance so that research could proceed cautiously. Still other commentators found the guidelines too onerous and restrictive in light of the potential benefits of this research for medicine, argriculture, and industry. Some felt that the guidelines are perhaps more stringent than necessary given the available evidence on the likelihood of hazards, but supported them as a compromise that would best serve the scientific community and the public at large. Many commentators urged that the guidelines be adopted as soon as possible to afford more specific direction to this research area.

I understand and appreciate the concerns of those who urge that this research proceed because of the benefits and of those who urge caution because of potential hazards. The guidelines issued today allow the research to go forward in a manner responsive and appropriate to hazards that may be realized in the future.

The object of these guidelines is to ensure that experimental DNA recombination will have no ill effects on those engaged in the work, on the general public, or on the environment. The essence of their construction is subdivision of potential experiments by class, decision as to which experiments should be permitted at present, and assignment to these of certain procedures for containment of recombinant organisms.

Containment is defined as physical and biological. Physical containment involves the isolation of the research by procedures which have evolved over many years of experience in laboratories studying infectious micro-organisms. P1 containment--the first physical containment level--is that used in most routine bacteriology laboratories. P2 and P3 afford increasing isolation of the research from the environment. P4 represents the most extreme measures used for containing virulent pathogens, and permits no escape of contaminated air, wastes, or untreated materials. Biological containment is the use of vectors or hosts that are crippled by mutation so that the recombinant DNA is incapable of surviving under natural conditions.

The experiments now permitted under the guidelines involve no known additional hazard to the workers or the environment beyond the relatively low risk known to be associated with the source materials. The additional hazards are speculative and therefore not quantifiable. In a real sense they are considerably less certain than are the benefits now clearly derivable from the projected research.

For example, the ability to produce, through "molecular cloning", relatively large amounts of pure DNA from the chromosomes of any living organism will have a profound effect in many areas of biology. No other procedure, not even chemical synthesis, can provide pure material corresponding to particular genes. DNA "probes," prepared from the clones will yield precise evidence on the presence or absence, the organization, and the expression of genes in health and disease.

Potential medical advances were outlined by scientists active in this research area who were present at the meeting of the Director's Advisory Committee. Of enormous importance, for example, is the opportunity to explore the malfunctioning of cells in complicated diseases. Our ability to understand a variety of hereditary defects may be significantly enhanced, with amelioration of their expression a real possibility. There is the potential to elucidate mechanisms in certain cancers, particularly those that might be caused by viruses.

Instead of mere propagation of foreign DNA, the expression of the genes of one organism by the cell machinery of another may alter the new host and open opportunities for manipulating the biological properties of cells. In certain prokaryotes (organisms with a poorly developed nucleus, like bacteria), this exchange of genetic information occurs in nature. Such exchange explains, for instance, an important mechanism for the changing and spreading of resistance to antibiotics in bacteria. Beneficial effects of this mechanism might be the production of medically important compounds for the treatment and control of disease. Examples frequently cited are the production of insulin, growth hormone, specific antibodies, and clotting factors absent in victims of hemophilia.

Aside from the potential medical benefits, a whole host of other applications in science and technology have been envisioned. Examples are the largescale production of enzymes for industrial use and the development of bacteria that could ingest and destroy oil spills in the sea. Potential benefits in agriculture include the enhancement of nitrogen fixation in certain plants, permitting increased food production.

While the projected research offers the possibility of many benefits, it must proceed only with assurance that potential hazards can be controlled or prevented. Some commentators are concerned that nature may maintain a barrier to the exchange of DNA between prokaryotes and eukaryotes (higher organisms, with a well-formed nucleus)--a barrier that can now be crossed by experimentalists. They further argue that expression of the foreign DNA may alter the host in unpredictable and undesirable ways. Conceivable harm could result if the altered host has a competitive advantage that would foster its survival in some niche within the ecosystem. Other commentators believe that the endless experiments in recombination of DNA which nature has conducted since the beginning of life on the earth, and which have accounted in part for the evoluaiton of species, have most likely involved exchange of DNA between widely disparate species. They argue that prokaryotes such as bacteria in the intestines of man do exchange DNA with this eukaryotic host and that the failure of the altered prokaryotes to be detected attests to a sharply limited capacity of such recombinants to survive. Thus nature, this argument runs, has already tested the probabilities of harmful recombination and any survivors of such are already in the ecosystem. The fact is that we do not know which of the above-stated propositions is correct.

The international scientific community, as exemplified by the Asilomar conference and the deliberations attendant upon preparation of the present guidelines, has indicated a desire to proceed with research in a conservative manner. And most of the considerable public commentary on the subject, while urging caution, has also favored proceeding. Three European groups have independently arrived at the opinion that recombinant DNA research should proceed with caution. These are the Working Party on Experimental Manipulation of the Genetic Composition of Micro-Organisms, whose "Ashby Report" was presented to Parliament in the United Kingdom by the Secretary of State for Education and Science in January 1975; the Advisory Committee on Medical Research of the World Health Organization, which issued a press release in July 1975; and the European Molecular Biology Organization Standing Committee on Recombinant DNA, meeting in February 1976.

There is no means for a flat proscription of such research throughout the world community of science. There is also no need to attempt it. It is likely that the evaluation engendered in the preparation and application of these guidelines will lead to beneficial review of some of the containment practices in other work that is not technically defined as recombinant DNA research.

Recombinant DNA research with which these guidelines are concerned involves micro-organisms such as bacteria or viruses or cells of higher organisms growing in tissue culture. It is extremely important for the public to be aware that this research is not directed to altering of genes in humans although some of the techniques developed in this research may have relevance if this is attempted in the future.

NIH recognizes its responsibility to conduct and support research designed to determine the extent to which certain potentially harmful effects from recombinant DNA molecules may occur. Among these are experiments, to be conducted under maximum containment, that explore the capability of foreign genes to alter the character of host or vector, rendering it harmful, as through the production of toxic products.

Given the general desire that no rare and unexpected event arising from this research shall cause irreversible damage, it is obvious that merely to establish conservative rules of conduct for one group of scientists is not enough. The precautions must be uniformly and unanimously observed. Second, there must be full and timely exchange of experiences so that guidelines can be altered on the basis of new knowledge. The guidelines must also be implemented in a manner that protects all concerned--the scientific workers most likely to encounter unexpected hazards and all forms of life within our bio-sphere. The responsibility of the scientists involved is as inescapable and extreme as is their opportunity to beneficially enrich our understanding.

B. Environmental Policy Considerations. A number of commentators urged NIH to consider preparing an environmental impact statement on recombinant DNA research activity. They evoked the possibility that organisms containing recombinant DNA molecules might escape and affect the environment in potentially harmful ways.

I am in full agreement that the potentially harmful effects of this research on the environment should be assessed. As discussed throughout this paper, the guidelines are premised on physical and biological containment to prevent the release or propagation of DNA recombinants outside the laboratory. Deliberate release of organisms into the environment is prohibited. In my view, the stipulated physical and biological containment ensures that this research will proceed with a high degree of safety and precaution. But I recognize the legitimate concern of those urging than an evironmental impact assessment be done. In view of this concern and ensuing public debate, I have reviewed the appropriateness of such an assessment and have directed that one be undertaken.

The purpose of this assessment will be to review the environmental effects, if any, of research that may be conducted under the guidelines. The assessment will provide further opportunity for all concerned to address the potential benefits and hazards of this most important research activity. . . .

It should be noted that the development of the guidelines was in large part tantamount to conducting an environmental impact assessment. For example, the objectives of recombinant DNA research, and alternate approaches to reach those objectives, have been considered. The potential hazards and risks have been analyzed. Alternative approaches have been thoroughly considered, to maximize safety and minimize potential risk. And an elaborate review structure has been created to achieve these safety objectives. From a public policy viewpoint, however, the environmental impact assessment will be yet another review that will provide further opportunity for the public to participate and comment on the conduct of this research.

National Institutes of Health Guidelines for Research Involving
Recombinant DNA Molecules--June, 1976

I.  Introduction.

The purpose of these guidelines is to recommend safeguards for research on recombinant DNA molecules to the National Institutes of Health and to other institutions that support such research. In this context we define recombinant DNA's as molecules that consist of different segments of DNA which have been joined together in cell-free systems, and which have the capacity to infect and replicate in some host cell, either autonomously or as an integrated part of the host's genome.

This is the first attempt to provide a detailed set of guidelines for use by study sections as well as practicing scientists for evaluating research on recombinant DNA molecules. We cannot hope to anticipate all possible lines of imaginative research that are possible with this powerful new methodology. Nevertheless, a considerable volume of written and verbal contributions from scientists in a variety of disciplines has been received. In many instances the views presented to us were contradictory. At present, the hazards may be guessed at, speculated about, or voted upon, but they cannot be known absolutely in the absence of firm experimental data -- and, unfortunately, the needed data were, more often than not, unavailable. Our problem then has been to construct guidelines that allow the promise of the methodology to be realized while advocating the considerable caution that is demanded by what we and others view as potential hazards.

In designing these guidelines we have adopted the following principles, which are consistent with the general conclusions that were formulated at the International Conference on Recombinant DNA Molecules held at Asilomar Conference Center, Pacific Grove, California, February 1975 (3):  (i) There are certain experiments for which the assessed potential hazard is so serious that they are not to be attempted at the present time.  (ii) The remainder can be undertaken at the present time provided that the experiment is justifiable on the basis that new knowledge or benefits to humankind will accrue that cannot readily be obtained by use of conventional methodology and that appropriate safeguards are incorporated into the design and execution of the experiment. In addition to an insistence on the practice of good microbiological techniques, these safeguards consist of providing both physical and biological barriers to the dissemination of the potentially hazardous agents.  (iii) The level of containment provided by these barriers is to match the estimated potential hazard for each of the different classes of recombinants. For projects in a given class, this level is to be highest at initiation and modified subsequently only if there is a substantiated change in the assessed risk or in the applied methodology.  (iv) The guidelines will be subjected to periodic review (at least annually) and modified to reflect improvements in our knowledge of the potential biohazards and of the available safeguards.

In constructing these guidelines it has been necessary to define boundary conditions for the different levels of physical and biological containment and for the classes of experiments to which they apply. We recognize that these definitions do not take into account existing and anticipated special procedures and information that will allow particular experiments to be carried out under different conditions than indicated here without sacrifice of safety. Indeed, we urge that individual investigators devise simple and more effective containment procedures and that study sections give consideration to such procedures which may allow change in the containment levels recommended here.

It is recommended that all publications dealing with recombinant DNA work include a description of the physical and biological containment procedures practiced, to aid and forewarn others who might consider repeating the work.

II. Containment.

Effective biological safety programs have been operative in a variety of laboratories for years. Considerable information therefore already exists for the design of physical containment facilities and the selection of laboratory procedures applicable to organisms carrying recombinant DNAs. The existing programs rely upon mechanisms that, for convenience, can be divided into two categories: (i) a set of standard practices that are generally used in microbiological laboratories, and (ii) special procedures, equipment, and laboratory installations that provide physical barriers which are applied in varying degrees according to the estimated biohazard.

Experiments on recombinant DNAs by their very nature lend themselves to a third containment mechanism -- namely, the application of highly specific biological barriers. In fact, natural barriers do exist which either limit the infectivity of a vector or vehicle (plasmid, bacteriophage or virus) to specific hosts, or its dissemination and survival in the environment. The vectors that provide the means for replication of the recombinant DNAs and/or the host cells in which they replicate can be genetically designed to decrease by many orders of magnitude the probability of dissemination of recombinant DNAs outside the laboratory.

As these three means of containment are complementary, different levels of containment appropriate for experiments with different recombinants can be established by applying different combinations of the physical and biological barriers to a constant use of the standard practices. We consider these categories of containment separately here in order that such combinations can be conveniently expressed in the guidelines for research on the different kinds of recombinant DNAs.

A. Standard practices and training. The first principle of containment is a strict adherence to good microbiological practices. Consequently, all personnel directly or indirectly involved in experiments on recombinant DNAs must receive adequate instruction. This should include at least training in aseptic techniques and instruction in the biology of the organisms used in the experiments so that the potential biohazards can be understood and appreciated.

Any research group working with agents with a known or potential biohazard should have an emergency plan which describes the procedures to be followed if an accident contaminates personnel or environment. The principal investigator must ensure that everyone in the laboratory is familiar with both the potential hazards of work and the emergency plan. If a research group is working with a known pathogen for which an effective vaccine is available, all workers should be immunized. Serological monitoring, where appropriate, should be provided.

B. Physical containment levels. A variety of combinations (levels) of special practices, equipment, and laboratory installations that provide additional physical barriers can be formed. For example, 31 combinations are listed in "Laboratory Safety at the Center for Disease Control"; four levels are associated with the "Classification of Etiologic Agents on the Basis of Hazard", four levels were recommended in the "Summary Statement of the Asilomar Conference on Recombinant DNA Molecules"; and the National Cancer Institute used three levels for research on oncogenic viruses. We emphasize that these are an aid to, and not a substitute for, good technique. Personnel must be competent in the effective use of all equipment needed for the requirement containment level as described below. We define only four levels of physical containment here, both because the accuracy with which one can presently assess the biohazards that may result from recombinant DNAs does not warrant a more detailed classification, and because additional flexibility can be obtained by combination of the physical with the biological barriers. Though different in detail, these four levels (P1, P2, P3, P4) approximate those given for human etiologic agents by the Center for Disease Control. (i.e., classes 1 through 4), in the Asilomar summary statement (i.e., minimal, low, moderate, and high), and by the National Cancer Institute for oncogenic viruses (i.E., low, moderate, and high), as is indicated by the P-number or adjective in the following headings. It should be emphasized that the descriptions and assignments of physical containment detailed below are based on existing approaches to containment of hazardous organisms.

We anticipate, and indeed already know of, procedures which enhance physical containment capability in novel ways. For example, miniaturization of screening, handling, and analytical procedures provides substantial containment of a given host-vector system. Thus, such procedures should reduce the need for the standard types of physical containment, and such innovations will be considered by the Recombinant DNA Molecule Program Advisory Committee.

The special practices, equipment and facility installations indicated for each level of physical containment are required for the safety of laboratory workers, other persons, and for the protection of the environment. Optional items have been excluded; only those items deemed absolutely necessary for safety are presented. Thus, the listed requirements present basic safety criteria for each level of physical containment. Other microbiological practices and laboratory techniques which promote safety are to be encouraged. Additional information giving further guidance on physical containment is provided in a supplement to the guidelines.

P1 Level (Minimal). A laboratory suitable for experiments involving recombinant DNA molecules requiring physical containment at the P1 level is a laboratory that possesses no special engineering design features. It is a laboratory commonly used for microorganisms of no or minimal biohazard under ordinary conditions of handling. Work in this laboratory is generally conducted on open bench tops. Special containment equipment is neither required nor generally available in this laboratory. The laboratory is not separated from the general traffic patterns of the building. Public access is permitted.

The control of biohazards at the P1 level is provided by standard microbiological practices of which the following are examples: (i) Laboratory doors should be kept closed while experiments are in progress. (ii) Work surfaces should be decontaminated daily and following spills of recombinant DNA materials. (iii) Liquid wastes containing recombinant DNA materials should be decontaminated before disposal. (iv) Solid wastes contaminated with recombinant DNA materials should be decontaminated or packaged in a durable leak-proof container before removal from the laboratory. (v) Although pipetting by mouth is permitted, it is preferable that mechanical pipetting devices be used. When pipetting by mouth, cotton-plugged pipettes shall be employed. (vi) Eating, drinking, smoking, and storage of food in the working area should be discouraged. (vii) Facilities to wash hands should be available. (viii) An insect and rodent control program should be provided. (ix) The use of laboratory gowns, coats, or uniforms is discretionary with the laboratory supervisor.

P2 Leval (Low). A laboratory suitable for experiments involving recombinant DNA molecules requiring physical containment at the P2 level is similar in construction and design to the P1 laboratory. The P2 laboratory must have access to an autoclave within the building; it may have a Biological Safety Cabinet. Work which does not produce a considerable aerosol is conducted on the open bench. Although this laboratory is not separated from the general traffic pattern of the building, access to the laboratory is limited when experiments requiring P2 level physical containment are being conducted. Experiments of lesser biohazard potential can be carried out concurrently in carefully demarcated areas of the same laboratory.

The P2 laboratory is commonly used for experiments involving microorganisms of low biohazard such as those which have been classified by the Center for Disease Control as Class 2 agents (5).

The following practices shall apply to all experiments requiring P2 level physical containment: (i) Laboratory doors shall be kept closed while experiments are in progress. (ii) Only persons who have been advised of the potential biohazard shall enter the laboratory. (iii) Children under 12 years of age shall not enter the laboratory. (iv) Work surfaces shall be decontaminated daily and immediately following spills of recombinant DNA materials. (v) Liquid wastes of recombinant DNA materials shall be decontaminated before disposal. (vi) Solid wastes contaminated with recombinant DNA materials shall be decontaminated or packaged in a durable leak-proof container before removal from the laboratory. Packaged materials shall be disposed of by incineration or sterilized before disposal by other methods. Contaminated materials that are to be processed and reused (i.e., glassware) shall be decontaminated before removal from the laboratory. (vii) Pipetting by mouth is prohibited; mechanical pipetting devices shall be used. (viii) Eating, drinking, smoking, and storage of food are not permitted in the working area. (ix) Facilities to wash hands shall be available within the laboratory. Persons handling recombinant DNA materials should be encouraged to wash their hands frequently and when they leave the laboratory. (x) An insect and rodent control program shall be provided. (xi) The use of laboratory gowns, coats, or uniforms is required. Such clothing shall not be worn to the lunch room or outside the building. (xii) Animals not related to the experiment shall not be permitted in the laboratory. (xiii) Biological Safety Cabinets and/or other physical containment equipment shall be used to minimize the hazard of aerosolization of recombinant DNA materials from operations or devices that produce a considerable aerosol (e.g., blender, lyophilizer, sonicator, shaking machine, etc.). (xiv) Use of the hypodermic needle and syringe shall be avoided when alternative methods are available.

P3 Level (Moderate). A laboratory suitable for experiments involving recombinant DNA molecules requiring physical containment at the P3 level has special engineering design features and physical containment equipment. The laboratory is separated from areas which are open to the general public. Separation is generally achieved by controlled access corridors, air locks, locker rooms or other double-doored facilities which are not available for use by the general public. Access to the laboratory is controlled. Biological Safety Cabinets are available within the controlled laboratory area. An autoclave shall be available within the building and preferably within the controlled laboratory area. The surfaces of walls, floors, bench tops, and ceilings area easily cleanable to facilitate housekeeping and space decontamination.

Directional air flow is provided within the controlled laboratory area. The ventilation system is balanced to provide for an inflow of supply air from the access corridor into the laboratory. The general exhaust air from the laboratory is discharged outdoors and so dispensed to the atmosphere as to prevent reentry into the building. No recirculation of the exhaust air shall be permitted without appropriate treatment.

No work in open vessels involving hosts or vectors containing recombinant DNA molecules requiring P3 physical containment is conducted on the open bench. All such procedures are confined to Biological Safety Cabinets.

The following practices shall apply to all experiments requiring P3 level physical containment: (i) The universal biohazard sign is required on all laboratory access doors. Only persons whose entry into the laboratory is required on the basis of program or support needs shall be authorized to enter. Such persons shall be advised of the potential hazards before entry and they shall comply with posted entry and exit procedures. Children under 12 years of age shall not enter the laboratory. (ii) Laboratory doors shall be kept closed while experiments are in progress. (iii) Biological Safety Cabinets and other physical containment equipment shall be used for all procedures that produce aerosols of recombinant DNA materials (e.g., pipetting, plating, flaming, transfer operations, grinding, blending, drying, sonicating, shaking, etc.). (iv) The work surfaces of Biological Safety Cabinets and other equipment shall be decontaminated following the completion of the experimental activity contained within them. (v) Liquid wastes containing recombinant DNA materials shall be decontaminated before disposal. Solid wastes contaminated with recombinant DNA materials shall be decontaminated or packaged in a durable leak-proof container before removal from the laboratory. Packaged material shall be sterilized before disposal. Contaminated materials that are to be processed and reused (i.e., glassware) shall be sterilized in the controlled laboratory area or placed in a durable leak-proof container before removal from the controlled laboratory area. This container shall be sterilized before the materials are processed. (vii) Pipetting by mouth is prohibited; mechanical pipetting devices shall be used. (viii) Eating, drinking, smoking, and storage of food are not permitted in the laboratory. (ix) Facilities to wash hands shall be available within the laboratory. Persons shall wash hands after experiments involving recombinant DNA materials and before leaving the laboratory. (x) An insect and rodent control program shall be provided. (xi) Laboratory clothing that protects street clothing (i.e., long sleeve solid-front or wrap-around gowns, no-button or slipover jackets, etc.) shall be worn in the laboratory. FRONT-BUTTON LABORATORY COATS ARE UNSUITABLE. Gloves shall be worn when handling recombinant DNA materials. Provision for laboratory shoes is recommended. Laboratory clothing shall not be worn outside the laboratory and shall be decontaminated before it is sent to the laundry. (xii) Raincoats, overcoats, topcoats, coats, hats, caps, and such street outerwear shall not be kept in the laboratory. (xiii) Animals and plants not related to the experiment shall not be permitted in the laboratory. (xiv) Vacuum lines shall be protected by filters and liquid traps. (xv) Use of the hypodermic needle and syringe shall be avoided when alternate methods are available. (xvi) If experiments of lesser biohazard potential are to be conducted in the same laboratory concurrently with experiments requiring P3 level physical containment they shall be conducted only in accordance with all P3 level requirements. (xvii) Experiments requiring P3 level physical containment can be conducted in laboratories where the directional air flow and general exhaust air condition described above cannot be achieved, provided that this work is conducted in accordance with all other requirements listed and is contained in a Biological Safety Cabinet with attached glove ports and gloves. All material before removal from the Biological Safety Cabinets shall be sterilized or transferred to a non-breakable, sealed container, which is then removed from the cabinet through a chemical decontamination tank, autoclave, ultraviolet air lock, or after the entire cabinet has been decontaminated.

P4 Level (High). Experiments involving recombinant DNA molecules requiring physical containment at the P4 level shall be confined to work areas in a facility of the type designed to contain microorganisms that are extremely hazardous to man or may cause serious epidemic disease. The facility is either a separate building or it is a controlled area, within a building which is completely isolated from all other areas of the building. Access to the facility is under strict control. A specific facility operations manual is available. Class III Biological Safety Cabinets are available within work areas of the facility.

A P4 facility has engineering features which are designed to prevent the escape of microorganisms to the environment. These features include: (i) monolithic walls, floors, and ceilings in which all penetrations such as for air ducts, electrical conduits, and utility pipes are selaed to assure the physical isolation of the work area and to facilitate housekeeping and space decontamination; (ii) air locks through which supplies and materials can be brought safely into the facility; (iii) contiguous clothing change and shower rooms through which personnel enter into and exit from the facility; (iv) double-door autoclaves to sterilize and safely remove wastes and other materials from the facility; (v) a biowaste treatment system to sterilize liquid effluents if facility drains are installed; (vi) a separate ventiliation system which maintains negative air pressures and directional air flow within the facility; and (vii) a treatment system to decontaminate exhaust air before it is dispersed to the atmosphere. A central vacuum utility system is not encouraged; if one is installed, each branch line leading to a laboratory shall be protected by a high efficiency particulate air filter.

The following practices shall apply to all experiments requiring P4 level physical containment: (i) The universal biohazard sign is required on all facility access doors and all interior doors to individual laboratory rooms where experiments are conducted. Only persons whose entry into the facility or individual laboratory rooms is required on the basis of program or support needs shall be authorized to enter. Such persons shall be advised of the potential biohazards and instructed as to the appropriate safeguards to ensure their safety before entry. Such persons shall comply with the instructions and all other posted entry and exit procedures. Under no condition shall children under 15 years of age be allowed entry. (ii) Personnel shall enter into and exit from the facility only through the clothing change and shower rooms. Personnel shall shower at each exit from the facility. The air locks shall not be used for personnel entry or exist except for emergencies. (iii) Street clothing shall be removed in the outer facility side of the clothing change area and kept there. Complete laboratory clothing including undergarments, pants and shirts or jumpsuits, shoes, head cover, and gloves shall be provided and used by all persons who enter into the facility. Upon exit, this clothing shall be stored in lockers provided for this purpose or discarded into collection hampers before personnel enter into the shower area. (iv) Supplies and materials to be taken into the facility shall be placed in an entry air lock. After the outer door (opening to the corridor outside of facility) has been secured, personnel occupying the facility shall retrieve the supplies and materials by opening the interior air lock door. This door shall be secured after supplies and materials are brought into the facility. (v) Doors to laboratory rooms within the facility shall be kept closed while experiments are in progress. (vi) Experimental procedures requiring P4 level physical containment shall be confined to Class III Biological Safety Cabinets. All materials, before removal from these cabinets, shall be sterilized or transferred to a non-breakable sealed container, which is then removed from the system through a chemical decontaminated tank, autoclave, or after the entire eyetem has been decontaminated. (vii) No materials shall be removed from the facility unless they have been sterilized or decontaminated in a manner to prevent the release of agents requiring P4 physical containment. All wastes and other materials and equipment not damaged by high temperature or steam shall be sterilized in the double-door autoclave. Biological materials to be removed from the facility shall be transferred to a non-breakable sealed container which is then removed from the facility through a chemical decontamination tank or a chamber designed for gas sterilization. Other materials which may be damaged by temperature or steam shall be sterilized by gaseous or vapor methods in an air lock or chamber designed for this purpose. (viii) Eating, drinking, smoking, and storage of food are not permitted in the facility. Foot-operated water fountains located in the facility corridors are permitted. Separate potable water piping shall be provided for these water fountains. (ix) Facilities to wash hands shall be available within the facility. Persons shall wash hands after experiments. (x) An insect and rodent control program shall be provided. (xi) Animals and plants not related to the experiment shall not be permitted in the facility. (xii) If a central vacuum system is provided, each vacuum outlet shall be protected by a filter and liquid trap in addition to the branch line HEPA filter mentioned above. (xiii) Use of the hypodermic needle and syringe shall be avoided when alternative methods are available. (iv) If experiments of lesser biohazard potential are to be conducted in the facility concurrently with experiments requiring P4 level containment, they shall be confined in Class I or Class II Biological Safety Cabinets or isolated by other physical containment equipment. Work surfaces of Biological Safety Cabinets and other equipment shall be contaminated following the completion of the experimental activity contained within them. Mechanical pipetting devises shall be used. All other practices listed above with the exception of (vi) shall apply.

C. Shipment. To protect product, personnel, and the environment, all recombinant DNA material will be shipped in containers that meet the requirement issued by the U.S. Public Health Service (Section 72.25 of Part 72, Title 42, Code of Federal Regulations), Department of Transportation (Section 173.387 (b) of Part 173, Title 49, Code of Federal Regulations) and the Civil Aeronautics Board (C.A.B. No. 82, Official Air Transport Restricted Articles Tariff No. 6-D) for shipment of etiologic agents. Labeling requirements specified in these Federal regulations and tariffs will apply to all viable recombinant DNA materials in which any portion of the material is derived from an etiologic agent listed in paragraph (c) of 42 CFR 7225. Additional information on packing and shipping is given in a supplement to the guidelines.

D. Biological containment levels. Biological barriers are specific to each host-vector system. Hence the criteria for this mechanism of containment cannot be generalized to the same extent as for physical containment. This is particularly true at the present time when our experience with existing host-vector systems and our predictive knowledge about projected systems are sparse. The classification of experiments with recombinant DNAs that is necessary for the construction of the experimental guidelines can be accomplished with least confusion if we use the host-vector system as the primary element and the source of the inserted DNA as the secondary element in the classification. It is therefore convenient to specify the nature of the biological containment under host-vector headings such as those given below for Escherichia coli K-12.

III. Experimental Guidelines.

A general rule that, though obvious, deserves statement is that the level of containment required for any experiment on DNA recombinants shall never be less than that required for the most hazardous component used to construct and clone the recombinant DNA (i.e., vector, hosts, and inserted DNA). In most cases the level of containment will be greater, particularly when the recombinant DNA is formed from species that ordinarily do not exchange genetic information. Handling the purified DNA will generally require less stringent precautions than will propagating the DNA. However, the DNA itself should be handled at least as carefully as one would handle the most dangerous of the DNAs used to make it.

The above rule by itself effectively precludes certain experiments -- namely, those in which one of the components is in Class 5 of the "Classification of Etiologic Agents on the Basis of Hazard" as these are excluded from the United States by law and USDA administrative policy. There are additional experiments which may engender such serious biohazards that they are not to be performed at this time. These are considered prior to presentation of the containment guidelines for permissible experiments.

A. Experiments that are not to be performed. We recognize that it can be argued that certain of the recombinants placed in this category could be adequately contained at this time. Nonetheless, our estimates of the possible dangers that may ensue if that containment fails are of such a magnitude that we consider it the wisest policy to at least defer experiments on these recombinant DNAs until there is more information to accurately assess that danger and to allow the construction of more effective biological barriers. In this respect, these guidelines are more stringent than those initially recommended.

The following experiments are not to be initiated at the present time: (i) Cloning of recombinant DNAs derived from the pathogenic organisms in Classes 3, 4, and 5 of "Classification of Etiologic Agents on the Basis of Hazard" (5), or oncogenic viruses classified by NCI as moderate risk (6), or cells known to be infected with such agents, regardless of the host-vector system used. (ii) Deliberate formation of recombinant DNAs containing genes for the biosynthesis of potent toxins (e.g., botulinum or diphtheria toxins; venoms from insects, snakes, etc.). (ii) Deliberate creation from plant pathogens of recombinant DNAs that are likely to increase virulence and host range. (iv) Deliberate release into the environment of any organism containing a recombinant DNA molecule. (v) Transfer of a drug resistance trait to microorganisms that are not known to acquire it naturally if such acquisition could compromise the use of a drug to control disease agents in human or veterinary medicine or agriculture.

In addition, at this time large-scale experiments (e.g., more than 10 liters of culture) with recombinant DNAs known to make harmful products are not to be carried out. We differentiate between small- and large-scale experiments with such DNAs because the probability of escape from containment barriers normally increases with increasing scale. However, specific experiments in this category that are of direct societal benefit may be excepted from this rule if special biological containment precautions and equipment designed for large-scale operations are used, and provided that these experiments are expressly approved by the Recombinant DNA Molecule Program Advisory Committee of NIH.

B. Containment guidelines for permissible experiments. It is anticipated that most recombinant DNA experiments initiated before these guidelines are next reviewed (i.e., within the year) will employ E. coli K-12 host-vector biohazards committee 7 to: (i) advise the institution on policies, (ii) create and maintain a central reference file and library of catalogs, books, articles, newsletters, and other communications as a source of advise and reference regarding, for example, the availability and quality of the safety equipment, the availability and level of biological containment for various host-vector systems, suitable training of personnel and data on the potential biohazards associated with certain recombinant DNAs, (iii) develop a safety and operations manual for any P4 facility maintained by the institution and used in support of recombinant DNA research, (iv) certify to the NIH on applications for research support and annually thereafter, that facilities, procedures, and practices and the training and expertise of the personnel involved have been reviewed and approved by the institutional biohazards committee.

The biohazards committee must be sufficiently qualified through the experience and expertise of its membership and the diversity of its membership to ensure respect for its advice and counsel. Its membership should include individuals from the institution or consultants, selected so as to provide a diversity of discipline relevant to recombinant DNA technology, biological safety, and engineering. In addition to possessing the professional competence necessary to assess and review specific activities and facilities, the committee should possess or have available to it, the competence to determine the acceptability of its findings in terms of applicable laws, regulations, standards of practices, community attitudes, and health and environmental considerations. Minutes of the meetings should be kept and made available for public inspection. The institution is responsible for reporting names of and relevant background information on the members of its biohazards committee to the NIH.

The following roles and responsibilities define an administrative framework in which safety is an essential and integrated function of research involving recombinant DNA molecules.

A. Principal Investigator. The principal investigator has the primary responsibility for: (i) determining the real and potential biohazards of the proposed research, (ii) determining the appropriate level of biological and physical containment, (iii) selecting the microbiological practices and laboratory techniques for handling recombinant DNA materials, (iv) preparing procedures for dealing with accidental spills and overt personnel contamination, (v) determining the applicability of various precautionary medical practices, serological monitoring, and immunization, when available, (vi) securing approval of the proposed research prior to initiation of work, (vii) submitting information on purported EK2 and EK3 systems to the NIH Recombinant DNA Molecule Program Advisory Committee and making the strains available to others, (viii) reporting to the institutional biohazards committee and the NIH Office of Recombinant DNA Activities new information bearing on the guidelines, such as technical information relating to hazards and new safety procedures or innovations, (ix) applying for approval from the NIH Recombinant DNA Molecule Program Advisory Committee for large scale experiments with recombinant DNAs known to make harmful products (i.e., more than 10 liters of culture), and (x) applying to NIH for approval to lower containment levels when a cloned DNA recombinant derived from a shotgun experiment has been rigorously characterized and there is sufficient evidence that it is free of harmful genes.

Before work is begun, the principal investigator is responsible for: (i) making available to program and support staff, copies of those portions of the approved grant application that describe the biohazards and the precautions to be taken, (ii) advising the program and support staff of the nature and assessment of the real and potential biohazards, (iii) instructing and training this staff in the practices and techniques required to ensure safety, and in the procedures for dealing with accidentally created biohazards, and (iv) informing the staff of the reasons and provisions for any advised or requested precautionary medical practices, vaccinations, or serum collection.

During the conduct of research, the principal investigator is responsible for: (i) supervising the safety performance of the staff to ensure that the required safety practices and techniques are employed, (ii) investigating and reporting in writing to the NIH Office of Recombinant DNA Activities and the institutional biohazards committee any serious or extended illness of a worker or any accident that results in (a) inoculation of recombinant DNA materials through cutaneous penetration, (b) ingestion of recombinant DNA materials, (c) probably inhalation of recombinant DNA materials following gross aerosolization, or (d) any incident causing serious exposure to personnel or danger of environmental contamination, (iii) investigating and reporting in writing to the NIH Office of Recombinant DNA Activities and the institutional biohazards committee any problems pertaining to operation and implementation of biological and physical containment safety practices and procedures, or equipment or facility failure, (iv) correcting work errors and conditions that may result in the release of recombinant DNA materials, and (v) ensuring the integrity of the physical containment (e.g., biological safety cabinets) and the biological containment (e.g., genotypic and phenotypic characteristics, purity, etc.).

B. Institution. Since in almost all cases, NIH grants are made to institutions rather than to individuals, all the responsibilities of the principal investigator listed above are the responsibilities of the institution under the grant, fulfilled on its behalf by the principal investigator. In addition, the institution is responsible for establishing an institutional systems. These are also the systems for which we have the most experience and knowledge regarding the effectiveness of the containment provided by existing hosts and vectors necessary for the construction of more effective biological barriers.

For these reasons, E. coli K-12 appears to be the system of choice at this time, although we have carefully considered arguments that many of the potential dangers are compounded by using an organism as intimately connected with a man as is E. coli. Thus, while proceeding cautiously with E. coli, serious efforts should be made toward developing alternate host-vector systems. . . .

IV. Roles and Responsibilities.

Safety in research involving recombinant DNA molecules depends upon how the research team applies these guidelines. Motivation and critical judgment are necessary, in addition to specific safety knowledge, to ensure protection of personnel, the public, and the environment.

The guidelines given here are to help the principal investigator determine the nature of the safeguards that should be implemented. These guidelines will be incomplete in some respects because all conceivable experiments with recombinant DNAs cannot now be anticipated. Therefore, they cannot substitute for the investigator's own knowledgeable and discriminating evaluation. Whenever this evaluation calls for an increase in containment over that indicated in the guidelines, the investigator has a responsibility to institute such an increase. In contrast, the containmentconditions called for in the guidelines should not be decreased without review and approval at the institutional and NIH levels.

C. NIH Initial Review Groups (Study Sections). The NIH Study Sections, in addition to reviewing the scientific merit of each grant application involving recombinant DNA molecules, are responsible for: (i) making an independent evaluation of the real and potential biohazards of the proposed research on the basis of these guidelines, (ii) determining whether the proposed physical containment safeguards certified by the institutional biohazards committee are appropriate for control of these biohazards, (iii) determining whether the proposed biological containment safeguards are appropriate, (iv) referring to the NIH Recombinant DNA Molecule Program Advisory Committee or the NIH Office of Recombinant DNA Activities those problems pertaining to assessment of biohazards or safeguard determination that cannot be resolved by the Study Sections.

The membership of the Study Sections will be selected in the usual manner. Biological safety expertise, however, will be available to the Study Sections for consultation and guidance.

D. NIH Recombinant DNA Molecule Program Advisory Committee. The Recombinant DNA Molecule Program Advisory Committee advises the Secretary, Department of Health, Education, and Welfare, the Assistant Secretary for Health, Department of Health, Education, and Welfare, and the Director, National Institutes of Health, on a program for the evaluation of potential biological and ecological hazards of recombinat DNAs (molecules resulting from different segments of DNA that have been joined together in cell-free systems, and which have the capacity to infect and replicate in some host cell, either autonomously or as an integrated part of their host's genome), on the development of procedures which are designed to prevent the spread of such molecules within human and other populations, and on guidelines to be followed by investigators working with potentially hazardous recombinants.

The NIH Recombinant DNA Molecule Program Advisory Committee has responsibility for: (i) revising and updating guidelines to be followed by investigators working with DNA recombinants, (ii) for the time being, receiving information on purported EK2 and EK3 systems and evaluating and certifying that host-vector systems meet EK2 and EK3 criteria, (iii) resolving questions concerning potential biohazard and adequacy of containment capability if NIH staff or NIH Initial Review Group so request, and (iv) reviewing and approving large scale experiments with recombinant DNAs known to make harmful products (e.g., more than 10 liters of culture).

E. NIH Staff. NIH Staff has responsibility for: (i) assuring that no NIH grants or contracts are awarded for DNA recombinant research unless they (a) conform to these guidelines, (b) have been properly reviewed and recommended for approval, and (c) include a properly executed Memorandum of Understanding and Agreement, (ii) reviewing and responding to questions or problems or reports submitted by institutional biohazards committees or principal investigators, and disseminating findings, as appropriate, (iii) receiving and reviewing applications for approval to lower containment levels when a cloned DNA recombinant derived from a shotgun experiment has been rigorously characterized and there is sufficient evidence that it is free of harmful genes, (iv) referring items covered under (ii) and (iii) above to the NIH Recombinant DNA Molecule Program Advisory Committee, as deemed necessary, and (v) performing site inspections of all P4 physical containment facilities, engaged in DNA recombinant research, and of other facilities as deemed necessary.

Statement By Joseph A. Califano, Jr., Secretary Of Health, Education And Welfare Relative To The Promulgation Of New National Guidelines For The Conduct Of Research On Recombinant DNA, December 17, 1978. The Revised DNA Guidelines Are Found at 43 Fed. Reg. 60108-60131 (December 22, 1978).

I am announcing today several actions affecting the conduct of recombinant DNA research in this country.

In taking these steps, I have been guided by my responsibility to allow the maximum freedom of scientific inquiry consistent with the protection of the public health and the environment and with respect for the important ethical concerns surrounding genetic research in general.

The research techniques used to produce recombined molecules of deoxyribonucleic acid, the complex chemical that codes genetic information for all living cells, hold great promise for significantly advancing our understanding of fundamental biological processes. Moreover, this research may also hold potential for the commercial production of needed biological materials and agricultural products.

From the pioneering days of this research, many of this nation's leading scientists expressed concern that the insertion of foreign genes into micro-organisms could carry the potential for harm by yielding new disease-producing organisms. Although no harm has resulted from recombinant DNA research to date, there has been widespread uncertainty as to the degree of risk involved.

We must always recognize that scientific knowledge is not immutable; it is constantly changing as research generates additional information and understanding. Public policy in the field of science must therefore be flexible--to allow change as knowledge and understanding increase. The requirements that we impose must constantly be revised and updated to reflect new knowledge. Today the experience and insights that we have gained provide the basis for relaxing some of the restrictions the National Institutes of Health first imposed in 1976 on recombinant DNA research it funds.

The actions I am announcing today strive to allow the greatest freedom of scientific inquiry possible. At the same time, they provide the protections necessary to safeguard the public health and environment and also provides the opportunity for those concerned to raise any ethical issues posed by recombinant DNA research.

Specifically, I am today

. Approving final guidelines prepared by the National Institutes of Health that significantly revise the sfaety requirements for conducting recombinant DNA research;

. Taking immediate steps to require that research conducted by private companies complies with the NIH guidelines, primarily through use of the regulatory authority of the Food and Drug Administration;

. Requesting the Environmental Protection Agency to review its authority and to take all action it can to require compliance with the NIH guidelines by companies that carry out DNA research but whose products are not regulated by the Food and Drug Administration;

. Directing NIH to increase its research designed to determine the extent of risk associated with recombinant DNA research;

. Broadening substantially the public representation on the HEW advisory committee that will assist NIH in administering the revised guidelines;

. Increasing significantly public access to information about recombinant DNA research activities and increasing public participation in the administration of the guidelines in local communities.

Revised Guidelines.

The revised final guidelines that NIH has developed and that I am approving today set new directions for regulation of future recombinant DNA research. These final guidelines retain much of the guidelines that NIH published in proposed form last July. But NIH has made many revisions based on public comment and on the review conducted by a Departmental committee.

The final guidelines relax some of the restrictions under which recombinant DNA research has been conducted since 1976, and at the same time increase the role of the public in approving and monitoring recombinant DNA experiments.

In particular, these final guidelines relax in two major respects the guidelines that were placed in effect in 1976.

. The revisions exempt altogether five categories of experiments from the guidelines' restrictions. NIH has concluded that these experiments present no known health risk. Approximately one-third of research convered under the existing guidelines would be exempted under the revised standards. The revised guidelines continue to ban all six categories of potentially hazardous research that the 1976 guidelines prohibited. They will now, however, permit the Director of NIH to grant--following public notice and comment--case by case exceptions to these prohibitions with appropriate safeguards.

. The revised guidelines will ease restrictions on other permissible experiments. Depending on the potential risk of an experiment, both the 1976 guidelines and today's revised guidelines require a researcher to comply with one of four levels of protective laboratory proceudres and one of three levels of restrictions on the type of organism that may be used in the research. The revised guidelines assign almost all categories of research physcial containment and/or biological containment levels at least one step lower than in the 1976 guidelines. Since the likelihood of harm now appears more remote than was once anticipated, the scientific community has now concluded that this downgrading is appropriate. The four levels of physical containment and three levels of "biological containment"--the use of weakened organisms that cannot survive outside the laboratory--set by the 1976 guidelines would remain the same.

Based on the review and public hearing conducted by a Departmental committee, the guidelines have been significantly rewritten from the July version to increase public participation at both the local and national level:

. Twenty percent of the members of local Institutional Biosafety Committees (IBC's) must represent the general public, and have no connection to the institution. The 1976 guidelines had no such requirements for public participation.

. Important records must be made public. The bulk of IBC records must be made available to the public and problems, violations, illnesses and accidents must be reported to NIH.

. At the national level, major actions cannot be taken without advice of the Recombinant DNA Advisory Committee (RAC) with public and Federal agency comment. Major actions include decisions to approve on a case by case basis experiments that are generally prohibited, to exempt additional categories of research from the guidelines, to permit the insertion of genes in new types of bacteria, and to approve changes in the guidelines themselves.

Finally, today's revised guidelines provide more explicit guidance both for local institutions and for NIH to follow in implementing the guidelines.

. Institutions must develop emergency plans covering accidental spills and personnel contamination; health surveillance programs for projects needing such safeguards; and training programs for IBC members, researchers, and other laboratory staff.

. Under the revised guidelines, the NIH Director cannot approve proposed actions unless he determines that they present no significant risk to health or the environment.

Guideline Coverage.

The revised guidelines apply to all recombinant DNA research conducted at any institution which receives NIH funds for recombinant DNA research. At these institutions, even research conducted without NIH support must comply with the guidelines. Other research agencies of the Federal government have assured us that they will require compliance with the NIH guidelines for all recombinant DNA research that they conduct or support.

We are also taking action to assure that the guidelines apply, to the greatest extent possible, to research conducted in the private sector.

. At my direction, the Food and Drug Administration is today announcing its intent to propose that any recombinant DNA research submitted to satisfy FDA's regulatory requirements must have been conducted in compliance with the NIH guidelines.

. I have also written to Douglas Costle, the Administrator of the Environmental Protection Agency (EPA) and asked him to review EPA's regulatory authority to determine whether EPA can regulate recombinant DNA research conducted privately that is not submitted to the FDA. I have asked him to take all action he can.

If both FDA and EPA act to regulate privately conducted recombinant DNA research, virtually all recombinant DNA research in this country would be brought under the requirements of the revised guidelines.

Broadened Committee Membership.

I will announce shortly the names of fourteen new members of HEW's recombinant DNA Advisory Committee. In addition to scientists who are experts in molecular biology and other disciplines, the Committee will be expanded to include persons knowledgeable in a wide variety of fields such as law, public policy, ethics, the environment and public health. The Committee will serve as the principal advisory body to the Director of NIH and to the Secretary of HEW on recombinant DNA policy.

Increased Risk Assessment Research.

While our knowledge about the risks of recombinant DNA has increased dramatically, much remains unknown. The scientific community must continue to assess the extent of the risks posed by recombinant DNA research. I am therefore directing the Assistant Secretary for Health and the Director of the National Institutes of Health to formulate a plan for carrying out a balanced program of additional risk assessment experiments. In my view, the more risk assessment experiments NIH conducts or supports, the better we can judge whether the guidelines --and actions taken under them--afford appropriate protection for health and environment.

Today's action represents the culmination of a long and thorough process that has sought at each step to balance the important concerns involved in recombinant DNA research. The National Institutes of Health in 1976 published guidelines to govern research which it funds.

The 1976 guidelines:

. Prohibited six categories of recombinant DNA experiments which experts felt posed significant hazards.

. Defined degrees of physican and biological containment necessary to prevent recombinant DNA organisms from escaping into the environment and surviving.

. Described permissible categories of recombinant DNA research and assigned levels of physical and biological containment for each.

. Described specific roles and responsibilities for principal investigators, research institutions, institutional biohazard committees, and the NIH.

Since issuance of the 1976 guidelines, recombinant DNA techniques have become much more widely used in research, and more has been learned about the limits of potential risks in using this technology.

In light of this new knowledge, the Director, NIH, on July 28, 1978 proposed substantial modification and relaxation of the guidelines. At that time, I named a Departmental review committee consisting of Peter Libassi, the Department's General Counsel, as Chairperson; Dr. Donald Frederickson, the Director of NIH, as Vice Chaiperson; Dr. Julius Richmond, Assistant Secretary for Health; and Dr. Henry Aaron, then Assistant Secretary for Planning and Evaluation. I asked the Committee to examine the proposed guidelines and to hold a public hearing on the guidelines.

In reviewing the guidelines, the committee solicited and heard comments from representatives of environmental groups, unions, pharmaceutical companies, institutional bio-safety committees and Congressional staff members. The committee reviewed more than 170 letters from the public commenting on the revisions. The committee played a vital role in the process which led to the revised guidelines and unanimously recommended that the revised guidelines be approved.

These revised guidelines provide for a flexible, open system that can accommodate new scientific information that may warrant change, either to relax or to increase safety requirements.

I applaud all who have labored to developed these guidelines: the scientific community, the public, and workers at the Federal, State and local levels. This research holds promise for adding to our understanding about basic biological processes. These guidelines should permit that promise to be realized without presenting any significant risk to public health or the environment.

[See also, Schmeck, "New Guides Issued on Gene Research," N. Y. Times, Dec. 17, 1978, at 32, col. 1.]

. . . .

Revised NIH Guidelines became effective January 29, 1980, and were obtained by incorporating into the December 1978 Guidelines all the changes made following the February 16-17, 1979, May 21-23, 1979, September 6-7, 1979, and December 6-7, 1979, meetings of the NIH Recombinant DNA Advisory Committee (RAC). The Complete Resivsed Guidelines are to be found in 45 Federal Register 6724 (January 29, 1980). Other pertinent historical citations concerning the guidelines are to be found at 45 Federal Register 6718 (January 29, 1980) and 45 Federal Register 7182 (January 31, 1980).

These revised NIH guidelines are viewed largely as being quite liberal in that their effect is to ease or--in some specific instances-- remove most of the existing rules concerning DNA experimentaion. These guidelines are expected to double the nearly eight hundred current experiments in the field being conducted in various universities and research centers and, at the same time, encourage increased industrial efforts to use genetic engineering, develop brain hormones, etc. The NIH rules continue to limit experiments to 10 liters (2.6 gallons) of any batch of genes growing in bacteria--commonly modified versions of the intestinal bacteria, $E.$ coli.

As previously noted in an earlier part of this Appendix, the 1976 NIH guidelines required each DNA experiment to be performed in one of four types of laboratories depending on the risk involved. The new rules downgrade the requirements for 80 per cent of all current recombinant DNA research. Nearly half of this percentage may be undertaken in the simplest laboratory. The very basic and uncomplicated experiments using modified E. coli need no longer even be reported to NIH--although they are required to still be screened by a local safety committee.

Legally, these NIH guidelines apply only to research centers obtaining federal funds. President Carter has urged Congress--to no avail--to require that industry comply with these guidelines. Within the new guidelines, however, is a procedure which allows firms to register their research plans on a voluntary basis with NIH--who in turn agrees to keep the plans confidential and away from commercial competitors. [See Cohn, "U.S. Lowers Bars to Gene Engineering," Wash. Post, Jan. 30, 1980, at 1, col. 4].

## APPENDIX H

Recommendations--Research Involving Children, September 6, 1977

The National Commission for the Protection of Human Subjects of Biomedical and Behavioral Research makes the following recommendations for research involving children to:

The Secretary of Health, Education, and Welfare, with respect to research that is subject to his regulation, i.e., research conducted or supported under programs administered by him and research reported to him in fulfillment of regulatory requirements; and

The Congress, with respect to research that is not subject to regulation by the Secretary of Health, Education, and Welfare.

RECOMMENDATION (1). SINCE THE COMMISSION FINDS THAT RESEARCH INVOLVING CHILDREN IS IMPORTANT FOR THE HEALTH AND WELL-BEING OF ALL CHILDREN AND CAN BE CONDUCTED IN AN ETHICAL MANNER, THE COMMISSION RECOMMENDS THAT SUCH RESEARCH BE CONDUCTED AND SUPPORTED, SUBJECT TO THE CONDITIONS SET FORTH IN THE FOLLOWING RECOMMENDATIONS.

Comment: The Commission recognizes the importance of safeguarding and improving the health and well-being of children, because they deserve the best care that society can reasonably provide. It is necessary to learn more about normal development as well as disease states in order to develop methods of diagnosis, treatment and prevention of conditions that jeopardize the health of children, interfere with optimal development, or adversely affect well-being in later years. Accepted practices must be studied as well, for although infants cannot survive without continual support, the effects of many routine practices are unknown and some have been shown to be harmful.

Much research on childhood disorders or conditions necessarily involves children as subjects. The benefits of this research may accrue to the subjects directly or to children as a class. The Commission considers, therefore, that the participation of children in research related to their conditions should receive the encouragement and support of the federal government.

The Commission recognizes, however, that the vulnerability of children, which arises out of their dependence and immaturity, raises questions about the ethical acceptability of involving them in research. Such ethical problems can be offset, the Commission believes, by establishing conditions that research must satisfy to be appropriate for the involvement of children. Such conditions are set forth in the following recommendations.

RECOMMENDATION (2). RESEARCH INVOLVING CHILDREN MAY BE CONDUCTED OR SUPPORTED PROVIDED AN INSTITUTIONAL REVIEW BOARD HAS DETERMINED THAT: (A) THE RESEARCH IS SCIENTIFICALLY SOUND AND SIGNIFICANT; (B) WHERE APPROPRIATE, STUDIES HAVE BEEN CONDUCTED FIRST ON ANIMALS AND ADULT HUMANS, THEN ON OLDER CHILDREN, PRIOR TO INVOLVING INFANTS; (C) RISKS ARE MINIMIZED BY USING THE SAFEST PROCEDURES CONSISTENT WITH SOUND RESEARCH DESIGN AND BY USING PROCEDURES PERFORMED FOR DIAGNOSTIC OR TREATMENT PURPOSES WHENEVER FEASIBLE; (D) ADEQUATE PROVISIONS ARE MADE TO PROTECT THE PRIVACY OF CHILDREN AND THEIR PARENTS, AND TO MAINTAIN CONFIDENTIALITY OF DATA; (E) SUBJECTS WILL BE SELECTED IN AN EQUITABLE MANNER; AND (F) THE CONDITIONS OF ALL APPLICABLE SUBSEQUENT RECOMMENDATIONS ARE MET.

Comment: This recommendation sets forth general conditions that should apply to all research involving children. Such research must also satisfy the conditions of one or more of Recommendations (3)

through (6), as applicable; Recommendation (7); Recommendation (8), if permission of parents or guardians is not a reasonable requirement; Recommendation (9), if the subjects are wards of the state; and Recommendation (10), if the subjects are institutionalized.

Respect for human subjects requires the use of sound methodology appropriate to the discipline. The time and inconvenience requested of subjects should be justified by the soundness of the research and its design, even if no more than minimal risk is involved. In addition, research involving children should satisfy a standard of scientific significance, since these subjects are less capable than adults of determining for themselves whether to participate. If necessary, the IRB should obtain the advice of consultants to assist in determining scientific soundness and significance.

Whenever possible, research involving risk should be conducted first on animals and adult humans in order to ascertain the degree of risk and the likelihood of generating useful knowledge. Sometimes this is not relevant or possible, as when the research is designed to study disorders or functions that have no parallel in animals or adults. In such cases, studies involving risk should be initiated on older children to the extent feasible prior to including infants, because older children are less vulnerable and they are better able to understand and to assent to participation. In addition, they are more able to communicate about any physical or psychological effects of such participation.

In order to minimize risk, investigators should use the safest procedures consistent with good research design and should make use of information or materials obtained for diagnostic or treatment purposes whenever feasible. For example, if a blood sample is needed, it should be obtained from samples drawn for diagnostic purposes whenever it is consistent with research requirements to do so.

Adequate measures should be taken to protect the privacy of children and their families, and to maintain the confidentiality of data. The adequacy of procedures for protecting confidentiality should be considered in light of the sensitivity of the data to be collected (i.e., the extent to which disclosure could reasonably be expected to be harmful or embarrassing).

Subjects should be selected in an equitable manner, avoiding overutilization of any one group of children based solely upon administrative convenience or availability of a population living in conditions of social or economic deprivation. The burdens of participation in research should be equitably distributed among the segments of our society, no matter how large or small those burdens may be.

In addition to the foregoing requirements, research must satisfy the conditions of the following recommendations. as applicable.

RECOMMENDATION (3). RESEARCH THAT DOES NOT INVOLVE GREATER THAN MINIMAL RISK TO CHILDREN MAY BE CONDUCTED OR SUPPORTED PROVIDED AN INSTITUTIONAL REVIEW BOARD HAS DETERMINED THAT: (A) THE CONDITIONS OF RECOMMENDATION (2) ARE MET; AND (B) ADEQUATE PROVISIONS ARE MADE FOR ASSENT OF THE CHILDREN AND PERMISSION OF THEIR PARENTS OR GUARDIANS, AS SET FORTH IN RECOMMENDATIONS (7) AND (8).

Comment: If the IRB determines that proposed research will present no more than minimal risk to children, the research may be conducted or supported provided the conditions of Recommendation (2) are met and appropriate provisions are made for parental permission and the children's assent, as described in Recommendations (7) and (8) below. If the IRB is unable to determine that the proposed research will present no more than minimal risk to children, the research should be reviewed under Recommendations (4), (5) and (6), as applicable.

RECOMMENDATION (4). RESEARCH IN WHICH MORE THAN MINIMAL RISK TO CHILDREN IS PRESENTED BY AN INTERVENTION THAT HOLDS OUT THE PROSPECT OF DIRECT BENEFIT FOR THE INDIVIDUAL SUBJECTS, OR BY A MONITORING PROCEDURE REQUIRED FOR THE WELL-BEING OF THE SUBJECTS, MAY BE CONDUCTED OR SUPPORTED PROVIDED AN INSTITUTIONAL REVIEW BOARD HAS DETERMINED THAT:

> (A) SUCH RISK IS JUSTIFIED BY THE ANTICIPATED BENEFIT TO THE SUBJECTS:
> (B) THE RELATION OF ANTICIPATED BENEFIT TO SUCH RISK IS AT LEAST AS FAVORABLE TO THE SUBJECTS AS THAT PRESENTED BY AVAILABLE ALTERNATIVE APPROACHES:
> (C) THE CONDITIONS OF RECOMMENDATION (2) ARE MET: AND
> (D) ADEQUATE PROVISIONS ARE MADE FOR ASSENT OF THE CHILDREN AND PERMISSION OF THEIR PARENTS OR GUARDIANS, AS SET FORTH IN RECOMMENDATIONS (7) and (8).

Comment: The Commission emphasizes that the purely investigative procedures in research encompassed by Recommendation (4) should entail no more than minimal risk to children. Greater risk is permissible under this recommendation only if it is presented by an intervention that holds out the prospect of direct benefit to the individual subjects or by a procedure necessary to monitor the effects of such intervention in order to maintain the well-being of these subjects (e.g., obtaining samples of blood or spinal fluid in order to determine drug levels that are safe and effective for the subjects). Such risk is acceptable, for example, when all available treatments for a serious illness or disability have been tried without success, and the remaining option is a new intervention under investigation. The expectation of success should be scientifically sound to justify undertaking whatever risk is involved. It is also appropriate to involve children in research when accepted therapeutic, diagnostic or preventive methods involve risk or are not entirely successful, and new biomedical or behavioral procedures under investigation present at least an equally favorable risk-benefit ratio. The IRB should evaluate research protocols of this sort in the same way that comparable decisions are made in clinical practice. It should compare the risk and anticipated benefit of the intervention under investigation (including the monitoring procedures necessary for care of the child) with those of available alternative methods for achieving the same goal, and should also consider the risk and possible benefit of attempting no intervention whatsoever.

To determine the overall acceptability of the research, the risk and anticipated benefit of activities described in a protocol must be evaluated individually as well as collectively, as is done in clinical practice. Research protocols meeting the criteria regarding risk and benefit may be conducted or supported provided the conditions of Recommendation (2) are fulfilled and the requirements for assent of the children and for permission and participation of their parents or guardians, as set forth in Recommendations (7) and (8), will be met. If the research also includes a purely investigative procedure presenting more than minimal risk, the research should be reviewed under Recommendation (5) with respect to such procedure.

RECOMMENDATION (5). RESEARCH IN WHICH MORE THAN MINIMAL RISK TO CHILDREN IS PRESENTED BY AN INTERVENTION THAT DOES NOT HOLD OUT THE PROSPECT OF DIRECT BENEFIT FOR THE INDIVIDUAL SUBJECTS, OR BY A MONITORING PROCEDURE NOT REQUIRED FOR THE WELL-BEING OF THE SUBJECTS, MAY BE CONDUCTED OR SUPPORTED PROVIDED AN INSTITUTIONAL REVIEW BOARD HAS

DETERMINED THAT:

    (A) SUCH RISK REPRESENTS A MINOR INCREASE OVER MINIMAL RISK:
    (B) SUCH INTERVENTION OR PROCEDURE PRESENTS EXPERIENCES TO SUBJECTS THAT ARE REASONABLY COMMENSURATE WITH THOSE INHERENT IN THEIR ACTUAL OR EXPECTED MEDICAL, PSYCHOLOGICAL OR SOCIAL SITUATIONS, AND IS LIKELY TO YIELD GENERALIZABLE KNOWLEDGE ABOUT THE SUBJECTS' DISORDER OR CONDITION;
    (C) THE ANTICIPATED KNOWLEDGE IS OF VITAL IMPORTANCE FOR UNDERSTANDING OR AMELIORATION OF THE SUBJECTS' DISORDER OR CONDITION;
    (D) THE CONDITIONS OF RECOMMENDATION (2) ARE MET; AND
    (E) ADEQUATE PROVISIONS ARE MADE FOR ASSENT OF THE CHILDREN AND PERMISSION OF THEIR PARENTS OR GUARDIANS, AS SET FORTH IN RECOMMENDATIONS (7) AND (8).

Comment: An IRB must determine that three special criteria are met in order to approve research presenting more than minimal risk but no direct benefit to the individual subjects. First, the increment in risk must be no more than a minor increase over minimal risk. The IRB should consider the degree of risk presented by the research from at least the following four perspectives: a common-sense estimation of the risk; an estimation based upon investigators' experience with similar interventions or procedures; any statistical information that is available regarding such interventions or procedures; and the situation of the proposed subjects. Second, the research activity must be commensurate with (i.e., reasonably similar to) procedures that the prospective subjects and others with the specific disorder or condition ordinarily experience (by virtue of having or being treated for that disorder or condition). Finally, the research must hold out the promise of significant benefit in the future to children suffering from or at risk for the disorder or condition (including, possibly, the subjects themselves). If necessary, the advice of scientific consultants should be obtained to assist in determining whether the research is likely to provide knowledge of vital importance to understanding the etiology or pathogenesis, or developing methods for the prevention, diagnosis or treatment, or the disorder or condition affecting the subjects.

The requirement of commensurability of experience should assist children who can assent to make a knowledgeable decision about their participation in research, based on some familiarity with the intervention or procedure and its effects. More generally, commensurability is intended to assure that participation in research will be closer to the ordinary experience of the subjects. The use of procedures that are familiar or similar to those used in treatment of the subjects should not, however, be used as a major justification for their participation in research, but rather as one of several criteria regarding the acceptability of such participation.

In addition to these special criteria, the IRB should assure that the conditions of Recommendation (2) are fulfilled and the requirements for assent of the children and permission and participation of their parents or guardians, as set forth in Recommendations (7) and (8), will be met. If the proposed research includes an intervention or procedure from which the subjects may derive direct benefit, it should also be reviewed under Recommendation (4), with respect to that intervention or procedure.

    RECOMMENDATION (6). RESEARCH THAT CANNOT BE APPROVED BY AN INSTITUTIONAL REVIEW BOARD UNDER RECOMMENDATIONS (3), (4) AND (5), AS APPLICABLE, MAY BE CONDUCTED OR SUPPORTED PROVIDED THE INSTITUTIONAL REVIEW BOARD HAS DETERMINED THAT THE RESEARCH PRESENTS AN OPPORTUNITY TO UNDERSTAND, PREVENT OR ALLEVIATE A SERIOUS PROBLEM AFFECTING THE

HEALTH OR WELFARE OF CHILDREN AND, IN ADDITION, A NATIONAL ETHICAL ADVISORY BOARD AND, FOLLOWING OPPORTUNITY FOR PUBLIC REVIEW AND COMMENT, THE SECRETARY OF THE RESPONSIBLE FEDERAL DEPARTMENT (OR HIGHEST OFFICIAL OF THE RESPONSIBLE FEDERAL AGENCY) HAVE DETERMINED EITHER (A) THAT THE RESEARCH SATISFIED THE CONDITIONS OF RECOMMENDATIONS (3), (4) AND (5), AS APPLICABLE, OR (B) THE FOLLOWING:

>  (I) THE RESEARCH PRESENTS AN OPPORTUNITY TO UNDERSTAND, PREVENT OR ALLEVIATE A SERIOUS PROBLEM AFFECTION THE HEALTH OR WELFARE OF CHILDREN;
>  (II) THE CONDUCT OF THE RESEARCH WOULD NOT VIOLATE THE PRINCIPLES OF RESPECT FOR PERSONS, BENEFICENCE AND JUSTICE;
>  (III) THE CONDITIONS OF RECOMMENDATION (2) ARE MET; AND
>  (IV) ADEQUATE PROVISIONS ARE MADE FOR ASSENT OF THE CHILDREN AND PERMISSION OF PARENT(S) OR GUARDIANS, AS SET FORTH IN RECOMMENDATIONS (7) and (8).

Comment: If an IRB is unable for any reason to determine that proposed research satisfies the conditions of Recommendations (3), (4) and (5), as applicable, the IRB may nevertheless certify the research for review and possible approval by a national ethical advisory board and the Secretary of the responsible department. Such review is contingent upon an IRB's determination that the research presents an opportunity to understand, prevent or alleviate a serious problem affecting the health or welfare of children. Thereafter, the research should be reviewed by the national board and Secretary, with opportunity for public comment, to determine whether the conditions of Recommendations (3), (4) and (5), as applicable, are satisfied, or, alternatively, the research is justified by the importance of the knowledge sought and would not contravene principles of respect for persons, beneficence and justice that underlie these recommendations. In the latter instance, commencement of the research should be delayed pending Congressional notification and a reasonable opportunity for Congress to take action regarding the proposed research.

The provision for national review and approval under Recommendations (3), (4) and (5) is intended to fit the situation where an IRB has difficulty in applying those recommendations but considers the research of sufficient importance to warrent national review. Such difficulty may be resolved by a determination on the national level pursuant to Recommendation (6)(A) that the research does satisfy the conditions of the applicable earlier recommendations. Alternatively, the national review may determine either that the research satisfies the conditions of Recommendation (6)(B) or that it should not be conducted.

The Commission believes that only research of major significance, in the presence of a serious health problem, would justify the approval of research under Recommendation (6)(B). The problem addressed must be a grave one, the expected benefit should be significant, the hypothesis regarding the expected benefit must be scientifically sound, and an equitable method should be used for selecting subjects who will be invited to participate. Finally, appropriate provisions should be made for assent of the subjects and permission and participation of parents or guardians.

RECOMMENDATION (7). IN ADDITION TO THE DETERMINATIONS REQUIRED UNDER THE FOREGOING RECOMMENDATIONS, AS APPLICABLE, THE INSTITUTIONAL REVIEW BOARD SHOULD DETERMINE THAT ADEQUATE PROVISIONS ARE MADE FOR; (A) SOLICITING THE ASSENT OF THE CHILDREN (WHEN CAPABLE) AND THE PERMISSION OF THEIR PARENTS OR GUARDIANS; AND, WHEN APPROPRIATE, (B) MONITORING THE SOLICITATION OF ASSENT AND PERMISSION, AND INVOLVING AT LEAST ONE PARENT OR GUARDIAN IN THE CONDUCT OF THE RESEARCH. A CHILD'S OBJECTION TO PARTICIPATION IN RESEARCH SHOULD BE BINDING UN-

LESS THE INTERVENTION HOLDS OUT A PROSPECT OF DIRECT BENEFIT THAT IS IMPORTANT TO THE HEALTH OR WELL-BEING OF THE CHILD AND IS AVAILABLE ONLY IN THE CONTEXT OF THE RESEARCH.

Comment: The Commission uses the term parental or guardian "permission," rather than "consent," in order to distinguish what a person may do autonomously (consent) from what one may do on behalf of another (grant permission). Parental permission normally will be required for the participation of children in research. In addition, assent of the children should be required when they are seven years of age or older. The Commission uses the term "assent" rather than "consent" in this context, to distinguish a child's agreement from a legally valid consent.

Parental or guardian permission, as used in this recommendation, refers to the permission of parents, legally appointed guardians, and others who care for a child in a reasonably normal family setting. The last category might include, for example, step-parents or relatives such as aunts, uncles or grandparents who have established a continuing, close relationship with the child. Recommendation (8) describes circumstances in which the IRB may determine that the permission of parents or guardians is not appropriate because of the nature of the subject under investigation (e.g., contraception, drug abuse) or because of a failure in the relationship with the child (e.g., child abuse, neglect).

Parental or guardian permission should reflect the collective judgment of the family that an infant or child may participate in research. There are some research projects for which documented permission of one parent or guardian should be sufficient, such as research involving no more than minimal risk (as described in Recommendation (3)), or research in which risks or discomforts are related to a therapeutic, diagnostic or preventive intervention (as described in Recommendation (4)). In such cases, it may be assumed that the person giving formal permission is reflecting a family consensus. For research that is described in Recommendations (5) and (6), the permission of both parents should be documented unless one parent is deceased, unknown, incompetent or not reasonably available, or the child has a guardian or belongs to a single-parent family (i.e., when only one person has legal responsibility for the care, custody and financial support of the child). The IRB should determine for each project whether permission of one or both parents should be required, a substitute mechanism may be used, or the provision may be waived. In making such determination, the IRB should consider the nature of the activities described in the research protocol and the age, status and condition of the subjects.

The IRB should assure that children who will be asked to participate in research described in Recommendation (5) are those with good relationships with their parents or guardians and their physician, and who are receiving care in supportive surroundings. Projects approved under Recommendations (4) and (6) may also require scrutiny of this sort. The IRB may wish to appoint someone to assist in the selection of subjects and to review the quality of interaction between parents or guardian and child. A member of the board or a consultant such as the child's pediatrician, a psychologist, a social worker, a pediatric nurse, or other experienced and perceptive person would be appropriate. The IRB should be particularly sensitive to the difficulties surrounding permission when the investigator is the treating physician to whom the parents or guardian may feel an obligation.

Because of the dependence of infants, the traditional role of parents as protectors, and the general authority of parents to determine the care and upbringing of their children, the IRB may determine

that small children should participate in certain research only if
the parents or guardians participate themselves by being present during some or all of the conduct of the research. This role will vary
according to the nature of the research, the risk involved, the extent
to which the research entails possibly disturbing deviations from
normal routine, and the age and condition of the children. As a general rule, when infants participate in research that may cause physical discomfort or emotional stress and involves a significant departure from normal routine, a parent or guardian should be present.
However, if discomfort arises only as a result of therapeutic interventions that must continue over a considerable period of time, the
continual presence of parents need not be required. Parental presence
during the conduct of much behavioral research may not be feasible or
warranted, especially with older children. Generally, parents or
guardians should be sufficiently involved in the research to understand its effects on their children and be able to intervene, if necessary.

The Commission believes that children who are seven years of age
or older are generally capable of understanding the procedures and
general purpose of research and of indicating their wishes regarding
participation. Their assent should be required in addition to parental permission. However, if any child over six years of age is incapacitated so that he or she cannot reasonably be consulted, then parental permission should be sufficient, as it is for infants. The
objection of a child of any age to participation in research should be
binding except as noted below.

If the research protocol includes an intervention from which the
subjects might derive significant benefit to their health or welfare,
and that intervention is available only in a research context, the
objection of a small child may be overridden. Such would be the case,
for example, with a new drug that is not approved by the Food and Drug
Administration for general distribution until safety and efficacy have
been demonstrated in controlled clinical trials. Access to a drug
under investigation generally requires participation in the research.
Similar restrictions may be placed on other innovative therapies as a
precaution. As children mature, their ability to perceive and act in
their own best interest increases; thus, their wishes with respect to
such research should carry increasingly more weight. When school-age
children disagree with their parents regarding participation in such
research, the IRB may wish to have a third party discuss the matter
with all concerned and be present during the consent process. Although parents may legally override the objections of school-age
children in such cases, the burden of that decision becomes heavier in
relation to the maturity of the particular child.

Disclosure requirements for assent and permission are the same as
those for legally valid informed consent. Similarly, children and
parents or guardians should be free from duress. In order to assure
full understanding and freedom of choice, the IRB may determine that
there is a need for an advocate to be present during the decision-
making process. The need for third-party involvement in this process
will vary according to the risk presented by the research, and the
autonomy of the subjects. The advocate should be an individual who
has the experience and perceptiveness to fulfill such a role and who
is not related in any way (except in the role as advocate or member of
the IRB) to the research or the investigators.

Finally, the IRB should pay particular attention to the explanation and consent form, if any, to assure that appropriate language is
used.

RECOMMENDATION (8). IF THE INSTITUTIONAL REVIEW BOARD DETERMINES

THAT A RESEARCH PROTOCOL IS DESIGNED FOR CONDITIONS OR A SUBJECT POPULATION FOR WHICH PARENTAL OR GUARDIAN PERMISSION IS NOT A REASONABLE REQUIREMENT TO PROTECT THE SUBJECTS, IT MAY WAIVE SUCH REQUIREMENT PROVIDED AN APPROPRIATE MECHANISM FOR PROTECTING THE CHILDREN WHO WILL PARTICIPATE AS SUBJECTS IN THE RESEARCH IS SUBSTITUTED. THE CHOICE OF AN APPROPRIATE MECHANISM SHOULD DEPEND UPON THE NATURE AND PURPOSE OF THE ACTIVITIES DESCRIBED IN THE PROTOCOL, THE RISK AND ANTICIPATED BENEFIT TO THE RESEARCH SUBJECTS, AND THEIR AGE, STATUS AND CONDITION.

Comment: Circumstances that would justify modification or waiver of the requirement for parental or guardian permission include: (1) research designed to identify factors related to the incidence or treatment of certain conditions in adolescents for which, in certain jurisdictions, they legally may receive treatment without parental consent; (2) research in which the subjects are "mature minors" and the procedures involved entail essentially no more than minimal risk that such individuals might reasonably assume on their own; (3) research designed to understand and meet the needs of neglected or abused children, or children designated by their parents as "in need of supervision"; and (4) research involving children whose parents are legally or functionally incompetent.

There is no single mechanism that can be substituted for parental permission in every instance. In some cases the consent of mature minors should be sufficient. In other cases court approval may be required. The mechanism invoked will vary with the research and the age, status and condition of the prospective subjects.

A number of states have specific legislation permitting minors to consent to treatment for certain conditions (e.g., pregnancy, drug addiction, venereal diseases) without the permission (or knowledge) of their parents. If parental permission were required for research about such conditions, it would be difficult to develop improved methods of prevention and therapy that meet the special needs of adolescents. Therefore, assent of such mature minors should be considered sufficient with respect to research about conditions for which they have legal authority to consent on their own to treatment. An appropriate mechanism for protecting such subjects might be to require that a clinic nurse or physician, unrelated to the research, explain the nature and the purpose of the research to prospective subjects, emphasizing that participation is unrelated to provision of care.

Another alternative might be to appoint a social worker, pediatric nurse, or physician to act as surrogate parent when the research is designed, for example, to study neglected or battered children. Such surrogate parents would be expected to participate not only in the process of soliciting the children's cooperation but also in the conduct of the research, in order to provide reassurance for the subjects and to intervene or support their desires to withdraw if participation becomes too stressful.

RECOMMENDATION (9). CHILDREN WHO ARE WARDS OF THE STATE SHOULD NOT BE INCLUDED IN RESEARCH APPROVED UNDER RECOMMENDATIONS (5) OR (6) UNLESS SUCH RESEARCH IS: (A) RELATED TO THEIR STATUS AS ORPHANS, ABANDONED CHILDREN, AND THE LIKE; OR (B) CONDUCTED IN A SCHOOL OR SIMILAR GROUP SETTING IN WHICH THE MAJORITY OF CHILDREN INVOLVED AS SUBJECTS ARE NOT WARDS OF THE STATE. IF SUCH RESEARCH IS APPROVED, THE INSTITUTIONAL REVIEW BOARD SHOULD REQUIRE THAT AN ADVOCATE FOR EACH CHILD BE APPOINTED, WITH AN OPPORTUNITY TO INTERCEDE THAT WOULD NORMALLY BE PROVIDED BY PARENTS.

Comment: It is important to learn more about the effects of various settings in which children who are wards of the state may be placed, as well as about the circumstances surrounding child abuse and neglect, in order to improve the care that is provided for such children by the community. Also, it is important to avoid embarrassment or psychological harm that might result from excluding wards of the state from research projects in which their peers in a school, camp or other group setting will be participating. Provision must be made to permit the conduct of such studies in ways that will protect the children involved, even though no parents or guardian are available to act in their behalf.

To this end, the IRB reviewing such research should evaluate the reasons for including wards of the state as research subjects and assure that such children are not the sole participants in a research project unless the research is related to their status as orphas, abandoned children, and the like. The IRB should require, as a minimum, that an advocate for each child be appointed to intercede, when appropriate, on the child's behalf. The IRB may also require additional protections, such as prior court approval.

RECOMMENDATION (10). CHILDREN WHO RESIDE IN INSTITUTIONS FOR THE MENTALLY INFIRM OR WHO ARE CONFINED IN CORRECTIONAL FACILITIES SHOULD PARTICIPATE IN RESEARCH ONLY IF THE CONDITIONS REGARDING RESEARCH ON THE INSTITUTIONALIZED MENTALLY INFORM OR ON PRISONERS (AS APPLICABLE) ARE FULFILLED IN ADDITION TO THE CONDITIONS SET FORTH HEREIN.

. . . .

APPENDIX I

United Nations General Assembly Declaration On The Rights of Mentally Retarded Persons, Adopted December 1971, Resolution 2856

The General Assembly,

Mindful of the pledge of the States Members of the United Nations under the Charter to take joint and separate action in co-operation with the Organization to promote higher standards of living, full employment and conditions of economic and social progress and development,

Reaffirming faith in human rights and fundamental freedoms and in the principles of peace, of the dignity and worth of the human person and of social justice proclaimed in the Charter,

Recalling the principles of the Universal Declaration of Human Rights, the International Covenants on Human Rights, 1 the Declaration of the Rights of the Child, 2 and the standards already set for social progress in the constitutions, conventions, recommendations and resolutions of the International Labour Organization, the United Nations Education, Scientific and Cultural Organization, the World Health Organization, the United Nations Children's Fund and of other organizations concerned,

Emphasizing that the Declaration on Social Progress and Development 3 has proclaimed the necessity of protecting the rights and assuring the welfare and rehabilitation of the physically and mentally disadvantaged,

Bearing in mind the necessity of assisting mentally retarded persons to develop their abilities in various fields of activities and of promoting their integration as far as possible in normal life,

Aware that certain countries, at their present state of development, can devote only limited efforts to this end,

Proclaims this Declaration on the Rights of Mentally Retarded Persons and calls for national and international action to ensure that it will be used as a common basis and frame of reference for the protection of these rights:

1. The mentally retarded person has, to the maximum degree of feasibility, the same rights as other human beings.

2. The mentally retarded person has a right to proper medical care and physical therapy and to such education, training, rehabilitation and guidance as will enable him to develop his ability and maximum potential.

3. The mentally retarded person has a right to economic security and to a decent standard of living. He has a right to perform productive work or to engage in any other meaningful occupation to the fullest possible extent of his capabilities.

4. Whenever possible, the mentally retarded person should live with his own family or with foster parents and participate in different forms of community life. The family with which he lives should receive assistance. If care in an institution becomes necessary, it should be provided in surroundings and other circumstances as close as possible to those of normal life.

5. The mentally retarded person has a right to a qualified

guardian when this is required to protect his personal well-being and interests.

6. The mentally retarded person has a right to protection from exploitation, abuse and degrading treatment. If prosecuted for any offense, he shall have a right to due process of law with full recognition being given to his degree of mental responsibility.

7. Whenever mentally retarded persons are unable, because of the severity of their handicap, to exercise all their rights in a meaningful way or it should become necessary to restrict or deny some or all of these rights, the procedure used for that restriction or denial or right must contain proper legal safeguards against every form of abuse. This procedure must be based on an evaluation of the social capability of the mentally retarded person by qualified experts and must be subject to period review and to the right of appeal to higher authorities.

2027th plenary meeting, 20 December 1971

---

[1] Resolution 2200 A (XXI).

[2] Resolution 1386 (XIV).

[3] Resolution 2542 (XXIV).

APPENDIX J

The Belmont Report
Ethical Principles And Guidelines For The Protection Of Human Subjects Of Research: Reports Of The National Commission For The Protection Of Human Subjects Of Biomedical and Behavioral Research. 44 Fed. Reg. 23192 April 19, 1979, Footnotes Omitted.

The Commission was mandated to identify the basic ethical principles that should underline the conduct of biomedical and behavioral research involving human subjects and to develop guidelines which should be followed in order to assure that such research is conducted in accordance with those principles; thus, the Commission was directed to consider: the boundaries between biomedical and behavioral research and the accepted and routine practice of medicine; the role of assessment of risk-benefit criteria in the determination of the appropriateness of research involving human subjects; appropriate guidelines for the selection of human subjects for participation in such research, and the nature and definition of informed consent in various research settings.

The Belmont Report attempts to summarize the basic ethical principles identified by the Commission in the course of its deliberations. Unlike most other reports of the Commission, the Report does not make specific recommendations for administrative action by the Secretary of Health, Education and Welfare. Rather, the Commission recommended that the Belmont Report be adopted in its entirety, as a statement of the Department's policy; and the Department accordingly requested public comments on the proposal until July 17, 1979. [No definitive action of any nature will be taken on the Report until sometime in 1981].

The Report

Ethical Principles And Guidelines For Research Involving Human Subjects

Scientific research has produced substantial social benefits. It has also posed some troubling ethical questions. Public attention was drawn to these questions by reported abuses of human subjects in biomedical experiments, especially during the Second World War. During the Nuremberg War Crime Trials, the Nuremberg Code was drafted as a set of standards for judging physicians and scientists who had conducted biomedical experiments on concentration camp prisoners. This code became the prototype of many later codes intended to assure that research involving human subjects would be carried out in an ethical manner.

The codes consist of rules, some general, others specific, that guide the investigators or the reviewers of research in their work. Such rules often are inadequate to cover complex situations; at times they come into conflict, and they are frequently difficult to interpret or apply. Broader ethical principles will provide a basis on which specific rules may be formulated, criticized and interpreted.

Three principles, or general prescriptive judgments, that are relevant to research involving human subjects are identified in this statement. Other principles may also be relevant. These three are comprehensive, however, and are stated at a level of generalization that should assist scientists, subjects, reviewers and interested citizens to understand the ethical issues inherent in research involving human subjects. These principles cannot always be applied so as to resolve beyond dispute particular ethical problems. The objective is to provide an analytical framework that will guide the resolution of ethical problems arising from research involving human subjects.

This statement consists of a distinction between research and practice, a discussion of the three basic ethical principles, and remarks about the application of these principles.

A. Boundaries Between Practice And Research.

It is important to distinguish between biomedical and behavioral research, on the one hand, and the practice of accepted therapy on the other, in order to know what activites ought to undergo review for the protection of human subjects of research. The distinction between research and practice is blurred partly because both often occur together (as in research designed to evaluate a therapy) and partly because of notable departures from standard practice are often called "experimental" when the terms "experimental" and "research" are not carefully defined.

For the most part, the term "practice" refers to interventions that are designed solely to enhance the well-being of an individual patient or client and that have a reasonable expectation of success. The purpose of medical or behavioral practice is to provide diagnosis, preventive treatment or therapy to particular individuals. By contrast, the term "research" designates an activity designed to test an hypothesis, permit conclusions to be drawn, and thereby to develop or contribute to generalizable knowledge (expressed, for example, in theories, principles, and statements of relationships). Research is usually described in a formal protocol that sets forth an objective and a set of procedures designed to reach that objective.

When a clinician departs in a significant way from standard or accepted practice, the innovation does not, in and of itself, constitute research. The fact that a procedure is "experimental," in the sense of new, untested or different, does not automatically place it in the category of research. Radically new procedures of this description should, however, be made the object of formal research at an early stage in order to determine whether they are safe and effective. Thus, it is the responsibility of medical practice committees, for example, to insist that a major innovation be incorporated into a formal research project.

Research and practice may be carried on together when research is designed to evaluate the safety and efficacy of a therapy. This need not cause any confusion regarding whether or not the activity requires review; the general rule is that if there is any element of research in an activity, that activity should undergo review for the protection of human subjects.

B. Basic Ethical Principles.

The expression "basic ethical principles" refers to those general judgments that serve as a basic justification for the many particular ethical prescriptions and evaluations of human actions. Three basic principles, among those generally accepted in our cultural tradition, are particularly relevant to the ethics of research involving human subjects; the principles of respect of persons, beneficencence and justice.

1. Respect for Persons.

Respect for persons incorporates at least two ethical convictions: first, that individuals should be treated as autonomous agencts, and second, that persons with diminished autonomy are entitled to protection. The principle of respect for persons thus divides into two separate moral requirements: the requirement to acknowledge autonomy and the requirement to protect those with diminished autonomy.

An autonomous person is an individual capable of deliberation about personal goals and of acting under the direction of such deliberation. To respect autonomy is to give weight to autonomous persons' considered opinions and choices while refraining from obstructing their actions unless they are clearly detrimental to others. To show lack of respect for an autonomous agent is to repudiate that persons' considered judgments, to deny an individual the freedom to act on those considered judgments, or to withhold information necessary to make a considered judgment, when there are no compelling reasons to do so.

However, not every human being is capable of self determination. The capacity for self determination matures during an individual's life, and some individuals lose this capacity wholly or in part because of illness, mental disability, or circumstances that severely restrict liberty. Respect for the immature and the incapacitated may require protecting them as they mature or while they are incapacitated.

Some persons are in need of extensive protection, even to the point of excluding them from activities which may harm them; other persons require little protection beyond making sure they undertake activities freely and with awareness of possible adverse consequences. The extent of protection afforded should depend upon the risk of harm and the likelihood of benefit. The judgment that any individual lacks autonomy should be periodically re-evaluated and will vary in different situations.

In most cases of research involving human subjects, respect for persons demands that subjects enter into the research voluntarily and with adequate information. In some situations, thowever, application of the principle is not obvious. The involvement of prisoners as subjects of research provides an instructive example. On the one hand, it would seem that the principle of respect for persons requires that prisoners not be deprived of the opportunity to volunteer for research. On the other hand, under prison conditions, they may be subtly coerced or unduly influenced to engage in research activities for which they would not otherwise volunteer. Respect for persons would then dictate that prisoners be protected. Whether to allow prisoners to "volunteer" or to "protect" them presents a dilemma. Respecting persons, in most hard cases, is often a matter of balancing competing claims urged by the principles of respect itself.

2. Beneficence.

Persons are treated in an ethical manner not only by respecting their decisions and protecting them from harm, but also by making efforts to secure their well-being. Such treatment falls under the principle of beneficence. The term "beneficence" is often understood to cover acts of kindness or charity that go beyond strict obligation. In this document, beneficence is understood in a stronger sense, as an obligation. Two general rules have been formulated as complementary expressions of beneficent actions in this sense: (1) do not harm and (2) maximize possible benefits and minimize possible harms.

The Hippocratic maxim "do no harm" has long been a fundamental principle of medical ethics. Calude Bernard extended it to the real of research, saying that one should not injure one person regardless of the benefits that might come to others. However, even avoiding harm requires learning what is harmful; and, in the process of obtaining this information, persons may be exposed to risk of harm. Further, the Hippocratic Oath requires physicians to benefit their patients "according to their best judgment." Learning what will in fact benefit may require exposing persons to risk. The problem posed by these imperatives is to decide when it is justifiable to seek certain benefits despite the risks involved, and when the benefits should be foregone because of the risks.

The obligations of beneficence affect both individual investigators and society at large, because they extend both to particular research projects and to the entire enterprise of research. In the case of particular projects, investigators and members of their institutions are obliged to give forethought to the maximization of benefits and the reduction of risk that might occur from the research investigation. In the case of scientific research in general, members of the larger society are obliged to recognize the longer term benefits and risks that may result from the improvement of knowledge and from the development of novel medical, psychotherapeutic, and social procedures.

The principle of beneficence often occupies a well-defined justifying role in many areas of research involving human subjects. An example is found in research involving children. Effective ways of treating childhood diseases and fostering healthy developments are benefits that serve to justify research involving children--even when individual research subjects are not direct beneficiaries. Research also makes it possible to avoid the harm that may result from the application of previously accepted routine practices that on closer investigation turn out to be dangerous. But the role of the principle of beneficence is not always so unambiguous. A difficult ethical problem remains, for example, about research that presents more than minimal risk without immediate prospect of direct benefit to the children involved. Some have argued that such research is inadmissible, while others have pointed out that this limit would rule out much research promising great benefit to children in the future. Here again, as with all hard cases, the different claims covered by the principle of beneficence may come into conflict and force difficult choices.

3. Justice.

Who ought to receive the benefits of research and bear its burdens? This is a question of justice, in the sense of "fairness in distribution" or "what is deserved." An injustice occurs when some benefit to which a person is entitled is denied without good reason or when some burden is imposed unduly. Another way of conceiving the principle of justice is that equals ought to be treated equally. However, this statement requires explication. Who is equal and who is unequal? What considerations justify departure from equal distribution? Almost all commentators allow that distinctions based on experience, age, deprivation, competence, merit and position do sometimes constitute criteria justifying differential treatment for certain purposes. It is necessary, then, to explain in what respects people should be treated equally. There are several widely accepted formulations of just ways to distribute burdens and benefits. Each formulation mentions some relevant property on the basis of which burdens and benefits should be distributed. These formulations are (1) to each person an equal share, (2) to each person according to individual need, (3) to each person according to individual effort, (4) to each person according to societal contribution, and (5) to each person according to merit.

Questions of justice have long been associated with social practices such as punishment, taxation and political representation. Until recently, these questions have not generally been associated with scientific research. However, they are foreshadowed even in the earliest reflections on the ethics of research involving human subjects. For example, during the 19th and early 20th centuries, the burdens of serving as research subjects fell largely upon poor ward patients, while the benefits of improved medical care flowed primarily to private patients. Subsequently, the exploitation of unwilling prisoners as research subjects in Nazi concentration camps was condemned as a particularly flagrant injustice. In this country, in the 1940's, the Tuskegee syphilis study used disadvantaged, rural black men to study the untreated course of a disease that is by no means confined to that population. These subjects were deprived of demonstrably effective treatment in order not to interrupt the project, long after such treatment became generally available.

Against this historical background, it can be seen how conceptions of justice are relevant to research involving human subjects. For example, the selection of research subjects needs to be scrutinized in order to determine whether some classes (e.g., welfare patients, particular racial and ethnic minorities, or persons confined to institutions) are being systematically selected simply because of their easy availability, their compromised position, or their manipulability, rather than for reasons directly related to the problem being studied. Finally, whenever research supported by public funds leads to the development of therapeutic devices and procedures, justice demands both that these not provide advantages only to those who can afford them and that such research should not unduly involve persons from groups unlikely to be among the beneficiaries of subsequent applications of the research.

C. Applications.

Applications of the general principles to the conduct of research leads to consideration of the following requirements: informed consent, risk/benefit assessment, and the selection of subjects of research.

1. Informed Consent.

Respect for persons requires that subjects, to the degree that they are capable, be given the opportunity to choose what shall or shall not happen to them. This opportunity is provided when adequate standards for informed consent are satisfied.

While the importance of informed consent is unquestioned, controversy prevails over the nature and possibility of an informed consent. Nonetheless, there is widespread agreement that the consent process can be analyzed as containing three elements: information, comprehension and voluntariness.

Information.

Most codes of research establish specific items for disclosure intended to assure that subjects are given sufficient information. These items generally include: the research procedure, their purposes, risks and anticipated benefits, alternative procedures (where therapy is involved), and a state offering the subject the opportunity to ask questions and to withdraw at any time from the research. Additional items have been proposed, including how subjects are selected, the person responsible for the research, etc.

However, a simple listing of items does not answer the question of what the standard should be for judging how much and what sort of information should be provided. One standard frequently invoked in medical practice, namely the information commonly provided by practitioners in the field or in the locale, is inadequate since research takes place precisely when a common understanding does not exist. Another standard, currently popular in malpractice law, requires the practitioner to reveal the information that reasonable persons would wish to know in order to make a decision regarding their care. This, too, seems insufficient since the research subject, being in essence a volunteer, may wish to know considerably more about risks gratuitously undertaken than do patients who deliver themselves into the hands of a clinician for needed care. It may be that a standard of "the reasonable volunteer" should be proposed: the extent and nature of information should be such that persons, knowing that the procedure is neither necessary for their care nor perhaps fully understood, can decide whether they wish to participate in the furthering of knowledge. Even when some direct benefit to them is anticipated, the subjects should understand clearly the range of risk and the voluntary nature of participation.

A special problem of consent arises where informing subjects of some pertinent aspect of the research is likely to impair the validity of the research. It many cases, it is sufficient to indicate to subjects that they are being invited to participate in research of which some features will not be revealed until the research is concluded. In all cases of research involving incomplete disclosure, such research is justified only if it is clear that (1) incomplete disclosure is truly necessary to accomplish the goals of the research, (2) there are no undisclosed risks to subjects that are more than minimal, and (3) there is an adequate plan for debriefing subjects, when appropriate, and for dissemination of research results to them. Information about risks should never be withheld for the purpose of eliciting the co-operation of subjects, and truthful answers should always be given to direct questions about the research. Care should be taken to distinguish cases in which disclosure would destroy or invalidate the research from cases in which disclosures would simply inconvenience the investigator.

Comprehension.

The manner and context in which information is conveyed is as important as the information itself. For example, presenting information in a disorganized and rapid fashion, allowing too little time for consideration or curtailing opportunities for questioning, all may adversely affect a subject's ability to make an informed choice.

Because the subject's ability to understand is a function of intelligence, rationality, maturity and language, it is necessary to adapt the presentation of the information to the subject's capacities. Investigators are responsible for ascertaining that the subject has comprehended the information. While there is always an obligation to ascertain that the information about risk to subjects is complete and adequately comprehended, when the risks are more serious, that obligation increases. On occasion, it may be suitable to give some oral or written tests of comprehension.

Special provision may need to be made when comprehension is severely limited--for example, by conditions of immaturity or mental disability. Each class of subjects that one might consider as incompetent (e.g., infants and young children, mentally disabled patients, the terminally ill and the comatose) should be considered on its own terms. Even for these persons, however, respect requires giving them the opportunity to choose to the extent they are able, whether or not to participate in research. The objections of these subjects to involvement should be honored, unless the research entails providing them a therapy unavailable elsewhere. Respect for persons also requires seeking the permission of other parties in order to protect the subjects from harm. Such persons are thus respected both by acknowledging their own wishes and by the use of third parties to protect them from harm.

The third parties chosen should be those who are most likely to understand the incompetent subject's situation and to act in that person's best interest. The person authorized to act on behalf of the subject should be given an opportunity to observe the research as it proceeds in order to be able to withdraw the subject from the research, if such action appears in the subject's best interest.

Voluntariness.

An agreement to participate in research constitutes a valid consent only if voluntarily given. This element of informed consent requires conditions free of coercion and undue influence. Coercing occurs when an overt threat of harm is intentionally presented by one person to another in order to obtain compliance. Undue influence, by contract, occurs through an offer of an excessive, unwarranted, inappropriate or improper reward or other overture in order to obtain compliance. Also, inducements that would ordinarily be acceptable may become undue influences if the subject is especially vulnerable.

Unjustifiable pressures usually occur when persons in positions of authority or commanding influence--especially where possible sanctions are involved--urge a course of action for a subject. A continuum of such influencing factors exists, however, and it is impossible to state precisely where justifiable persuasion ends and undue influence begins. But undue influence would include actions such as manipulating a person's choice through the controlling influence of a close relative and threatening to withdraw health services to which an individual would otherwise be entitled.

2. Assessment of Risks and Benefits.

The assessment of risks and benefits requires a careful arrayal of relevant data, including, in some cases, alternative ways of obtaining the benefits sought in the research. Thus, the assessment presents both an opportunity and a responsibility to gather systematic and comprehensive information about proposed research. For the investigator, it is a means to examine whether the proposed research is properly designed. For a review committe, it is a method for determining whether the risks that will be presented to subjects are justified. For prospective subjects, the assessment will assist the determination whether or not to participate.

The Nature and Scope of Risks and Benefits.

The requirement that research be justified on the basis of a favorable risk/benefit assessment bears a close relation to the principle of beneficence, just as the moral requirement that informed consent be obtained is derived primarily from the principle of respect for persons. The term "risk" refers to a possibility that harm may occur. However, when expressions such as "small risk" or "high risk" are used, they usually refer (often ambiguously) both to the change (probability) of experiencing a harm and the severity (magnitude) of the envisioned harm.

The term "benefit" is used in the research context to refer to something of positive value related to health or welfare. Unlike "risk," "benefit" is not a term that expresses probabilities. Risk is properly contrasted with harms rather than risks of harm  Accordingly, so-called risk/benefit assessments are concerned with the probabilities and magnitudes of possible harms and anticipated benefits. Many kinds of possible harms and benefits need to be taken into account. There are, for example, risks of psychological harm, physical harm, legal harm, social harm and economic harm and the corresponding benefits. While the most likely types of harms to research subjects are those of psychological or physical pain or injury, other possible kinds should not be overlooked.

Risks and benefits or research may affect the individual subjects, the families of the individual subjects, and society at large (or special groups of subjects in society). Previous codes and federal regulations have required that risks to subjects be outweighed by the sum of both the anticiapted benefit to the subject, if any, and the anticipated benefit to society in the form of knowledge to be gained from the research. In balancing these different elements, the risks and benefits affecting the immediate research subject will normally carry special weight. On the other hand, interest other than those of the subject may on some occasions be sufficient by themselves to justify the risks involved in the research, so long as the subjects' rights have been protected. Beneficence thus requires that we protect against risk of harm to subjects and also that we be concerned about the loss of the substantial benefits that might be gained from research.

The Systematic Assessment of Risks and Benefits.

It is commonly said that benefits and risks must be "balanced" and shown to be "in a favorable ratio." The metaphorical character of these terms draws attention to the difficulty of making precise judgments. Only on rare occasions will quantitative techniques be available for the scrutiny of research protocols. However, the idea of systematic, nonarbitrary analysis of risks and benefits should be emulated insofar as possible. This ideal requires those making decisions about the justifiability of research to be thorough in the accumulation and assessment of information about all aspects of the research, and to consider alternatives systematically. This procedure renders the assessment of research more rigorous and precise, while making communication between review board members and investigators less subject to misinterpretation, misinformation and conflicting judgments. Thus, there should first be a determination of the validity of the presuppositions of the research; then the nature, probability and magnitude of risk should be distinguished with as much clarity as possible. The method of ascertaining risks should be explicit, especially where there is no alternative to the use of such vague categories as small or slight risk. It should also be determined whether an investigator's estimates of the probability of harm or benefits are reasonable, as judged by known facts or other available studies.

Finally, assessment of the justifiability of research should reflect at least the following considerations: (i) Brutal or inhumane treatment of human subjects is never morally justified; (ii) Risks should be reduced to those necessary to achieve the research objective. It should be determined whether it is in fact necessary to use human subjects at all. Risk can perhaps never be entirely eliminated, but it can often be reduced by careful attention to alternative procedures. (iii) When research involves significant risk of serious impairment, review committees should be extraordinarily insistent on the justification of the risk (looking usually to the likelihood of benefit to the subject--or, in some rare cases, to the manifest voluntariness of the participation); (iv) Whe vulnerable populations are involved in research, the appropriateness of involving them should itself be demonstrated. A number of variables go into such judgments, including the nature and degree of risk, the condition of the particular population involved, and the nature and level of the anticipated benefits; (v)Relevant risks and benefits must be thoroughly arrayed in documents and procedures used in the informed consent process.

3. Selection of Subjects.

Just as the principle of respect for persons finds expression in the requirements for consent, and the principle of beneficence in risk/benefit assessment, the principles of justice gives rise to moral requirements that there be fair procedures and outcomes in the selection of research subjects.

Justice is relevant to the selection of subjects of research at two levels: the social and the individual. Individual justice in the selection of subjects would require that researchers exhibit fairness: thus, they should not offer potentially beneficial research on to some patients who are in their favor or select only "undesirable" persons for risky research. Social justice requires that a distinction be drawn between classes of subjects that ought, and ought not, to participate in any particular kind of research, based on the ability of members of that class to bear burdens and on the appropriateness of placing further burdens on already burdened persons. Thus, it can be considered a matter of social justice that there is an order of preference in the selection of classes of subjects (e.g., the institutionalized mentally infirm or prisoners) may be involved as research subjects, if at all, only on certain conditions.

Injustice may appear in the selection of subjects, even if individua subjects are selected fairly by investigators and treated fairly in the course of research. This injustice arises from social, racial, sexual and cultural biases institutionalized in society. Thus, even if individual researchers are treating their research subjects fiarly, and even if IRBs are taking caring to assure that subjects are selected fairl within a particular institution, unjust social patterns may nevertheless appear in the overall distribution of the burdens and benefits of research. Although individual institutions or investigators may not be able to resolve a problem that is pervasive in their social setting, they can consider distributive justice in selecting research subjects.

Some populations, especially institutionalized ones, are already burdened in many ways by their infirmities and environments. When research is proposed that involves risks and does not include a therapeutic component, other less burdened classes of persons should be called upon first to accept these risks of research, except where the research is directly related to the specific conditions of the class involved. Also, even though public funds for research may often flow in the same directions as public funds for health care, it seems unfair that populations dependent on public health care consitute a pool of preferred research subjects if more advantaged populations are likely to be the recipients of the benefits.

One special instance of injustice results from the involvement of vulnerable subjects. Certain groups, such as racial minorities, the economically disadvantaged, the very sick, and the institutionalized may continually be sought as research subjects, owing to their ready availability in settings where research is conducted. Given their dependent status and their frequently compromised capacity for free consent, they should be protected against the danger of being involved in research solely for administrative convenience, or because they are easy to manipulate as a result of their illness or socioeonomic condition.

TABLE OF PRINCIPAL CASES BY CHAPTER

## Chapter 1

In re Quinlan 9
Eichner v. Dillon 13
Superintendent, Belchertown State School et al v. Saikewicz 8, 12

## Chapter 2

Buck v. Bell 20, 21, 29
Eisenstadt v. Baird 27
Gleitman v. Cosgrove 24
Griswold v. Connecticut 27
Loving v. Virginia 27
Roe v. Wade 18, 27
Skinner v. Oklahoma 27
Stewart v. Long Island College Hospital 24

## Chapter 3

Aiken v. Clary 46
Bellotti v. Baird 60
Bravery v. Bravery 56
Canterbury v. Spence 25, 29, 47
Clonce v. Richardson 44
Dandridge v. Williams 38
Doe v. Bolton 54
Eichner v. Dillon 68
Eisenstadt v. Baird 54
Haven v. Randolph 50
In re Cavitt 37
In re Maida Yetter 67
In re M.K.R. 59
In re Quinlan 43
In re Simpson 59
In re Willoughby 38
Jehovah Witness v. King County Hospital 66
John F. Kennedy Memorial Hospital v. Heston 42
Johnson v. Indiana 55
Kaimowitz v. Dept. Mental Health 64
Keyishian v. Bd. of Regents 62
Mugler v. Kansas 54
Natanson v. Kline 33
New State Ice Co. v. Liebman 62
Nielsen v. Regents of University California 58
NLRB v. Jones-Laughlin 39
O'Connor v. Donaldson 53
Planned Parenthood v. Danforth 60
Prince v. Massachusetts 59, 66
Raleigh Fitkin-Paul Morgan Memorial Hospital v. Anderson 67
Roe v. Wade 54, 55, 60
Schloendorff v. New York Hospital 46

Singleton v. Wulff 61
Skinner v. Oklahoma 54
Slago v. Leland Stanford Jr. Univ. Bd. of Trustees 33, 48
Slater v. Baker & Stapleton, C.B. 32
Stump v. Sparkman 61
Strunk v. Strunk 59
Sweezy v. New Hampshire 62
Trogun v. Fruchtman 48
United States v. Rutherford 68
Wall v. Brim 52
Wyatt v. Stickney 69

## Chapter 4

Becker v. Schwartz 82-85, 101
Berman v. Allan 102
Betancourt v. Gaylor 79
Bonbrest v. Kotz 80
Curlender v. Bio-Science Laboratories 95, 103
Coleman v. Garrison 78, 96
Custodia v. Baer 77, 96
Dietrich v. Northampton 80
Dumer v. St. Michael's Hosp. 71
Gleitman v. Cosgrove 90, 102
Jacobs v. Theimer 79, 97, 98, 111
Jorgensen v. Meade-Johnson Laboratories 100
Korman v. Hagen 81
Park v. Chessin 70, 72-76, 82, 92, 93, 94, 101
Reick v. Med. Protective Company 71, 98
Roe v. Wade 79
Sherlock v. Stillwater Clinic 76
Speck v. Finegold 88-89
Stewart v. Long Island College Hospital 71, 79, 90, 91, 99
Story Parchment Co. v. Patterson Parchment Paper Company 71
Terrell v. Garcia 78, 87
Troppi v. Scarf 80, 87
Womach v. Buchhorn 81
Wilczynski v. Goodman 95
Williams v. State 71
Zepeda v. Zepeda 71, 86, 92
Ziemba v. Sternberg 79, 99

## Chapter 5

Del Zio v. Columbia Hospital 109
Metropolis Theater Co. v. Chicago 117
Shapiro v. Thompson 116
Skinner v. Oklahoma 116

Chapter 6

Becker v. Schwartz  122
Fitzgerald v. Rueckl  121, 127
Gursky v. Gursky  118
Park v. Chessin  122
People v. Sorensen  118
Tarasoff v. Regents of
Univ. California  127

Chapter 7

Griswold v. Connecticut  136
Kaimowitz v. Dept. Mental
Health  136
Meyer v. Nebrasks  136
Scientists' Institute for
Pub. Info., Inc. v. AEC  143
Whitney v. California  135

- - - -

In re Dinnerstein  17
In re Spring  17
Rennie v. Klein  69
Rogers v. Okin  69

# INDEX

Abortion   41, 54-55, 60, 65, 93,
    therapeutic   95, 97

Abortuses   140

Academic freedom   39, 62

Adoption Identification Act,   116

Albinism   15

Amaurotic   15

Amniocentesis   18, 22, 82

Artificial insemination   104, 118, 164
    adulterous nature   118, 125
    confidentiality   121
    donors   107
        identity   122, 123, 127
        liability   121-122
    status of issue   118, 125
    public acceptance   126
    religious attitudes   155
    state statutes   125
    use by unmarrieds   119

Artificial life support systems   68

Asexual reproduction   105

Assault and battery   47, 48

Atkinson, Richard C.   62

Bazelon, David   149

Bevis, Douglas   104

Bioethics   145-147

Bioethical creed   146, 150

Biological revolution   5

Biomedicine   145, 146

Bok, Sissela   13

Burt, Robert   141

Calabresi, Guido   12, 16, 51

Callahan, Dan   138

Case utilitarians   150

Chromosomes   14

Civil liberties   19

Civil Rights Act   41

Civil rights   67

Cleft palate   15

Cloning   105, 114
    methods   112, 113
    government control   114

Clubfoot   15

Colorado Medical Center   22

Commerce Clause   39

Consent   46
    Restatement of Torts   46

Cooley's anemia   24

Cystic fibrosis   15

Damages: computation and physicians liability   71, 90-93, 96-98, 100, 102

Darwin, Charles   145

Death, new definition   140

Death with dignity   68

deChardin, Teilhard   145

Dilation and curettage   42

Dominant gene   10, 14

Doctrine of therapeutic privilege   49

Down's syndrome   22, 30, 31, 82

Due Process   55

Dwarfism   15

Eighth Amendment   41

Embryo implants   104, 108, 110

Equal Protection Clause   39, 61

Ethical tribunals   132, 133, 144

Ethics Advisory Board, HEW,   108, 142

Ettinger, Robert   163

Etzioni, A.   45, 144

Eugenics   1, 5
    positive   1, 104-106, 119
    negative   1, 19

Eugenic sterilization statutes   29

Eugenic sterilization   35-37, 53
    involuntary   37
    voluntary   37, 70-71, 88, 96
        minors   38
    compulsory   58, 59
    Roman Catholic Church   155, 156, 161

Experimentation with prisoners   141

Fear   135

Federal Privacy Act   126

Female eggs, how obtained   108

First Amendment   41

Fletcher, Joseph   23, 148, 151, 166

Floodgate argument   72

Food & Drug Administration   68

Frankel, Charles   5, 6, 22, 26

Fraud   73

Freedom of Information Act   127

Freedom to gain knowledge   129

Fried, Charles   49

Fundamental values   8

Gaylin, Willard   110, 112, 124

Genetic code   1

Genetic counseling   19, 25, 26

Genetic counselor   26

Genetically defective   21

Genetic deficiencies   11

Genetic disease   1, 3, 15, 31
    lifetime costs   30, 31

Genetic engineering   2, 6, 116, 130

Genetic manipulation   2

Genetic screening   18, 19

Genetic superiority   106

Good life   6

Grad, Frank   124, 126, 149

Granfield, David   23, 65

Greenawalt, Kent   28

Harelip   15

Hellegers, Andre   3, 6

Hemophiliac   11

Hippocratic Oath   50

Hitler, Adolf   1

Hobson's choice   84, 89

Holmes, Oliver Wendell   21, 28, 124

Human experimentation   139-140, 142
    therapeutic   39-40, 49
    non therapeutic   39-40
    disclosure   40, 49

Human life   7

Humanism   2

Huntington's chorea   10, 11, 24, 120

Hyde Amendment   55

Idiocy   15

India   28

Informed consent   32-42, 45-47, 51
    absence of   32
    fetal   40, 41
    proxy   57, 58, 59
    substituted   57, 58, 59
    diagnosis   33
    disclosure   33, 34, 52

Informed outsider   38

Ingelfinger, Franz   152

International Code of Ethical Behavior   133, 134, 136

In vitro fertilization   104, 109, 110, 111, 119, 147

Jehovah Witness  42, 47, 66-67

Juvenile cataract  10

Kallikak, Martin  30

Kant, Immanuel  145

Kass, Leon  4, 7, 152

Katz, Jay  46, 48

Laetrile  68

Lappe, Marc  5

Lederberg, Joshua  114

Lorber, John  13

Malpractice  70-73, 75-77, 88-89, 101

McCormick, Richard A.  13, 23

Medical treatment: right to refuse  42-44, 67

Mendel's law  24

Mental incompetent: right to refuse treatment  67

Metaethics  146

Mill, John Stuart  45, 46, 53

Mongolism  100-101, 103

Mongoloid  15

Muller, Herman  116, 164-166

Nagel, Ernest  135, 136

Nat'l Commission for Protection of Human Subjects & Behavioral Research  131, 139

Nat'l Envt'l Policy Act  143

Negligence  48, 50, 70-74, 80-84, 88-89, 90, 99

New Biology  1, 119, 145

Non therapeutic fetal research  132, 137

Normalization  53

Omega Point  145

Parens patri(ae)  42, 66

Parental duty  18

Parthenogenesis  105, 115

Pauling, Linus  25, 114

Peel Report  142

Peer Review Group Committees  144

Pellegrino, Edmund  7

Penetrance  14

Phenylketonuria (PKU)  24, 25, 31

Phenylpyruvic amentia  15, 31

Plante, Marcus  52

Plato  1

Polycystic congenital  15

Pope Pius XII  56

Prenatal injuries  80

Pres. Comm. for Study of Ethical Problems in Medicine & Biomedical Behavioral Research  140

Prosser, William  23

Protection of human subjects  64

Public intelligence  137

Public morals  54

Pyloric stenosis  15

Quinlan, Karen  43, 68

Ramsey, Paul  55-56, 61, 65-66

Randomized clinical trial  40

Rape  72

Rawls, John  151

Recessive genes  16, 20

Recombinant DNA  4-5, 11, 63, 137-138

Reilly, Philip  126-128

Relf, May & Minnie  57

Religion  153-155, 157

Retinoblastoma  10

Right of privacy  67, 68

Right to die  8, 43, 67

Right to know  129

Robertson, John A.  22, 23

Rorvik, David  108-109, 114-115

Rubella  71, 91, 101

Rules of ethical conduct  150

Rule utilitarians  146

Russell, Betrand  117, 158-159, 166

Saikewicz, Joseph  8, 12

Science  153-155

Scientific inquiry  129
    judicial review of  130
    government control  130-133

Self-determination  2

Shaw, Margery W.  88, 98

Shettles, Landrum  104, 113

Sickle Cell Anemia  24

Society, loss of  73, 74

Spina bifidia  9, 15

Steele, M. W.  58, 61

Stephen, Sir James Fitzjames  124

Steptoe, Patrick  108

Sterilization  11
    mental incompetents  20

Stone, Alan A.  60, 141

Substantive Due Procee  48

Substituted consent: children and legally incompetent, 57, 58

Substituted judgment  9, 38, 43, 44

Szasz, Thomas  68

Surrogate mothering  110, 124, 125

Tay Sachs  24, 31, 89, 95

Thalidomide  100, 101

Theology  153-157
    Protestant  156-157
    Roman Catholic  155-156, 160
    Judiac  157, 162

Tort immunity: intra familial  87, 88

Tribe, Lawrence H.  16, 30, 66, 114

Uniform Parentage Act  123
    state adoptions  128

Vukowich, William T.  27, 29, 43, 116

Watson, James  112, 113

Whitehead, Alfred North  7

Willowbrook consent order  55

Womb renting  110

Wrongful birth  70-72, 77-80, 83, 90, 94, 102-103

Wrongful conception  76-77, 83

Wrongful life  18, 70-76, 81-83, 86 95, 99

X-disorders  11, 112

X-linked disorder  111

XYY chromosome  24

Zone of privacy  20, 71